CONTAMINANTS in the ENVIRONMENT

A Multidisciplinary Assessment of Risks to Man and Other Organisms

Edited by
Aristeo Renzoni, Niccolò Mattei,
Lorena Lari, and Maria Cristina Fossi

LEWIS PUBLISHERS
Boca Raton   Ann Arbor   London   Tokyo

Library of Congress Cataloging-in-Publication Data

Contaminants in the environment: a multidisciplinary assessment of risks
  to man and other organisms / editors, Aristeo Renzoni.
    p. cm.
  Includes bibliographical references and index.
  ISBN 0-87371-853-4
    1. Environmental risk assessment. 2. Pollutants—Environmental aspects.
  I. Renzoni, Aristeo.
  GE145.C66 1994
  363.73—dc20   93-45760
                                                                                     CIP

   This book contains information obtained from authentic and highly regarded sources. Reprinted material is quoted with permission, and sources are indicated. A wide variety of references are listed. Reasonable efforts have been made to publish reliable data and information, but the author and the publisher cannot assume responsibility for the validity of all materials or for the consequences of their use.

   Neither this book nor any part may be reproduced or transmitted in any form or by any means, electronic or mechanical, including photocopying, microfilming, and recording, or by any information storage or retrieval system, without prior permission in writing from the publisher.

   All rights reserved. Authorization to photocopy items for internal or personal use, or the personal or internal use of specific clients, may be granted by CRC Press, Inc., provided that $.50 per page photocopied is paid directly to Copyright Clearance Center, 27 Congress Street, Salem, MA 01970 USA. The fee code for users of the Transactional Reporting Service is ISBN 0-87371-853-4/94/$0.00+$.50. The fee is subject to change without notice. For organizations that have been granted a photocopy license by the CCC, a separate system of payment has been arranged.

   CRC Press, Inc.'s consent does not extend to copying for general distribution, for promotion, for creating new works, or for resale. Specific permission must be obtained in writing from CRC Press for such copying.

   Direct all inquiries to CRC Press, Inc., 2000 Corporate Blvd., N.W., Boca Raton, Florida 33431.

© 1994 by CRC Press, Inc.
Lewis Publishers is an imprint of CRC Press

No claim to original U.S. Government works
International Standard Book Number 0-87371-853-4
Library of Congress Card Number 93-45760
Printed in the United States of America 1 2 3 4 5 6 7 8 9 0
Printed on acid-free paper

Aristeo Renzoni was born in Siena on 10 August 1929. He obtained a degree in veterinary science from Perugia University in November 1951. He was temporary assistant in the Department of Anatomy, Perugia 1953–1958, Guest Assistant in the Department of Veterinary Pathologisches Anatomy in Hannover, Germany, from January 1958 to August 1958, Assistant and Acting Director of the Department of Zoology at the Zoological Station in Naples from August 1958 to December 1960, Associate Professor, Institute of Biology, University of Siena 1960–1970, and Full Professor of Hydrobiology from January 1970 to October 1989 and Full Professor of Ecology since 1989. He was Chairman of the Department of Environmental Biology from January 1984 to November 1989 and has been lecturer in Poultry Husbandry since January 1980.

As Visiting Scientist he has worked at the Marine Laboratory of Oyster and Mussel Culture in Wemeldinge (Holland, summer 1960), at the University of California in Berkeley (10 months in 1963), at the Marine Laboratory for Oyster Culture in Conway, North Wales (spring 1973), at the Marine Laboratory of NOAA in Milford USA (2 months in 1974), and at the Marine Laboratory (Mussel Watch) Bodega Bay, University of California (1982, 1984). From 1975 to 1980 he was director of the United Nations Environmental program (UNEP) project: Environmental Research in the Mediterranean (Tyrrhenian Section).

In the past 10 years as visiting scientist under various cultural exchanges and agreements he has collected material and/or has been doing research in Romania, China, Switzerland, Egypt, Zimbabwe, Canada, Argentina, Norway, and Iceland. He is coordinator of an Erasmus project with the Universities of Reading and Wageningen and a participant (1988, 1989, 1990) in the Erasmus Project for the Small Islands (from 1991 he is the coordinator). He is also active in the fields of marine biology, marine resources, environmental conservation, and landscape ecology.

He is president and/or member of the Scientific Committee of several laboratories and institutes including the Laboratory for the Study of Natural Resources of the Lesina Lagoon of National Research Council of Italy.

Maria Cristina Fossi is a Staff Member of the Department of Environmental Biology of Siena University. She graduated with a Ph.D. in Environmental Science from Genoa University and has done postdoctoral research in the Department of Environmental Biology of Siena University on the fate, bioaccumulation, and biological effects of xenobiotic compounds in the Mediterranean. Since 1981, she has worked on biochemical adaptation to polluted environments by pollution-tolerant species, toxicological effects of xenobiotics on sea-bird populations, the use of biochemical and metabolic biomarkers (MFO, esterases, porphyrins) for the evaluation of environmental pollution, and MFO detoxification activity in Antarctic organisms. Her recent research has been concerned with the development of a methodological approach for the assessment of toxicological risk in endangered vertebrate species by means of nondestructive biomarkers. Dr. Fossi is author or co-author of over 100 papers, reviews, articles, and chapters.

She is on the editorial board of the *International Journal of Ecotoxicology,* member of the European Science Foundation Group for Biological Impact Assessment, and member of the Panel of Consultants of UNEP for the Marine Mammal Action Plan. She is a senior expert to the Italian Foreign Affairs Ministry in matters related to ecotoxicological research in developing countries. In 1993, she edited, with Dr. Claudio Leonzio, a book entitled *Nondestructive Biomarkers in Vertebrates*; in May 1992 they organized an international workshop in Siena on the same subject.

She is a member of the American Chemical Society, the Society of Environmental Toxicology and Chemistry, the European Society for Comparative Physiology and Biochemistry, the Italian Society of Ecology, and the Italian Society of Marine Biology. She has close research collaborations with Lee Shugart (Oak Ridge International Laboratory, USA), Francesco De Matteis (MRC Toxicology Unit, UK), David Peakall (National Wildlife Service, Canada), Michael Moore and D. Livingstone (Plymouth Research Marine Institute, UK), and S. Tanabe (Department of Environmental Conservation, Ehime University, Japan).

She has been one of the organizers of the Summer School since 1987.

Lorena Lari is a Ph.D. student at the Department of Environmental Biology of the University of Siena. Her field of interest is ecotoxicology and the use of biochemical biomarkers to determine toxicological hazard to wildlife. Dr. Lari is author of about 20 articles in the scientific literature. She has been one of the organizers of the Summer School for several years.

Niccolò Mattei is on the staff of the Department of Environmental Biology of the University of Siena, Italy. His professional training is in environmental biology (Ph.D. in Environmental Science) specializing in aquatic organisms. His Ph.D. studies concerned analysis of DNA damage due to xenobiotics. He is active in the field of aquaculture and mollusc culture techniques in lagoons and the sea. He worked for three months with Prof. Lee Shugart at the Oak Ridge National Laboratory (Tennessee, USA). He is currently working in the field of aquatic toxicology, DNA damage on organisms of brackish water in collaboration with the University of Venice and the School of Animal and Microbial Sciences of the University of Reading (UK), and the Consejo Superior de Investigaciones Cientificas of Castellon (Spain). He has been one of the organizers of the Summer School for a "Multidisciplinary Assessment of Environmental Risks for Human Health" since 1987.

PREFACE

The Summer School for a "Multidisciplinary Assessment of Environmental Risks for Human Health" was made possible by the generous support of the University of Siena and the Italian Council of Research and by the unstinting efforts of members of the Department of Environmental Biology of the University. The aim was to give a broad, interdisciplinary view of the hazards presented by environmental contaminants to man, either directly or indirectly. The course was given by internationally recognized experts, both from Italy and abroad. As the presentations were given in English, it was possible to invite students from many different countries to attend the School. This international character was consistent with the basic aim of the School, which was to present a wide range of topics, including global issues, giving certain of them in reasonable depth and detail. The School was held for the first time in 1987, at the Certosa di Pontignano, on the outskirts of Siena. It lasted for 2 weeks and there were 54 presentations by visiting lecturers. It was attended by 30 advanced students, who held Ph.D. degrees or were of similar standing. Fifteen students came from Italy and 15 from abroad.

The School proved to be a great success and there were many applications for entry in subsequent years. Four Summer Schools were held each year between 1987 and 1991. With the support of fellowships, half of the students came from abroad. Thus, over the 5-year period, there were students from the following countries: Argentina, Australia, Bangladesh, Bulgaria, Brasil, Costa Rica, Chile, China, Croatia, Czechoslovakia, Ecuador, Egypt, Gambia, Great Britain, Holland, India, Japan, Mexico, New Zealand, Pakistan, Poland, Portugal, Russia, South Africa, Spain, Syria, Turkey, Venezuela, and Zimbabwe.

Apart from achieving its primary educational objectives, the School established many links between scientists from different countries that have led to collaborative scientific work. The completion of the first 5 years of the School has, however, left one piece of unfinished business. Throughout the running of the School, lecturers produced notes and reprints that were circulated to students but these were never drawn together into one integrated collection. The present publication brings together a collection of reviews based on lectures given at the School. It is hoped that they will be useful not only for students attending the School at Siena in future years, but also to a much wider audience of scientists and interested lay people who are concerned with the problems of Environmental Toxicology.

INTRODUCTION

The Summer School "Multidisciplinary Assessment of Environmental Risks for Human Health" focused on human health hazards presented by contaminants with the major emphasis on direct effects. However, there are also indirect effects of contaminants that are or may be detrimental to mankind. Contaminants can have damaging effects on aquatic organisms such as fish and small molluscs, which are important sources of food—especially in third-world countries. Also, the injudicious use of insecticides has sometimes led to outbreaks of pest infestation, causing damage to crops, as a consequence of the decimation of natural predators or parasites. Thus, the program of the Summer School has also included certain presentations on ecotoxicological issues that have relevance to the main theme of the course.

The School gave attention to disparate aspects of the problem of risk assessment—from the molecular mechanisms that underly toxicity to the practical and administrative issues of environmental management. The basis scientific issues of environmental toxicology apply to all animals. In considering them, no distinction has been made between man and other animals. Indeed, most of the scientific work has involved the use of laboratory animals acting as surrogates for man. Thus the first three sections of the book deal mainly with principles that are relevant to living organisms generally. The fourth section, however, deals specifically with the question of risks for human health. The final section deals with environmental management in a broad way, touching on issues relevant to both the "human environment" and the "natural environment." The individual sections will now be considered in a little more detail.

Section I is concerned with the distribution and fate of contaminants. It includes a chapter dealing with models that may be used to predict the distribution of chemicals in the global environment. Two other chapters discuss approaches to the problem of monitoring. A more specialized chapter deals with one of the most serious problems encountered in environmental toxicology—the biomagnification of persistent contaminants in marine food chains.

Section II focuses on the relationship between metabolism and toxicity. This is fundamental to an understanding of the scientific basis of selective toxicity. An important practical aspect of this is the suitability of animal models for humans in toxicity testing. A better understanding of the metabolic regulation of toxicity should lead to the employment of better models for predicting toxicity to man. Although metabolism of contaminants is usually associated with detoxification, there are important exceptions to this rule. Some carcinogens and some highly toxic organo-phosphorus insecticides are activated by metabolism. These issues are brought out by contributors to this section.

Section III deals with toxic effects of contaminants. Clearly in the present text, it would not be appropriate to attempt a comprehensive view of the subject. Instead only a few selected examples are given, and emphasis is placed on the question of biomarkers. Until now, much of the work on environmental contamination has been concerned only with the measurement of levels of chemicals. Seldom has it been possible to say anything about the consequent effects. Taken alone, data on residue levels are of little assistance in reaching decisions about the control of environmental chemicals. A major purpose of the development of biomarkers is to overcome this problem—to provide measures of harmful effects of chemicals in the environment, thus giving a firm scientific basis for policy decisions on pollution control.

Section IV is concerned with the question of effects of contaminants on human health. With eight chapters, this is the largest section in the book, in keeping with the particular focus of the School. General issues such as environmental epidemiology and fate of inhaled substances are dealt with here, but the main focus is on the human health risks presented by specific compounds.

Section V concludes the text by focusing on issues of environmental management. These include valuable contributions on the policy of International Organizations concerned with pollution.

Regarding the overall book, it should be stressed that this is a collection of articles by authorities in specific fields. It focuses on specific issues, within a wide framework. It is hoped that this will provide valuable reading to students attending the Summer School at Siena in the future, and also to others following courses in environmental toxicology elsewhere. Finally it may also be of value to interested lay people having a scientific background.

ACKNOWLEDGMENTS

Thanks are due to all the collaborators from the Department of Environmental Biology of the University of Siena who have actively participated in the organization of the Summer School for a "Multidisciplinary Assessment of Environmental Risks for Human Health" held in the Certosa of Pontignano, sited in the hills of "Chianti" in the countryside of Tuscany, Italy.

Particular thanks go to the Rector of the University of Siena, Prof. Luigi Berlinguer, who supported the school during these 5 years.

TABLE OF CONTENTS

Preface

Introduction

Section I—Distribution and Fate of Contaminants

Chapter 1 Value and limitations of evaluative models in field situations 3
Eros Bacci

Chapter 2 Fate of persistent organochlorines in the marine environment 19
Shinsuke Tanabe

Chapter 3 Biological monitoring 29
Lee R. Shugart

Chapter 4 Pollution biomonitoring: A summing up 37
Henri M. André

Section II—Metabolism and Toxicity

Chapter 5 Comparative metabolism and selective toxicity 65
Colin H. Walker

Chapter 6 Application of metabolic studies to the evaluation of the toxicity of
chemicals to man 69
L. Vittozzi

Chapter 7 Comparative aspects of the metabolism and toxicity of xenobiotics
in fish 75
L. Vittozzi, J. Kaizer, A. Iannelli, and S. Soldano

Chapter 8 The role of cytochrome P-450 in drug metabolism and toxicity 81
Francesco De Matteis

Chapter 9 Estimation of induction of enzymes that metabolize xenobiotics *in vitro*
and *in vivo* 93
John R. Bend

Section III—Toxic Effects of Contaminants

Chapter 10 Environmental impact of pollutants: Biochemical responses as indicators of
exposure and toxic action (biochemical biomarkers) 105
Colin H. Walker

Chapter 11 Reactions of molluscan lysosomes as biomarkers of pollutant-induced
cell injury 111
Michael N. Moore

Chapter 12 Biochemical approach to toxicity evaluation of the rodenticides, norbormide, and α naphthyl thiourea 125
R. Radakrishnamurty

Chapter 13 Detection of xenobiotic–protein adducts: Electrophoretic and immunochemical approaches 129
B. Magi, B. Marzocchi, C. Lazzeri, L. Bini, and V. Pallini

Chapter 14 Structural and biochemical alterations in the gills of copper-exposed mussels 135
A. Viarengo, N. Arena, L. Canesi, F. A. Alia, and M. Orunesu

Chapter 15 Alternative methods in ecotoxicological research and testing 145
Julia Fentem and Michael Balls

Section IV—Contaminants and Risks for Human Health

Chapter 16 Cancer risks from arsenic in drinking water 163
Allan H. Smith, Claudia Hopenhayn-Rich, Michael N. Bates, Helen M. Goeden, I. Hertz-Picciotto, Heather M. Duggan, R. Wood, M. J. Kosuett, and M. T. Smith

Chapter 17 Serum 2,3,7,8–tetrachlorodibenzo-p-dioxin levels of New Zealand pesticide applicators and their implication for cancer hypotheses 179
A. H. Smith, D. G. Patterson, Jr., M. L. Warner, R. Mackenzie, and L. L. Needham

Chapter 18 Recombinant human papilloma type 16 DNA induces progressive changes in mouse 3T3 cells and human epithelial cells 189
Joseph A. DiPaolo

Chapter 19 Spatial distribution of dose in the human lung after inhalation of poorly soluble radionuclides and its sequelae 193
H. Cottier, A. Burkhardt, R. Kraft, F. Meister and A. Zimmermann

Chapter 20 The fate of inhaled toxic substances 199
H. Cottier and R. Kraft

Chapter 21 The pathogenesis of pulmonary alveolitis 203
H. Cottier and R. Kraft

Chapter 22 Environmental epidemiology 207
Pietro Comba

Chapter 23 Exposure to phenoxy herbicides and chlorinated dioxins and cancer risks: An inconsistent pattern of facts and frauds? 213
Olav Axelson

Chapter 24   Converging epidemiologic findings on radon in mines and homes as
             a risk of lung cancer 221
             Olav Axelson

Section V—Environmental Management

Chapter 25   The role of microorganisms in the environment decontamination 235
             Enrica Galli

Chapter 26   The use of pesticides and their levels in food in Eastern Europe:
             The example of Poland 247
             Jerzy Falandysz

Chapter 27   Contributions of the World Health Organization to pesticide safety 257
             Gary J. Burin

Chapter 28   World policy in the new environmental age 263
             David E. Alexander

    Index   277

SECTION I

Distribution and Fate of Contaminants

CHAPTER 1

Value and Limitations of Evaluative Models in Field Situations

Eros Bacci

INTRODUCTION

The environmental fate of contaminants is determined by the joint action of several factors ascribable either to the nature of the substance or to the environment. So, on the one hand, the physical and chemical properties will define the potential for reactivity and mobility and, on the other, the environmental variables will control the degree to which the these potentials operate. Under field conditions, environmental variables (e.g., temperature, pH, light wavelength and irradiation intensity, water and air turnover and turbulence, living organisms) are quite complex to analyze and may produce significant changes in the environmental behavior of contaminants. This makes the physicomathematical models for environmental chemistry very complex and difficult to apply, especially when several details are considered with the aim of producing a simulation of a *real system*.

An alternative approach consists in developing simple models simulating *evaluative environments*,[1,2] in which the environmental variables are standardized and reduced to the essential (*evaluative models*). In this case the different behavior of various chemicals will essentially depend only on their physicochemical properties (*intrinsic properties*). The main limitation of evaluative models is that the results are not directly related to real situations and, consequently, the absolute data (e.g., an expected concentration in a given environmental compartment) cannot be transferred to the field. However, the results of a series of chemicals may be applied, for instance, to *rank* the potential of different substances to reach and contaminate water bodies or to rank their potential to generate vapor drift from the site of release. Data from evaluative models may also be *calibrated* by means of laboratory and field measurements of selected reference chemicals to produce information on the order of magnitude of the expected environmental levels.

A combination of physicomathematical evaluative models, laboratory models, and field studies may be helpful in understanding major processes characterizing the environmental fate of chemical contaminants. The aim of the present chapter is to show some possible applications of this approach to real cases:

- hexachlorobenzene (HCB) and polychlorinated biphenyls (PCBs) in the Mediterranean Sea;
- tributyltin (TBT) in a harbor area;
- vapor drift and bioconcentration in plant foliage.

More details on the use of evaluative models in ecotoxicology are reported in a book recently appeared.[4]

HCB AND PCBs IN THE MEDITERRANEAN SEA

Hexachlorobenzene (HCB) and polychlorinated biphenyls (PCBs) are *semivolatile* organochlorinated hydrocarbons (SOCs). The term "semivolatile" indicates that these substances, characterized by liquid or supercooled liquid vapor pressure values from 10^{-5} to 10^{-1} Pa (20°C) are of low volatility. However, these chemicals are at same time *conservative*, or recalcitrant to degradation, and their long life makes volatilization phenomena significant.

Concern about SOCs began in the 1960s, when DDT residues (the most famous SOCs) were detected almost everywhere: in soils never treated with insecticides,[17] in birds and seals that never leave the Antarctic,[33] in animal and human tissues,[34,49] in rain,[1] and in the air of remote parts of the world.[36] The "*spurious peaks*" frequently found in biological samples for DDT residue analysis were identified by Jensen[3] as components of the PCB mixtures mainly used in electrical capacitors and transformers. In the second half of the 1960s another widespread contaminant, hexachlorobenzene (HCB), originating from agricultural and nonagricultural sources,[23] was identified.

During the 1970s and 1980s, research revealed the significance of environmental contamination by other "DDT-like" compounds, originating from intentional applications (i.e., agricultural use, vector control operations) and unintentional dispersion (PCBs) or unintentional production.

All these facts prompted the Mediterranean countries to move against Mediterranean Sea contamination in cooperation with certain U.N. Agencies (FAO, WHO, IAEA) and under the coordination of UNEP. The Governing Council of UNEP indicated the Mediterranean as a priority area, and launched the Mediterranean Action Plan (MAP).[43] Under MAP, extensive monitoring data on DDT and related compounds (DDTs), PCBs and other chlorinated hydrocarbons in marine organisms was collected and published,[45,46] leading to assessment documents on the state of pollution of the Mediterranean Sea by these compounds.[5,47]

Starting from physicochemical properties of HCB and PCBs, and from available field data, mainly on the levels in different environmental matrices (air, water, fish, sediment), it is possible to assess present contamination and future trends by means of simple evaluative models.

Environmental Distribution and Fate

Because of their low chemical reactivity, only physical and partition properties will be considered in describing the main trends in the environmental fate of HCB and PCBs. As far as PCBs are concerned, the properties of mixture containing 60% chlorine (Aroclor 1260) were used (Table 1).

The SOCs discussed here are nonpolar, and consequently are not easily transported by water. The situation of transport by air is quite different: due to air turnover, vapor movements of trace contaminants are significant when the vapor pressure of the chemical is as low as 10^{-6} Pa.[6,9] In the case of chemicals that are solid at ambient temperature (20–25°C) properties such as water solubility and vapor pressure have to be considered with reference to the liquid state (supercooled liquid for solids). This is because the high dispersion in the environment impedes the formation of crystals (i.e., the solid state).

Once in the air, the partition between gaseous and particulate phases is regulated by the supercooled liquid vapor pressure of the chemical:[13,19,32] substances with a liquid vapor pressure of

Table 1. Selected Properties of HCB and of a PCB Mixture[a]

Name	Molar mass	S_L (mol/m$_3$)	P_L(Pa)	K_{aw}	log K_{ow}
PCBs (60% Cl)	361[b]	4.8×10^{-5}	8.4×10^{-4}	7.1×10^{-3}	6.9
HCB	284.8	1.0×10^{-3}	1.3×10^{-1}	5.4×10^{-2}	6.0

[a]From data reported in Bacci et al.[8] S_L and P_L indicate water solubility and vapor pressure of the liquid phase.
[b]Average of the components of the Aroclor 1260 mixture.

3×10^{-5} Pa (25°C), can be expected to be equally distributed between air and particulate in rural environments. In urban environments, due to a higher particulate concentration, 3×10^{-4} Pa is the vapor pressure value for equipartitioning. Higher vapor pressure values will lead to an increase of the fraction in the vapor phase.

So even p,p'-DDT, the vapor pressure of which is 1.5×10^{-4} Pa at 20°C (as supercooled liquid[41]), migrates slowly but continuously from soil to air (where it is essentially in the vapor phase), reaching remote sites.[48] However, soil can delay the movement of less volatile and less polar chemicals from contaminated sites to remote areas and seas. In this retardation phenomena, plant foliage may play a significant role, due to the high affinity of foliage for SOC vapors.[7,15]

Air has a high carrying potential, even for long distances: vapor movement from contaminated soils to the air is probably the main path of chemical displacement, in a sort of global gas chromatographic process.[8,35] Vapor movement, together with some dry and wet deposition, can transport significant quantities of SOCs from contaminated land to oceans.[25] Air–water exchange is regulated by the air–water partition coefficient, K_{aw}. It is important to remember that the K_{aw} of water is 1.7×10^{-5} (20°C). As a consequence, chemicals with K_{aw} values 2 or more orders of magnitude lower than water are not able to pass from water to air and, if the water evaporates, their concentration in the water will increase.[42] Chemicals with a K_{aw} above 10^{-5} are able to pass from water to air, and this is the case of the chemicals here discussed (Table 1). The water-to-air transfer is particularly easy for HCB, with a K_{aw} of 5.4×10^{-2}. However, it can occur only when the water is already contaminated with SOCs. When the air is more contaminated than the water, the net flow will be reversed.

The mechanisms discussed above, in addition to input by dry deposition (air-borne particulate matter) and wet deposition (precipitation), are responsible for the reversible transfer of contaminants from air masses to water bodies.

In water, other phases are able to reversibly capture SOCs. As these chemicals are nonpolar, they can be taken up by particulate matter (both living and nonliving) and reach the sediment compartment (essentially the organic carbon fraction). Like soil in terrestrial systems, sediments constitute the main reservoir of HCB and PCBs in aquatic systems.

The main partition property regulating the adsorption of chemicals on suspended particles and sediments, as well as the bioconcentration of hydrophobic substances, is the 1-octanol/water partition coefficient, K_{ow}.[24,29]

The simple analysis of properties shown in Table 1 indicates that SOC-contaminated soils can constitute a source of SOC capable of reaching and contaminating distant land and water bodies by air transport. These processes are theoretically reversible; however, once bathyal zone sediments are contaminated, the return is probably more difficult. Another important aspect is the time needed for the recovery of contaminated water bodies: the sediment reservoir may play the role of a secondary contamination source.

Sources of Contamination

Agricultural run-off, rivers, and direct discharge of industrial and municipal wastes have been estimated to contribute a total organochlorine pesticide load of about 90 t/a.[47] No information is available for PCBs. However, as recently shown,[21] atmospheric transport of SOCs from land to

Table 2. Partition, Concentration (Calculated and Measured Values), and Total Mass of SOCs in the Mediterranean System at Equilibrium (20°C)[a]

	Partition (%)	Concentration Calculated	Concentration Measured
		PCBs	
Air	0.066	0.74 ng/m^3	0.54 ng/m^3
Water	9.224	0.1 ng/L	0.1–0.2 ng/L
Sediment	90.358	20 ng/g	16 ng/g
Fish	0.352	39 ng/g	68 ng/g
Total quantity of PCBs: 5000 t			
		HCB	
Air	2.33	0.26 ng/m^3	0.1–0.2 ng/m^3
Water	43.64	0.0048 ng/L	0.005 ng/L
Sediment	53.82	0.120 ng/g	0.002–1 ng/g
Fish	0.21	0.233 ng/g	0.3 ng/g
Total quantity of HCB: 50 t			

[a] Accessible volumes (m^3): air = 4.5×10^{15}; water = 4.5×10^{15}; sediment = 1.5×10^{11}; fish = 4.5×10^8.

ocean is the major route (78–99% of the total input). Riverine input of SOCs are unlikely to be the main route to marine environments.

The main land-based sources of SOCs are soils treated for agricultural practices or vector control operations, landfills, production sites, and all contaminated material and apparatus in *contact with flowing air,* located in Mediterranean and nonMediterranean countries. Thus for this group of contaminants, the Mediterranean Sea is not the classical semiclosed system characterized by very low water turnover (as it is for some trace metals), but an open system due to the carrying potential of air masses flowing over it and connecting it with world oceans and lands. This Mediterranean Sea is essentially the upper, mixed layer, down to 200 m depth, able to exchange contaminants with the air phase and to eliminate contaminants to deeper waters. The deep-Mediterranean, below 200 m, is less known. It is a dark and almost closed system, receiving large quantities of contaminants, by deposition of particulate matter (including fecal pellets from marine organisms), and unable to rapidly release them to the air and to destroy them by photochemical reactions. This second Mediterranean is probably the weakest component of the system. Direct and indirect input of stable chemicals into the deep-sea environment may trigger irreversible destructive processes.

An Evaluative Model for the Assessment of the Present Status of Contamination in the Off-Shore Mediterranean

A great effort was made under MED POL III and MED POL Phase II, leading to a picture of the contamination of several SOCs, including HCB and PCBs.[47] The available information in condensed form was presented in an assessment document.[5] Data on average levels in air, water, fish, and sediment are shown in Table 2.

The Mediterranean was assumed to be a homogeneous system, reduced to its main components: air, water, sediments, and fish, with accessible (to the contaminants) volumes as follows (Figure 1):

- *air*: 4.5×10^{15} m^3 (surface 3×10^{12} m^2, height 1.5×10^3 m)
- *water*: as for air 4.5×10^{15} m^3 (surface 3×10^{12} m^2, mean Mediterranean depth 1.5×10^3 m)
- *sediment*: Mediterranean surface (3×10^{12} m^2) for an accessible depth of 0.05 m gives 1.5×10^{11} m^3
- *fish*: taking a biomass concentration of 0.1 g/m^3 of water and the density of fish to be 1 g/cm^3, with a fish vol/vol concentration in water of 0.1 cm^3/m^3, the fish volume for the Mediterranean is 4.5×10^8 m^3

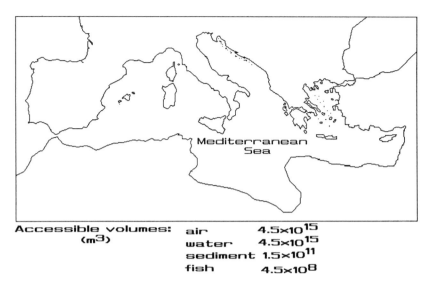

Figure 1. Mediterranean Sea. Main compartments and accessible volumes.

Air density is taken to be 1.19 kg/m³ and sediment density 1.5 kg/L. The organic carbon content of sediments is taken to be 4%.

To understand the present situation in the Mediterranean, let us consider a simple equilibrium model. Contamination sources are external and we assume that there are no significant concentration gradients in the system. Only air, water, sediments, and fish are considered and the system is assumed to be closed.

Once the system is defined, some basic properties of the chemicals in question are necessary to produce the simulation. These are molar mass, vapor pressure, water solubilty, and K_{ow}, as reported in Table 1.

Equilibrium fugacity models (Level I[28,31]) can now be used to simulate the distribution of HCB and PCBs in the Mediterranean. The load of each contaminant into the system was selected to obtain concentrations in the main environmental compartments approximating concentrations measured in the field (Table 2). In this way it is possible to obtain a reliable indication of the total quantity of each contaminant actually present in the system, as well as its distribution. The results are summarized in Table 2.

The data of Table 2 clearly indicate that the Mediterranean system is near to equilibrium. Thus any significant variation in concentration in a compartment will lead to a proportional variation in *all* the others.

Available estimates of organochlorine pesticide loads (no PCBs, and essentially only HCHs and DDTs) by direct input (industrial and municipal waste waters), run-off from treated fields and river transport, are of the order of 100 t/a.[44] The HCHs present in Mediterranean waters and sediments are about 3000 t, and DDT is of the order of 1000 t.[5] These data confirm the significance of air transport and indicate that if riverine input were to stop, the Mediterranean Sea would still be contaminated by long range air transport from Mediterranean and nonMediterranean areas, as in the case of open oceans and remote regions. The quantity of PCBs (5000 t) is in line with this interpretation: these chemicals are used in a far less dispersive manner than pesticides (lower run-off), nevertheless very large quantities reach the Mediterranean. This is not related to a substantially different environmental load, since the production and use of PCBs are of the same order of magnitude as that of DDTs and HCHs. The quantity of HCB, although small (50 t), produces significant contamination levels, particularly in fish, due to its high bioconcentration.

8   CONTAMINANTS IN THE ENVIRONMENT

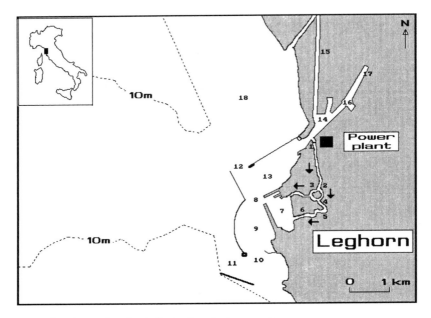

Figure 2. Location of sampling sites in the Leghorn harbor area.[10]

In this example, field data are in agreement with the picture derived from a simple fugacity-based equilibrium model. The model, together with a preliminary analysis of the properties of the chemicals, gives a general view of the behavior of these contaminants, in particular:

- the quantity of contaminants in the Mediterranean system;
- distribution among the main compartments;
- the main environmental reservoirs;
- the significance of vapor movements from contaminated inland sites and the importance of long range air transport.

The contamination of the Mediterranean Sea by SOCs reflects that of other seas and oceans and may be solved only by controlling the production of these chemicals on a global scale.

TRIBUTYLTIN (TBT) IN A HARBOR AREA

During the 1960s, triorganotin compounds, and particularly tributyltin (TBT), began to be used in marine antifouling treatments to protect cooling-water pipes of power stations and coastal industrial plants and in antifouling paints for boats, ships, and other marine structures. The rapid rise of the use of TBT was due to its very high biocidal activity against fouling organisms coupled with a relatively low toxicity to mammals. The high toxicity of TBT to nontarget species caused growing concern during the 1970s, particularly after the disasters of the French oyster industry, based on the Pacific oyster (*Crassostrea gigas*): oysters exposed to relatively low levels of TBT, released by marine antifouling paints, suffered reproductive failure, severe shell thickening, and malformations.[2]

Within the framework of the Mediterranean Action Plan (MAP)[43] a pilot survey was started in 1988 to evaluate the status of contamination by TBT in harbors and marinas in selected sites along the Mediterranean coast.[20] One of these sites was the Leghorn harbor area (Figure 2) where concentrations of TBT in harbor waters were of the order of 100–1000 ng/L, and ranged from

400 to 810 ng/L in sites representative of the main water mass. The TBT input into the system was of the order of 13.5 kg/day, with 10 kg/day originating from the antifouling treatment of a cooling pipe from a thermoelectric power plant and the remaining from the paint of pleasure boats and large ships.[10] Water turnover time in the *antiporto* area (Figure 2: sampling sites 8, 9, and 13) is of the order of 1 day. Field measurements indicated that the system was not far from a steady-state, concentrations of TBT in the water being constant within a factor of two in different period of the years 1988–1989.[11]

This chemical is rapidly degraded to inorganic tin by the progressive removal of the organic groups linked to the tin atom. The application of a thermodynamic partition model, as in the example of PCBs and HCB, is therefore inappropriate; besides, water turnover is large and should play a major role in TBT removal from the harbor. If degradation rates in sediments and water are known, a steady-state fugacity Level II model[30] with advection can be applied. The chemical is taken to be TBT chloride (TBT-Cl), the major species in marine environments.[26] The system is open; only water and sediment are considered, due to the negligible volatility of TBT-Cl from water. Water and sediment concentrations of TBT-Cl are assumed to be at equilibrium. A first-order degradation rate constant of 0.00413 h^{-1} (corresponding to a half-life of 7 days) is applied to the water compartment;[27] for sediment, a first-order degradation rate constant of 0.00018 h^{-1} (half-life of 160 days[40]) can be taken.

The physical properties of TBT-Cl are as follows:

Molar mass: 325.49 g/mol
Melting point = liquid
$K_{ow} = 5000$[26]
Vapor pressure, $P_L < 1 \times 10^{-6}$ Pa; taken to be 1×10^{-6} Pa
Solubility in water, $S_L = 3 \times 10^{-5}$ mol/m^3 [14]

Properties of the system:

Water volume, $V_W = 2.7 \times 10^7$ m^3 (surface area = 3×10^6 m^2, 9 m depth)
Water inflow, $I_W = 1.125 \times 10^6$ m^3/h
Water outflow, $O_W = 1.125 \times 10^6$ m^3/h
Sediment volume, $V_S = 1.5 \times 10^5$ m^3 (area as for water; depth 0.05 m)
Sediment density, $\rho = 1.5$ kg/liter
Organic carbon, mass fraction = 0.02

Input rate for TBT-Cl into the system, $E = 1.728$ mol/h (constant).

Reaction kinetics:

Degradation in water: first-order rate constant, $k_{degW} = 4.13 \times 10^{-3}$ h^{-1}
Degradation in sediment: first-order rate constant, $k_{degS} = 1.8 \times 10^{-4}$ h^{-1}

Advection:

Water advection rate constant, $k_A = I_W/V_W = 4.17 \times 10^{-2}$ h^{-1}

Calculating TBT-Cl input rate, I in mol/h:

$I = E = 1.728$ mol/h

Calculating D values:

For reaction in water, $D_{RW} = k_{degW} V_W Z_W = 3,345,942$ mol/(Pa-h)

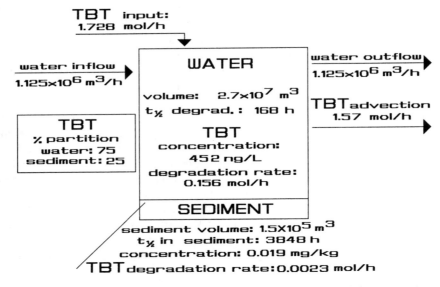

Figure 3. A Level II fugacity model for TBT in the Leghorn harbor area (steady-state and equilibrium).

For advection in water, $D_{AW} = kA\ V_W\ Z_W = 33{,}750{,}000$ mol/(Pa-h)
For reaction in sediment, $D_{RS} = k_{degS}\ V_S\ Z_S = 49{,}851$ mol/(Pa-h)

Calculating fugacity, f:

$f = I/\Sigma D_i = 4.65 \times 10^{-8}$ Pa

Calculating TBT-Cl concentrations and partition into the system:

Concentration in water, $C_W = Z_W f = 1.39 \times 10^{-6}$ mol/m³ = 452 ng/liter
Quantity of TBT-Cl in the water phase, $Q_{TBTW} = C_W\ V_W = 37.5$ mol (12.2 kg)
Concentration in sediment, $C_S = Z_S f = 8.58 \times 10^{-5}$ mol/m³ = 5.72×10^{-5} mol/t = 18.6 ng/g
Quantity of TBT-Cl in the sediment phase, $Q_{TBTS}\ C_S V_S = 12.9$ mol (4.2 kg)

The total quantity, Q_{TOT} of TBT-Cl at equilibrium will be about 50 mol, 25% in the sediment and 75% in the water.

Persistence, T, of TBT-Cl in this system appears to be very short:

$T = Q_{TOT}/I = 50/1.728 = 29$ h

The total reaction rate is subdivided as follows:

Water, $D_{RW} f = 0.156$ mol/h
Sediment, $D_{RS} f = 0.0023$ mol/h

Advection:

Water, $D_{AW} f = 1.57$ mol/h

The reaction will occur mainly in the water phase, which is the main reservoir of the chemical; advection contributes to more than 90% of the chemical output from the system (Figure 3).

The concentration of TBT in the water resulting from the model is similar to the field data. Of course this does not necessarily mean that the model perfectly reflects a real situation. However, main trends in partition and the relative weight of reaction and advection probably are not far from reality. Furthermore, the model may be essential in indicating research needs such as the measurement of:

- the water/sediment partition coefficient
- degradation rates in water and sediment
- contamination levels outside the harbor due to the advection of TBT

One of the critical points of the model may seem to be the high water turnover selected (turnover time = 1 day). It could be said that the significance of TBT advection by water is only due to this high turnover. Analyzing the sensitivity of the model to this parameter (water inflow) it can easily be seen that if water turnover is reduced by one order of magnitude (i.e., from 1.125×10^6 to 0.1125×10^6 m³/h, corresponding to a turnover time of 10 days), the persistence of TBT in the system will increase to 160 h, concentrations in water and sediments will rise by a factor of 5, and disappearance from the system will be negligible via degradation in sediments (0.7% of the total), reaching 49.4 and 49.9% of the total with, respectively, degradation in water and advection with water renewal. Even under these conditions, water advection is still the main way out for TBT.

VAPOR DRIFT AND BIOCONCENTRATION IN PLANT FOLIAGE

A relatively new problem arising from the application of pesticides in agriculture is that of *inadvertent residues in crops where they are not allowed*. These may be produced by inadvertent spraying of cultures by aerosol drift, and by vapor drift from treated fields.

In the second case, two processes are involved:

- volatilization of active ingredients
- vapor bioconcentration in plant foliage.

Evaluative models may be applied to rank the volatilization potential of different pesticides and potential for bioconcentration in plant tissues.

Volatilization Potential

This can be evaluated by means of the simple approach[22] for inert surfaces:

$$J = AP(M)^{1/2}$$

where A is a proportionality constant, J the vapor flux, P vapor pressure, and M the molar mass of evaporating substance.

In semiclosed systems, with constant temperature and air turnover (e.g., a greenhouse), this relationship can be written as follows:

$$C_A = BJ$$

and

$$C_A = const\, P(M)^{1/2}$$

Table 3. Examples of Calculation of the 0.5% Saturation Concentration in Air, $C_{A0.5\% sat}$, of Some Pesticides (25°C)

Active ingredient	Vapor pressure, P (Pa)	$C_{A0.5\%sat}$ (25°C) (mol/m³)
Dichlorvos	7000 × 10⁻³ᵃ	1.41 × 10⁻⁵
Butylate	100 × 10⁻³ᵃ	2.02 × 10⁻⁷
Aldicarb	13 × 10⁻³ᵇ	2.62 × 10⁻⁸
Alachlor	2.9 × 10⁻³ᵇ	5.85 × 10⁻⁹
Chlorthal-dimethyl (DCPA)	0.33 × 10⁻³ᶜ	6.66 × 10⁻¹⁰
Cymoxanil	0.08 × 10⁻³ᵇ	1.61 × 10⁻¹⁰
Terbutryn	0.0004 × 10⁻³ᵃ	8.07 × 10⁻¹³
Cypermethrin	0.0002 × 10⁻³ᵇ	4.03 × 10⁻¹³

ᵃSuntio et al.[41]
ᵇWorthing and Hance.[50]
ᶜDePablo.[18]

where *const* is a new proportionality constant (e.g., the greenhouse constant). When C_A is in mol/m³, P is in Pa and M is the molar mass, in g/mol. In small glass greenhouses (vol = 0.2 m³), with an air turnover of 150 L/h, and sand contaminated with about 1 mmol of selected chemicals and kept moist, Bacci et al.[9] found the following value for the constant: 10^{-7} mol/(m³ Pa) at 20–25°C.

For pesticides the molar mass is in the range 100–400 g/mol. Consequently, the concentration in air, C_A, corresponds to a first approximation to *0.5% of the chemical saturation concentration.*

Levels over treated fields may vary greatly due to uncontrolled temperature, wind speed and turbulence, water content in surface soil, and vegetation density. However, with moist soils and low wind at 1–2 m over the ground, 0.2–2% saturation can be expected. For instance, after an application of 1.5 kg/ha of *nitrapyrin,* Majewski et al.[33] found an average concentration of 0.4 mg/m³ or *1.5% saturation,* at a height of 1.5 m.

With different chemicals in the same environment, the relative concentrations in the air would probably be a fraction of the respective saturation values. This implies that molar mass differences are not of primary importance: in the case of pesticides, as stated before, they range from 100 to 400 and the relative square roots lie within a factor of 2, while vapor pressure is spread over several orders of magnitude.

To calculate the saturation concentration in air, or the *solubility in air,* the gas law may be applied:

$$PV = nRT$$

where P is the pressure in Pa, V the volume in m³, n the number of moles (mol), R the gas constant = 8.314 Pa·m³/(mol K), and T the absolute temperature, K.

From the gas law:

$$n/V = P/RT$$

If $V = 1$ m³ and P is the *vapor pressure, P/RT is the number of moles, n, corresponding to air saturation by the vapor of the chemical.*

Examples of calculations of the 0.5% saturation with different pesticides are shown in Table 3.

Concentrations such as those in Table 3 represent a *simple evaluative* model for ranking the potential to generate *vapor drift* from contaminated sites or agricultural fields. In the example, for solids, the vapor pressure of the solid state (at 25°C) was used. This seems appropriate when the formation or presence of crystals (i.e., the solid state) is likely, as in the case of fields just treated with pesticides.

More in general, with finely dispersed substances the supercooled vapor pressure values are more suitable. These can be calculated as previously indicated. For solids, the supercooled liquid vapor pressure, P_L, is generally higher than P_S. From P_S, the value of P_L for the same compound can be calculated by means of the *fugacity ratio, F,* which is $F = P_S/P_L$[41]:

$$F = P_S/P_L = \exp\{-\Delta S_f [(mp/T) - 1]/R\}$$

where R is the gas constant, 8.314 Pa-m³/(mol K), T the absolute temperature in K, mp the melting point, and ΔS_f the *entropy of fusion,* in J/(mol K). The entropy of fusion is the ratio $\Delta H_f/mp$, where ΔH_f is the heat of fusion.

In the absence of an experimental or calculated value of ΔS_f, a standard value of 56 J/(mol K) can be adopted, to a first approximation.[41] Average ΔS_f values of 56.5 J/mol K have been reported[51] for rigid aromatic molecules. From the previous equation, F values at 20°C can be obtained as follows:

$$F = \exp[-0.023\,(mp - 293)]$$

The vapor pressure of 10^{-6} Pa seems to be a cutoff value for volatilization from soils: for P_L $<10^{-6}$ Pa, vapor movements from contaminated soils probably become negligible. The temperature dependence of vapor pressure means that volatilization from soil, on the one hand, and adsorption onto solid particles and inclusion in water droplets, on the other, are related to environmental temperature. Cold climates will thus favor deposition, and warm climates will enhance volatilization.

Another remark concerns the possibility that an apparently nonvolatile chemical, if solid and with a high melting point, may become a substance that is relatively mobile in air when finely dispersed (supercooled liquid). In the case of refractory compounds, this soil-to-air transfer may be a key process in generating *global contamination.*[15,25]

Vapor drift concerns the displacement of a chemical from one site to another by volatilization and transport via air. The air, in the field, is the most mobile environmental phase and *though concentrations attained are low, the fluxes may be important,* due to high turbulence and advection. In general, air is the typical environmental compartment to which the famous statement *dilution is solution to pollution* may be applied. However, the presence of low concentrations may hide significant mass transfer from one phase to another, particularly when reconcentration phenomena occur.

Vapor movements from treated fields may generate air concentrations of the order of 0.5% of saturation values and plant tissues (mainly "green" tissues, i.e., leaves) are able to take up pesticide vapors as a function of their leaf/air partition coefficient (see below). Considering that present regulations in force in several countries do not allow commercialization of crops containing detectable amounts of pesticides not registered for use with that crop (with only a few exceptions for known soil carryover effects in rotational crops and for global contaminants, such as certain organochlorinated pesticides), the presence of a pesticide in a nontarget species may lead to economic damage to farmers. This possibility is not so terribly remote: as reported,[37] the California Department of Food and Agriculture found detectable residues of the herbicide 2,3,5,6-tetracloroterephthalate, DCPA, on a variety of produce samples from Monterey County and from the central valley in San Joaquin County to which it had not been applied. The herbicide was found in more than 10% of the samples analyzed of daikon (*Raphanus sativus* L.), dill (*Anethum graveolens* L.), kohlrabi (*Brassica oleracea* L.), and parsley (*Petroselinum crispum* Mill.), at levels of the order of 10–100 ng/g wet weight. To find an explanation for these facts, Ross et al.[37] carried out the following exeperiment: parsley plants were grown in pots around a circular field (1 ha) planted with onions (*Allium cepa* L.) and treated with DCPA at a rate of 7.08 kg/ha (DCPA is a herbicide currently applied to onion crops). Residues in the onions were well below

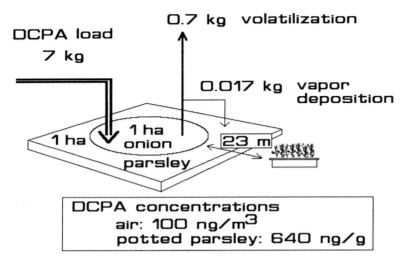

Figure 4. Mass balance 21 days after application of DCPA. Concentrations in air 30 m downwind; potted parsley plants 23 m downwind, time of exposure: 10 days. Data from Ross et al.[37]

the tolerance level (1 mg/kg)[16] and parsley planted back in the same soil, after the onion harvest, did not contain detectable levels of DCPA (detection limit = 20 ng/g) when sampled 217 and 336 days after the last application. However, the plants in pots installed around the onion field after the treatment (to avoid particulate drift) contained residue concentrations ranging from 58 to 640 ng/g 23 m from the edge of the onion field, 10 days after the DCPA application.

Wind speed ranged from 1 to 6.9 m/s and temperatures were as follows: average maximum 29°C and average minimum 9°C; average relative humidity ranged from 85 and 21%. Air samples, collected 30 m downwind from the edge of the circular onion field, at a height of 1.5 m, showed DCPA vapor concentrations during the first 21 days after the treatment, ranging from 22 to 910 ng/m³ with no relation with the time elapsed from treatment and with a median value of 100 ng/m³. The field was regularly irrigated to enhance DCPA volatilizaton.[39] Mass balance 21 days after the application indicated that about 10% of DCPA was dissipated by volatilization (Figure 4).

The 0.5% saturation concentration of DCPA vapors in the air (20°C) is 220 ng/m³. At a distance of 30 m from the treated field a dilution is likely but, due to the approximation of the proposed approach, it is not taken into consideration. The concentration of 220 ng/m³ is not far from the measured values. Levels in parsley exposed to a DCPA concentration in air of 220 ng/m³ can be calculated by the leaf/air bioconcentration factor, K_{LA}[7,8]:

$$K_{LA} = 0.022\ K_{OW}/K_{AW}$$

where K_{OW} and K_{AW} are the 1-octanol/water and the air/water partition coefficients, respectively (K_{AW} is the ratio of the solubilities in air and water, in mass/volume units). In Figure 5 field data are illustrated and compared with calculations based on a combination of the 0.5% approach and K_{LA}.

The data of Figure 5 are in good agreement with field measurements. The concentration of DCPA in parsley is very near the measured value. The approximations intrinsic to the applied approach, however, indicate that the good correspondence found in this case cannot be taken as a rule. These findings therefore represent a call for research, rather than a well-established approach to be applied in the field. A combination of the *0.5% approach* and the leaf/air bioconcentration factor, K_{LA}, however may be tentatively applied to evaluate the potential of pesticides

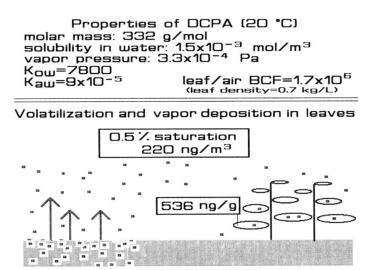

Figure 5. Evaluative model for off-target deposition of DCPA vapors on parsley.

to volatilize from soils and concentrates in plant foliage. This rough and simple estimation method could be used to rank the potential of different substances to generate *inadvertent residues* in crops by vapor drift.

CONCLUSIONS

The interpretative and predictive potentials of evaluative models have greatly improved in recent years. Designed to produce rankings, rather than data to be directly transferred to the field, they may approach real situations if correctly calibrated. The integration of these models with well-addressed field measurements and selected laboratory experiments seems, at present, to be a fruitful method of studying the environmental fate of trace contaminants.

REFERENCES

1. **Abbott, D. C., Harrison, R. B., Tatton, J. O'G., Thomson, J.** Organochlorine pesticides in the atmospheric environment, *Nature (London)*, 208, 1317–1318, 1965.
2. **Alzieu, C.,** TBT detrimental effects on oyster culture in France, in *Oceans '86 Proceedings, Vol. 4, International Organotin Symposium,* The Institute of Electrical and Electronics Engineers, New York, 1986, 1130–1134.
3. Anon., Report of a new chemical hazard, *New Scientist* December, 612, 1966.
4. **Bacci, E.,** *Ecotoxicology of Organic Contaminants,* Lewis Publishers, Chelsea, MI, 1994.
5. **Bacci, E.,** *Semivolatile Organochlorinated Hydrocarbons in the Mediterranean Sea: Sources, Fate, Hazard Assessment and Proposed Control Measures,* UNEP/MAP Agreement 042/90, Athens, 1991.
6. **Bacci, E., and Calamari, D.,** *Learning from field work and modelling,* Proceedings of the Workshop on Chemical Exposure Prediction, Trois Epis (Colmar, France), 11–13 June 1990. European Science Foundation, Strasbourg, 1990, 198–208.
7. **Bacci E., Calamari, D., Gaggi, C., and Vighi, M.,** Bioconcentration of organic chemical vapors in plant leaves: Experimental measurements and correlation, *Environ. Sci. Technol.,* 24, 885–889, 1990.
8. **Bacci, E., Cerejeira, M. J., Gaggi, C., Chemello, G., Calamari, D., and Vighi, M.,** Bioconcentration of organic chemical vapors in plant leaves: The azalea model, *Chemosphere,* 21, 525–535, 1990.

9. **Bacci, E., Cerejeira, M. J., Gaggi, C., Chemello, G., Calamari, D., and Vighi, M.,** Chlorinated dioxins: Volatilization from soils and bioconcentration in plant leaves, *Bull. Environ. Contam. Toxicol.,* 48, 401–408, 1992.
10. **Bacci, E., and Gaggi, C.,** Organotin compounds in harbour and marina waters from the northern Tyrrhenian Sea, *Mar. Pollut. Bull.,* 20, 290–292, 1989.
11. **Bacci, E., and Gaggi, C.,** Tributyltin in the Leghorn harbour area (Tuscany, Italy): A tentative hazard assessment to aquatic life, Proceedings of the 3rd International Organotin Symposium, Monaco, 17–20 1990, 140–146.
12. **Baughman, G. L., and Lassiter, R. R.,** Prediction of environmental pollutant concentration, in *Estimating the Hazard of Chemical Substances to Aquatic Life,* Cairns, J., Jr., Dickson, K. L., and Maki, A. W., Eds., American Society for Testing and Materials, Tech. Publ. 657, Philadelphia, PA, 1978, 35–54.
13. **Bidleman, T. F., Billings, W. N., and Foreman, W. T.,** Vapor-particle partitioning of semivolatile organic compounds: Estimates from field collections, *Environ. Sci. Technol.,* 20, 1038–1043, 1986.
14. **Blunden, S. J., Hobbs, L. A., and Smith, P. J.,** The environmnental chemistry of organotin compounds, in *Environmental Chemistry,* Bowen, H. J. M., Ed., The Royal Society of Chemistry, London, 1984, 49–77.
15. **Calamari, D., Bacci, E., Focardi, S., Gaggi, C., Morosini, M., and Vighi, M.,** The role of plant biomass in the global partition of chlorinated hydrocarbons, *Environ. Sci. Technol.,* 25, 1489–1495, 1991.
16. Code of Federal Regulations, Tolerances and exemptions from tolerance for pesticide chemicals in or on raw commodities, in *Protection of Environment,* 40CFR, Chapter 1, Part 180. The Office of the Federal Register, National Archives and Records Administration, Washington, D.C., 1988, 357.
17. **Cole, H., Barry, D., Frear, D. E. H., and Bradford, A.,** DDT levels in fish, streams, stream sediments, and soil before and after aerial spray application for fall cankerworm in northern Pennsylvania, *Bull. Environ. Contam. Toxicol.,* 2, 127–146, 1967.
18. **De Pablo, R. S.,** Vapor pressure of dimethyl tetrachloroterephthalate, *J. Chem. Eng. Data,* 26, 237–238, 1981.
19. **Foreman, W. T., and Bidleman, T. F.,** An experimental system for investigating vapor-particle partitioning of trace organic pollutants, *Environ. Sci. Technol.,* 21, 869–875, 1987.
20. **Gabrielides, G. P., Alzieu, C., Readman, J. W., Bacci, E., Aboul Dahab, O., Salihoglu, I.,** MED POL survey of organotins in the Mediterranean, *Mar. Pollut. Bull.,* 21, 233–237, 1900.
21. GESAMP, Atmospheric input of trace species to the world ocean, GESAMP Working Group No. 14, XIX/4, March 1989, Athens (WMO, 1989).
22. **Hartley, G. S.,** Evaporation of pesticides, *Adv. Chem. Ser.,* 86, 115–134, 1969.
23. **Heinisch, E.,** Input of pesticide agents from nonagricultural sources into environment, *Ekològia (CSSR),* 4, 97–109, 1985.
24. **Karickhoff, S. W., Brown, D. S., and Scott, T. A.,** Sorption of hydrophobic pollutants on natural sediments, *Water Res.,* 13, 241–248, 1979.
25. **Kurtz, D. A., Ed.,** *Long Range Transport of Pesticides,* Lewis Publishers, Chelsea, MI, 1990.
26. **Laughlin, R. B., Jr., Guard, H. E., and Coleman, W. M., III,** Tributyltin in seawater: Speciation and octanol-water partition coefficient, *Environ. Sci. Technol.,* 20, 201–204, 1986.
27. **Lee, R. F., Valkirs, A. O., and Selingman, P. F.,** Fate of tributyltin in estuarine waters, in *Oceans '87 Proceedings, Vol. 4, International Organotin Symposium,* The Institute of Electrical and Electronics Engineers, New York, 1987, 1411–1415.
28. **Mackay, D.,** Finding fugacity feasible, *Environ. Sci. Technol.,* 13, 1218–1223, 1979.
29. **Mackay, D.,** Correlation of bioconcentration factors, *Environ. Sci. Technol.,* 16, 274–278, 1982.
30. **Mackay, D.,** *Multimedia Environmental Models. The Fugacity Approach,* Lewis Publishers, Chelsea, MI, 1991.
31. **Mackay, D., and Paterson, S.,** Fugacity revisited, *Environ. Sci. Technol.,* 16, 654–660, 1982.
32. **Mackay, D., Paterson, S., and Schroeder, W. H.,** Model describing the rates of transfer processes of organic chemicals between the atmosphere and water, *Environ. Sci. Technol.,* 20, 810–816, 1986.
33. **Majewski, M. S., Glotfelty, D. E., Pau, K. T., and Seiber, J. N.,** A field comparison of several methods for measuring pesticide evaporation rates from soil, *J. Environ. Qual.,* 24, 1490–1497, 1990.

34. **Quinby, G. E., Hayes, W. J., Armstrong, J. F., and Durham, W. F.,** DDT storage in U.S. population, *J. Am. Med. Assoc.,* 191, 175–179, 1965.
35. **Risebrough, R. W.,** Beyond long-range transport: A model of a global gas chromatographic system, in *Long Range Transport of Pesticides,* Kurtz, D. A., Ed., Lewis Publishers, Chelsea, MI, 1990, 417–426.
36. **Risebrough, R. W., Huggett, R. J., Griffin, J. J., and Goldberg, E. D.,** Pesticides: Transatlantic movements in the northeast trades, *Science,* 159, 1233–1236, 1968.
37. **Ross, L. J., Nicosia, S., McChesney, M. M., Hefner, K. L., Gonzalez, D. A., and Seiber, J. N.,** Volatilizaton, off-site deposition, and dissipation of DCPA in the field, *J. Environ. Qual.,* 19, 715–722, 1990.
38. **Sladen, W. J. L., Menzie, C. M., and Reichel, W. L.,** DDT residues in Adelie penguins and crabeater seal from Antarctica: Ecological implications, *Nature (London),* 210, 670–673, 1966.
39. **Spencer, W. F., and Cliath, M. M.,** Pesticide volatilization as related to water loss from soil, *J. Environ. Qual.,* 2, 284–289, 1973.
40. **Stang, P. M., and Selingman, P. F.,** Distribution and fate of butyltin compounds in the sediment of San Diego Bay, in *Oceans '86 Proceedings, Vol. 4, International Organotin Symposium,* The Institute of Electrical and Electronics Engineers, New York, 1986, 1256–1261.
41. **Suntio, L. R., Shiu, W. Y., Mackay, D., Seiber, J. N., and Glotfelty, D.,** Critical review of Henry's law constants for pesticides, *Rev. Environ. Contam. Toxicol.,* 103, 1–59, 1988.
42. **Thomas, R. G.,** Volatilization from water, in *Handbook of Chemical Property Estimation Methods,* Lyman, W. J., Reehl, W. F., and Rosenblatt, D. H., Eds., American Chemical Society, Washington, D.C., 1990, chap. 15, 1–34.
43. **UNEP,** Convention for the protection of the Mediterranean Sea against pollution and its related protocols, United Nations, New York, 1982.
44. **UNEP,** Pollutants from land-based sources in the Mediterranean, UNEP Regional Seas Reports and Studies No. 32, 1984.
45. **UNEP,** Co-ordinated Mediterranean pollution monitoring and research programme (MED POL-Phase I). Final Report. 1975–1980, *MAP Technical Reports Series,* No. 9, UNEP, Athens, 1986, 276.
46. **UNEP/FAO,** Baseline studies and monitoring of DDT, PCBs and other chlorinated hydrocarbons in marine organisms (MED POL III), *MAP Technical Reports Series* No. 3, UNEP, Athens, 1986, 128.
47. **UNEP/FAO,** Assessment of the state of pollution of the Mediterranean Sea by organohalogen compounds, *MAP Technical Reports Series* No. 39, UNEP, Athens, 1990, 224.
48. **WHO,** DDT and its derivatives—Environmental aspects, *Environmental Health Criteria 83,* World Health Organization, Geneva, 1989.
49. **Woodwell, G. M., Craig, P. P., and Johnson, H. A.,** DDT in the biosphere: Where does it go?, *Science,* 174, 1101–1107, 1971.
50. **Worthing, C. R., Hance, R. J.,** *The Pesticide Manual,* 9th ed., The British Crop Protection Council, Farnham, UK, 1991.
51. **Yalkowsky, S. H.,** Estimation of entropies of fusion of organic compounds, *Ind. Eng.Chem. Fund.,* 18, 108–111, 1979.

CHAPTER 2

Fate of Persistent Organochlorines in the Marine Environment

Shinsuke Tanabe

INTRODUCTION

The present study concerns the fate of persistent organochlorines such as PCBs (polychlorinated biphenyls) in the marine environment, DDTs, HCHs (hexachlorocyclohexanes), and CHLs (chlordanes) in the marine environment, based on the data of their atmospheric and hydrospheric residue levels in the open ocean worldwide as well as in estuarine and coastal regions of Asia and Oceania.

According to the data on the distribution of these toxic contaminants in air and surface water from the western Pacific and the Indian Oceans during 1975 to 1981, higher concentrations were observed in the mid-latitudes of the northern hemisphere.[1] However, recent studies pointed out that there has been a temporal variation in organochlorine distribution in open ocean atmosphere and hydrosphere.[2] In this regard, the accelerating contamination by persistent organochlorines in the Arctic region was of great concern.[3] As a matter of fact, no significant point-source of contamination is present in the polar region and, therefore, this finding causes concern globally, particularly regarding long-range transport from the contaminant source of tropical areas. In fact, some tropical countries are still using persistent organochlorines in agriculture and public health, although many developed nations prohibited their production and usage more than a decade ago.[4-6] Taking into account the geographic shift of organochlorine usage, the present study also attempts to describe the global transport of these toxic chemicals from new point-source areas of contamination in the tropics.

BEHAVIOR AND FATE IN THE COASTAL REGION

According to the monitoring data on the distribution of persistent organochlorines in riverine and estuarine waters from urban areas of Asia and Oceania, apparently higher levels were observed in the tropical regions than in the temperate and colder ones (Figure 1). Such a prominent contamination in the tropics was also found in air samples from the same regions. Besides air and

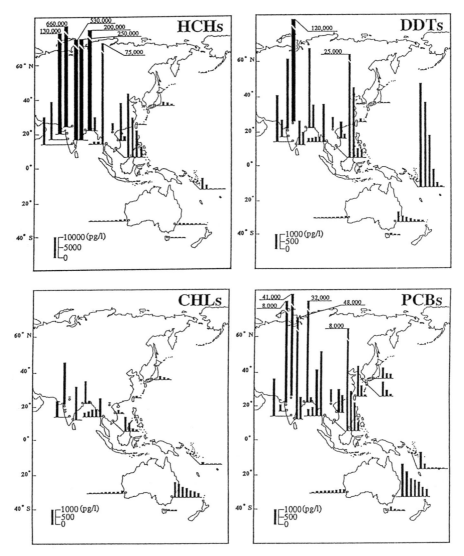

Figure 1. Distribution of persistent organochlorine residues in riverine and estuarine waters from urban areas of Asia and Oceania (1989–1990).[7]

water, biological samples like human breast milk[8] and birds[9] revealed a similar pattern of contamination, with tropical regions showing high levels. These results strongly support the suggestion that these are major emission source in tropical countries contributing to current global contamination by persistent organochlorines.

In this context, it should be remembered that most of the developing countries are located in the tropical belt where high temperature and heavy rainfall are rather common. The tropical agroecosystem characterized by the above climatic features may facilitate the rapid dissipation of toxic contaminants through air and water. In fact, a typical case was observed in Nigeria where approximately 98% of the DDT applied to the cowpea crop was volatilized during a period of 4 years.[10] The present investigations on the fate of the insecticide HCH in the tropical paddy ecosystems in South India provide more evidence of this conclusion. In this study, concentrations of HCH isomers in air, water, paddy soil, and sediments from South India (1987–1989) were compared with those from Japan (during their maximum usage period in the late 1960s and early

Table 1. Comparison of Residue Levels of HCHs in Air, Water, Paddy Soil, and Sediments from India and Japan

Country	Technical HCH applied per unit area (kg/ha)	Concentrations			
		Air (ng/m^3)	Water (ng/L)	Paddy Soil (ng/g dry wt.)	Sediment (ng/g dry wt.)
India	2.5	3.3–130	14–2,000	1.1–1,000	0.9–27
Japan	3.1	10–100	100–1,000	10–18,000	0.4–100

1970s) in order to clarify the specific behavior of toxic chemicals in tropical and temperate climates (Table 1). As a result, HCH residue levels in air and water were comparable or higher in India than in Japan.[11,12] On the other hand, soil and sediment samples had much lower levels in India than in Japan, although the HCH consumption per unit area is almost the same in both these countries.[13] Moreover, in the case of international comparisons of HCH residues in breast milk[8] and birds,[9] Indian samples revealed the highest levels in the world, reflecting the usage of this insecticide in large quantities in India at present. On the other hand, similar comparison in green mussels[14] and fish[9] showed that HCH residue levels in India were comparable to those in other countries.

These observations prompted us to further investigate whether the tropical agroecosystem characterized by high temperature differed in regard to pollutant dispersal, compared to that of temperate areas. Accordingly, an HCH application test was conducted in an experimental paddy field in the Vellar River watershed.[15] Interestingly, most of the HCH applied to the field was found to volatilize rapidly with only low residue levels being recorded in water, paddy plant, and soil. This fact suggested that applied insecticides were transported to the atmosphere to a large extent. More evidence concerning the environmental fate of insecticide residues in the tropics was provided by modeling their mass transfer in coastal environment. An attempt was made to assess the role of estuaries in the Vellar River watershed regarding behavior and fate of insecticide residues. Using hydrokinetic parameters of this river and the data on residue levels in air and water, the HCH flux in estuarine and coastal areas was calculated.[16] Consequently, it was estimated that about 99.6% of the applied HCH (41,830 kg, out of an approximate quantity of 42,000 kg used per year in this area) in the paddy areas of this watershed escaped to the air and only 0.4% drained to the estuary (Figure 2). Moreover, about 75% of the flux to the estuary was removed to the air. Thus only 0.1% of the applied HCH was ultimately drained to the sea. This means that HCH flux to the coastal water bodies through water is less significant and its residence time in the coastal aquatic environment is rather short whereas transfer to the atmosphere is quite large in the tropical areas.

TRANSPORT AND FATE IN THE OPEN OCEAN

Having followed the behavior and fate of persistent organochlorines in a representative tropical environment, the present study was extended to make clear where and how the contaminants are transported around the globe. The monitoring survey was conducted in various seas and oceans worldwide during the period of 1989 to 1990. As a result, in the case of HCHs, extensive contamination was still found in the northern hemisphere with high residue levels in tropical regions and colder waters near the Arctic (Figure 3). In contrast, DDT showed a different pattern of distribution, with high concentrations recorded only in the seas around tropical Asia. An interesting pattern was also observed for PCBs (Figure 4) and CHLs, which showed rather a uniform distribution.

In order to explain such patterns of global distribution, an attempt was made to estimate the mass transfer of these contaminants between air and water. Consequently, HCHs fluxes revealed

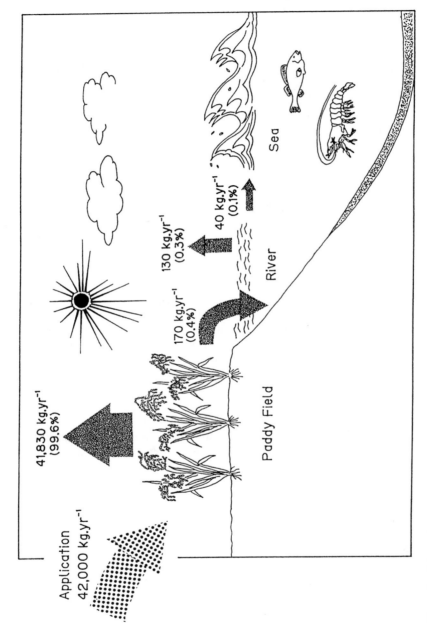

Figure 2. Schematic representation of the flux of HCHs Vellar River watershed, South India.[16]

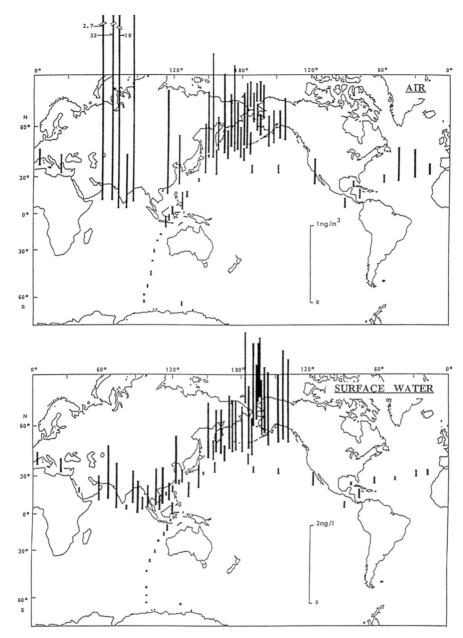

Figure 3. Distribution of HCH residues in air and surface seawater in open ocean (1989–1990).[17]

greater negative values in the East and South China Seas and the Bay of Bengal, which mean their active transfer from air to surface seawater (Figure 5). These seas play a role as a sink of HCHs due to their quite close proximity to the extensive HCH usage areas like India and China. Interestingly, colder waters near Arctic regions also showed negative values, indicating significant sinks for HCHs. Similar to HCHs, DDT mass transfer also showed greater negative values in tropical waters, which are also a sink for DDTs. In the case of PCBs (Figure 6) and CHLs, colder waters had greater negative values of flux, showing a sink for these contaminants as well.

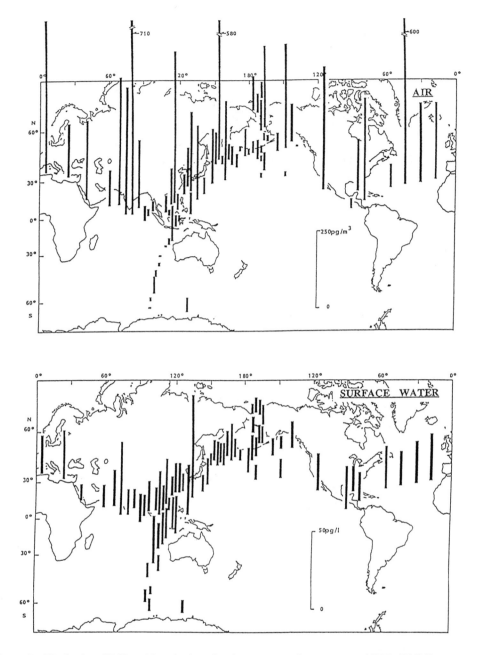

Figure 4. Distribution of PCB residues in air and surface seawater in open ocean (1989–1990).[17]

On the whole, these findings lend credence to the assumption that the large quantities of persistent organochlorines used in the tropics are released into the atmosphere and they disperse through the "long-range atmospheric transport" globally. In this context, HCHs and DDTs transported through the atmosphere are absorbed and deposited in the water bodies of seas and oceans nearby point-source contamination of land. Whereas, PCBs as well as CHLs are likely to be more transportable over the oceans than HCHs and DDTs. Regarding the geographical viewpoint, the

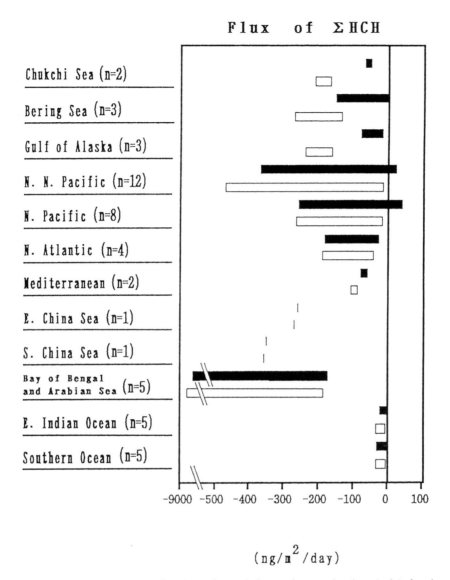

Figure 5. Fluxes by gas exchange of HCHs (sum of α- and γ-isomers) across the air–water interface in various seas and oceans.[17] Figures in parentheses are the number of pairing for air and surface water samples. Length of a bar shows the range of flux. Solid and open bars represent the flux for cases 1 (dissolved fraction) and 2 (particulate fraction), respectively.

Arctic seas and nearby oceans serve as one of the significant sinks for toxic and bioaccumulative contaminants.

ECOTOXICOLOGICAL IMPLICATION

Low residue levels and shorter residence time of toxic contaminants in the tropical coastal water bodies might be favorable from the viewpoint of environmental quality and human health. However, this may have wider implications for the global environment as a whole. Unfortunately,

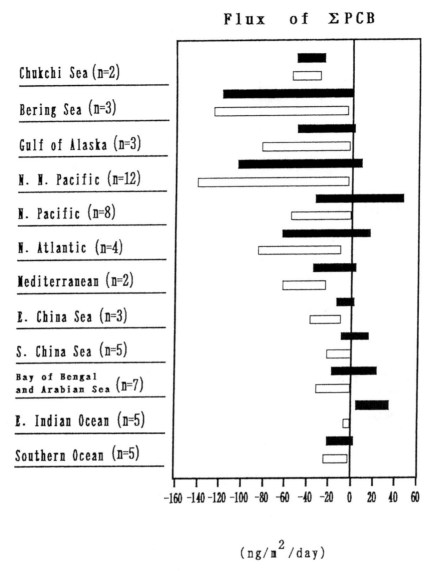

Figure 6. Fluxes by gas exchange of PCBs across the air–water interface in various seas and oceans.[17] Descriptions same as in Figure 5.

volatilized residues from the tropics disperse through the atmosphere globally and ultimately deposit into the sink-like open ocean environment including North Pole waters.

Such a sink should also be considered in an ecotoxicological perspective because organochlorine contaminants are extremely bioaccumulative and impose a toxic threat to resident and migratory marine organisms. In particular, marine mammals such as whales, dolphins, and seals that are at the top of foodchain are most likely to be vulnerable to their toxic effects since these animals are believed to be genetically sensitive to reproductive impairment by contaminants as exemplified in the case of PCBs.[18,19]

There is quite serious concern that the increasing usage and disposal of toxic chemicals in the tropical regions have a greater impact on the environment. However, this is not an intrinsic problem except in low latitude areas. The developed nations that are exporting many chemical products and providing funds and technology to the tropical countries should also share the responsibility for preventing global contamination from a tropical point source and thereby achieve the safety measure of chemical use in developing countries.

SUMMARY

Organochlorine residues in the estuarine and coastal air and water of Asia and Oceania showed apparent higher levels in tropical regions than in temperate and colder ones. Interestingly, such a trend was less pronounced in sediment and fish samples. These results can be explained by the fact that the usage of organochlorine insecticides is still continuing in developing countries of the tropics in spite of their restriction in developed nations. Moreover, the tropical agroecosystem characterized by high temperature indicates the rapid volatilization of applied insecticides to the atmosphere in large quantities and hence shorter residence time in the coastal aquatic environment.

At the same time, considerable residue levels of insecticides like HCHs were recorded in the waters off the northern North Pacific and the nearby Arctic regions as well as tropical waters. The mass transfer of some organochlorines by gas exchange between air and water revealed that they have a tendency to transfer from air to water, implying that polar waters and adjacent seas and oceans may play a role as a sink for such contaminants.

As a whole, it may be proposed that the tropical areas act as the major contamination source of some persistent organochlorines and the volatilized residues disperse through the atmosphere globally and ultimately deposit into the sink like Arctic waters.

REFERENCES

1. **Tatsukawa, R., Yamaguchi, Y., Kawano, M., Kannan, N., and Tanabe, S.,** Global transport of organochlorine insecticides—an 11-year case study (1975–1985) of HCHs and DDTs in the open ocean atmosphere and hydrosphere, in *Long Range Transport of Pesticides,* Kurtz, D. A., Ed., Lewis Publishers, Chelsea, MI, 1990, 127.
2. **Iwata, H., Tanabe, S., Sakai, N., and Tatsukawa, R.,** Distribution, behavior and fate of persistent organochlorines in air and water of western Pacific and Indian Oceans, *12th Annual Meeting of the Society of Environmental Toxicology and Chemistry,* Seattle, Washington, 1991, Abstract 2.
3. **Tanabe, S.,** Fate of toxic chemicals in the tropics, *Mar. Pollut. Bull.,* 22, 259, 1991.
4. **Mowbray, D. L.,** Pesticide control in the South Pacific, *Ambio,* 15, 22, 1986.
5. **Crick, H.,** Poisoned prey in the heart of Africa, *New Scientist,* 4, 29, 1990.
6. **Forget, G.,** Pesticides and the third world, *J. Toxicol. Environ. Health,* 32, 11, 1991.
7. **Iwata, H., Tanabe, S., Sakai, N., Nishimura, A., and Tatsukawa, R.,** Geographical distributions of persistent organochlorines in air, water and sediments from Asia and Oceania, and implications for their global redistribution from lower latitudes, *Environ. Pollut.,* 85, 15, 1994.
8. **Tanabe, S., Gondaira, F., Subramanian, A. N., Ramesh, A., Mohan, D., Kumaran, P., Venugopalan, V. K., and Tatsukawa, R.,** Specific pattern of persistent organochlorine residues in human breast milk from South India, *J. Agric. Food Chem.,* 38, 899, 1990.
9. **Ramesh, A., Tanabe, S., Kannan, K., Subramanian, A. N., Kumaran, P. L., and Tatsukawa, R.,** Characteristic trend of persistent organochlorine contamination in wildlife from a tropical agricultural watershed, South India, *Arch. Environ. Contam. Toxicol.,* 23, 26, 1992.
10. **Perfect, J.,** The environmental impact of DDT in a tropical agroecosystem. *Ambio,* 9, 16, 1980.
11. **Ramesh, A., Tanabe, S., Tatsukawa, R., Subramanian, A. N., Palanichamy, S., Mohan, D., and Venugopalan, V. K.,** Seasonal variations of organochlorine insecticide residues in air from Porto Novo, South India, *Environ. Pollut.,* 62, 213, 1989.
12. **Ramesh, A., Tanabe, S., Iwata, H., Tatsukawa, R., Subramanian, A. N., Mohan, D., and Venugopalan, V. K.,** Seasonal variation of persistent organochlorine insecticide residues in Vellar river waters in Tamil Nadu, South India, *Environ. Pollut.,* 67, 289, 1990.

13. **Ramesh, A., Tanabe, S., Murase, H., Subramanian, A. N., and Tatsukawa, R.,** Distribution and behavior of persistent organochlorine insecticides in paddy soil and sediments in the tropical environment: A case study in South India, *Environ. Pollut.,* 74, 293, 1991.
14. **Ramesh, A., Tanabe, S., Subramanian, A. N., Mohan, D., Venugopalan, V. K., and Tatsukawa, R.,** Persistent organochlorine residues in green mussels from coastal waters of South India, *Mar. Pollut. Bull.,* 21, 587, 1990.
15. **Tanabe, S., Ramesh, A., Sakashita, D., Iwata, H., Tatsukawa, R., Mohan, D., and Subramanian, A. N.,** Fate of HCH (BHC) in tropical paddy field: Application test in South India, *Int. J. Environ. Anal. Chem.,* 45, 45, 1991.
16. **Takeoka, H., Ramesh, A., Iwata, H., Tanabe, S., Subramanian, A. N., Mohan, D., Magendran, A., and Tatsukawa, R.,** Fate of HCH in the tropical coastal area, South India, *Mar. Pollut. Bull.,* 22, 209, 1991.
17. **Iwata, H., Tanabe, S., Sakai, N., and Tatsukawa, R.,** Distribution of persistent organochlorines in the oceanic air and surface seawater and the role of ocean on their global transport and fate, *Environ. Sci. Technol.,* 27, 1080, 1993.
18. **Tanabe, S.,** PCB problems in the future: Foresight from current knowledge. *Environ. Pollut.,* 50, 5, 1988.
19. **Cummins, J. E.,** Extinction: The PCB threat to marine mammals, *Ecologist,* 18, 193, 1988.

CHAPTER 3

Biological Monitoring

Lee R. Shugart

Biological monitoring has recently received considerable attention among ecologists and toxicologists as a potentially powerful approach for assessing environmental health, particularly in those environments impacted by pollution and in need of remediation. Specifically, biological monitoring refers to the use of biological organisms to determine environmental conditions. The underlying concept is that selected biological responses (biomarkers) measured in organisms and at various levels of biological complexity can provide sensitive indices of exposure and stress.[1-6] Often these indices correlate with subsequent deleterious consequences from such exposure.

Biomarkers can be measured at levels of biological organization ranging from the community to the molecular level, however, it is important to determine how information from the various levels can be most effectively used in a biological monitoring program. In this context, it is essential that field studies be designed to allow for the incorporation of laboratory testing.

ENVIRONMENTAL POLLUTION: A SOCIETAL DILEMMA

Human activities, e.g., agriculture, defense, energy production, manufacturing, recreation, and transportation, etc., although intended to be beneficial, often affect the physical environment in an adverse manner through the production and release of pollutants. Many of these pollutants are nonbiodegradable and therefore their persistence in the environment poses long-term threats to public health and indigenous wildlife. Even those activities intended to improve the quality of life can lead to problem situations that call for prevention or remediation. Thus environmental pollution has become a complex societal dilemma that must be addressed as there is a growing awareness that human health is inexorably linked to the health of the environment and it is becoming more difficult, if not impossible, to separate the two issues.[7,8] In this context, solutions are needed that define the problem of pollution early enough to preclude undesirable consequences and correct the situation without creating new and more troublesome problems, or stifle human initiative (i.e., within tolerable cost constraints). Implicit in this argument is an urgent need for practical, defensible strategies that provide information for establishing both priorities for remediation (environmental restoration) and endpoints for regulatory compliance.[6]

Effective environmental restoration will require knowledge of the fate and effect of pollution before, during, and after remediation. Often these substances are present in low concentrations and, unfortunately, under these circumstances it is difficult to evaluate risk. There are various reasons for this. First, a long latent period frequently exists between exposure and the subsequent expression of dysfunction, whether the concern is at the individual or the ecosystem level. Second, the effect(s) of exposure at the individual level will depend upon many factors, some of which are intrinsic, such as the age, sex, health, and nutritional status of the organism, and some of which are extrinsic, such as the dose, duration, and route of exposure. An approach that can help document whether or not a polluted environment constitutes a hazard to the health of the biota present is biological monitoring.

BIOLOGICAL MONITORING: THE CONCEPT

Biological monitoring refers to the use of living organisms to evaluate environmental conditions, and makes use of biological endpoints that are indicators (biomarkers) of environmental insults. The organism functions as an integrator of exposure, accounting for abiotic and physiological factors that modulate the dose of toxicant taken up from the environment. The subsequent magnitude of the response can then be used to estimate the severity of exposure, hopefully in time to take preventive or remedial measures. Monitoring organismal responses as a means of evaluating exposure and effect of contaminants and pollution is a recognized concept for use in environmental health research.[9-11] In the context of the discussion of biological monitoring in this chapter, the major emphasis is on environmental species, however, it should be obvious that the general principles covered are applicable to humans as well. Furthermore, it is well documented that environmental species can serve as sentinels for health effects in humans.[1,12]

The biological monitoring approach involves measuring a suite of selected responses at each of several levels of biological organization to assess the effects of exposure to pollutants on the individual. These responses in turn serve as early warning signals of stress and dysfunction, and provide insights into causal relationships between exposure and effects manifested at higher levels of biological organization. The types of biological responses measured with this approach range from the biomolecular/biochemical to the population and community levels and segregate along gradients both of toxicological and ecological relevance and of response time. Biomarkers that reflect the health status of biota at lower organization levels respond relatively rapidly to stress and have high toxicological relevance; those that reflect health conditions at higher organizational levels respond slowly to stress and have lower toxicological relevance but are more relevant ecologically. Conceptually this approach is represented in Figure 1, which illustrates the relationship between responses of different levels of biological organization and the relevance and time scales of the responses.[3,5]

BIOLOGICAL RESPONSES: BIOMARKERS

Introduction

Monitoring a biological response means the measurement and evaluation of a characteristic peculiar to a living organisms. A characteristic is any biological phenomenon (i.e., growth, reproduction, enzymic reaction, etc.) that can be detected. In a biological monitoring program, biological responses (biomarkers) are selected on the basis of their known susceptibility to perturbation by exogenous factors such as xenobiotics. The measured magnitude of change in the biomarker from its normal specified limits can be used to establish exposure and possibly to predict the anticipated change at other levels of biological organization.[2]

Classification

Biomarker classification usually adheres to some broad scheme that reflects levels of biological organization or function within living organisms. Some biomarkers are grouped under scientific

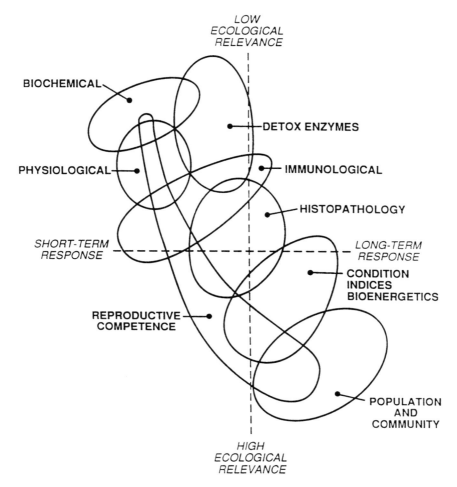

Figure 1. Relationship between levels of biological organization and the relevance and time scales of response (taken from Reference 20).

topics such as histopathology or immunology. Others are associated with physiological functions (i.e., detoxification), while others are categorized by the area of scientific investigation such as reproductive and developmental toxicology or environmental genotoxicology. Numerous examples of such classification can be found in the recent scientific literature.[2,3,5,10,13]

Advantages and Limitations

It should be clearly understood that biomarkers are an important component in environmental health research.[9-11,14,15] In this arena, the application of biomarkers for assessing human health is much further advanced than it is with defining the health of environmental species or various levels of environmental organization (i.e., ecosystems). For this reason many of the techniques and methodologies being applied to environmental species were taken from human health research activities.

Endpoints associated with the population-ecosystem level of effect such as density, species diversity, etc., although falling within the definition of a biomarker, have the disadvantage that they are slow to change, poorly understood, and display high natural variability. Furthermore, it is difficult to establish cause and effect relationships between these endpoints and the presence of environmental pollutants. Nevertheless, early warning signals of incipient ecological damage may be correlated with biological responses at the organismal level where important physiological process such as growth and reproduction are used to evaluate environmental insult.[16]

Biomarkers at lower levels of biological organization (molecular/biochemical) possess important features that are useful to biological monitoring programs. They are less species specific (highly conserved over evolution) and therefore applicable across phyla. They tend to be more sensitive to exposure and more readily amenable to detection. Some of these biomarkers are involved with changes to DNA integrity as a result of the metabolism of genotoxic compounds. Others comprised inductions of specific enzymes (mixed function oxidases) or other proteins (metallothionein) in response to certain chemical exposures. These induced responses are often considered to be adaptive, or protective responses by the exposed organisms. In addition, they are often associated with the mechanism of action, or metabolism, of the compounds producing the particular response. This proximity of the biochemical response to the mode of action or metabolism confers to these responses several advantages over many other types of biomarkers, primary of which is the potential to establish cause and effect relationships between exposure and biological impact. Of additional importance, particularly from the standpoint of monitoring environmental health, is the oftentimes inherent sensitivity of the molecular/biochemical biomarker; this characteristic often provides the first quantifiable change resulting from an environmental perturbation, including a chemical exposure. Furthermore, many are relatively specific to the compound or compound class evoking the response,[1,16,17] which is rarely true for other, nonmechanistic indices of organismal or ecological perturbation.

Selection and Prioritization

From both logistical and economical points of consideration, a biological monitoring program cannot include a large number of biomarkers. Selection can be fine-tuned without diminishing effectiveness if several general attributes and characteristics of biomarkers are recognized and considered.[18] Factors to be considered are the following:

1. Response: It should not be anticipated *a priori* that exposure of an organism to pollutants in its environment will elicit a biological response, or that responses will be observed through different levels of biological organization. Where possible, biomarker selection should be based on the known toxicological mechanism of action of the substances present, thus maximizing the probability of observing a biological response(s).

2. Temporal Occurrence: Some biomarkers are measurable early (hours) after exposure and are observable at the molecular/biochemical level of biological organization. These early responses may exhibit a temporal existence. Others appear much later (years) and are seen at higher levels of organization. Implicit in the concept of biomarker use is the potential for correlating responses among various levels of biological organization.

3. Variability: Sources of variability that may influence the measurement of a biological response generally fall into two categories: those that are inherent to the laboratory method, procedure, or assay selected, and those intrinsic to the species being sampled. Sample collection, preparation, and storage as well as reagent purity and instrument selection and calibration are examples of the first category. They are more easily controlled through adherence to quality assurance and quality control policies. Age, sex, and disease state of the organism being sampled, or environmental stresses such as climate or food availability are examples of factors that contribute variability in the second category. The effects of these latter factors on the biomarker assay are difficult to predict; however, they can be documented and accounted for by establishing baseline data from appropriate noncontaminated or reference sites.

4. Limitations: Specific limitations and restrictions apply to the use and interpretation of many of the biomarkers, and these factors should be verified by consulting the current scientific literature.[13]

The prioritization process involves the selection of biomarkers that are appropriate to the objectives of a given biological monitoring program. The suitability of a particular biomarker should be judged on the anticipated probability of obtaining information that will document either exposure, status of cellular compensatory mechanisms, or potential for harm. An in-depth survey of biomarkers was recently conducted.[13] From such a survey it is possible to identify a small subset (suite) of biomarkers with which to examine and evaluate both specific or general types of contamination problem. The identification of this subset and its supplementation represent the first step in the prioritization process.[18]

Interpretation

The interpretation of the biomarker data must be tempered by the current state of our scientific knowledge (i.e., specific limitations and restrictions). Some responses will be definitive indicators of exposure and even predictive of long-term adverse effects, whereas other responses will only be a signal of a potential problem.[1] The characterization of relationships between biomarker test results, toxic exposures, and environmental health effects requires special emphasis upon developing reference ranges and validating their sensitivity and specificity for biologic health endpoints. These factors are central to applying any screening test and should not be overlooked when examining populations that have been exposed to low levels of hazardous substances.[4,11,18] Therefore the collection of data from appropriate reference/control sites or populations cannot be overemphasized. Once adequate baseline information concerning the suite of biomarkers is established for a species or set of species within a defined area, one would be able to sample that species or species subset from selected sites and compare the results with established baselines. Such comparisons will help define the influence or contribution of intrinsic factors that contribute to variability.

The importance of integrating results from biomarker studies in the laboratory with those from the field (and vice versa) cannot be overemphasized.[4,18] The elucidation of relationships between known contaminants and biological response under controlled laboratory conditions will help in the selection of biomarkers for field studies. Conversely, biomarker data from field studies can direct the choice of biomarkers for more detailed laboratory studies.

AN INTEGRATED STUDY

The following is a synopsis of an integrated study in which biological monitoring played an important role not only in establishing the effects of remediation on a chemically polluted stream, but also in providing information applicable for regulatory compliance.[19]

Field studies concerned with effects of pollution on fish require that a variety of interacting environmental factors be considered relative to the evaluation of biological responses.[20] The homeostatic mechanisms of fish are continuously challenged by the demands of the aquatic environment, including exposure to contaminants, unfavorable temperatures, water velocity, sediment loads, hypoxia, reduced food availability, etc. These factors, individually or collectively, can impose considerable stress on physiological systems resulting in reduced growth, impaired reproduction, susceptibility to disease, and reduced capacity to tolerate subsequent stress. At the population level, effects of stress may be manifested as reduced recruitment and compensatory reserve. Since effects of toxicants on organisms may be manifested at one or several different levels of biological organization, a single stress response or responses at only one level of organization are not sufficient in themselves to fully evaluate and interpret the net effect of pollutant stress on fish. The biological monitoring approach, however, can be an effective technique to assess the integrative effects of stress on fish. This approach has been applied with redbreast sunfish (*Lepomis auritus*) in a stream that receives point source industrial discharges of mixed contaminants (PCBs, hydrocarbons, heavy metals, and chlorine).

In May 1985, a Biological Monitoring Program was developed for East Fork Poplar Creek (EFPC) in eastern Tennessee, United States. This stream originates within the Oak Ridge Y-12 Plant that produces nuclear weapons components for the Department of Energy. Water and sediment in the stream contain metals, organic chemicals, and radionuclides from releases that have occurred over the past 45 years. Effluents discharged from the Y-12 Plant enter the stream near its headwaters; further downstream the creek also receives urban and some agricultural runoff and effluent from the City of Oak Ridge's Wastewater Treatment Facility. Classified uses of EFPC, as designated by the Tennessee Department of Environment and Conservation, include growth and propagation of fish and aquatic life, and recreation, including fishing and swimming. Primarily because of elevated concentrations of mercury, fishing and swimming in EFPC have been prohibited since November 1982.

The biological monitoring program was developed under mandate of the National Pollutant Discharge Elimination System permit as an alternative approach to compliance with water quality standards. The monitoring program has two major objectives: first, to determine if the effluent limitations established for the Y-12 Plant, as stipulated in the NPDES permit, protect and maintain the classified uses of the stream; and second, to document environmental improvements from the implementation of a water pollution control program at the Y-12 Plant. This program seeks to eliminate direct discharges of wastewaters to EFPC and to reduce inadvertent release of pollutants.

At each of several sites (4 to 6) along the length of this stream and from a reference stream, 15 to 20 male sunfish of approximately the same size were collected by electroshocking over a 2-year period (October 1986–October 1988). Biomarkers such as the mixed-function oxidase enzymes and DNA damage have provided direct evidence of toxicant exposure, while condition indices and indicators related to lipid biochemistry and histopathology have reflected impaired lipid metabolism, immune and reproductive system dysfunction, and reduced growth potential.[19–21] At higher levels of organization, stress-mediated effects have included changes in the richness and biotic integrity of fish communities.

Overall the Biological Monitoring Program at EFPC has been effective in providing information on the spatial and temporal distribution and effects of pollution. Data from this study revealed a downstream gradient of increasing fish health. Fish from the lower reaches of EFPC showed improvement in health over the study period but the health of fish from the upper reaches has not improved. For example, results obtained over a 14-month period[21] indicate that the incidence of DNA strand breakage (a measure of genotoxic stress) in sunfish at several EFPC sampling stations was significantly higher than fish from the reference streams at the beginning of the study. However, the temporal and spatial decrease in DNA strand breakage in the fish population (Figure 2) suggested a decrease in genotoxic stress indicative of an improving aquatic environment during a period of intense remedial action. Furthermore, other biomarker responses indicated that decreased health of fish from EFPC was due primarily to toxicant exposure. Higher concentrations of detoxification enzyme, metallothionein, DNA damage, and liver somatic index were observed in EFPC fish. In addition, reproductive impairment was observed in fish collected near the Y-12 Plant but not in fish collected ≥4 km farther downstream.

SUMMARY

The main thrust of the thesis proposed here is that organisms dwelling in an environment impacted with anthropogenic pollutants exhibit detectable biological responses (biomarkers) that

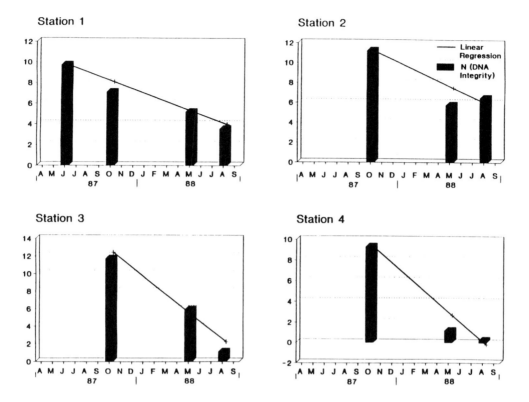

Figure 2. Temporal and spatial relationship of relative DNA strand breakage in bluegill sunfish (*Lepomis macrochirus*) from EFPC sampling stations compared to sunfish from a control, nonpolluted site from June 1987 through August 1988. Stations no. 1, 23.4 km; no. 2, 18.2 km; no. 3, 13.8 km; and no. 4, 6.3 km (taken from Reference 21).

reflect well understood biochemical and physiological mechanisms of toxicity. Biological monitoring is an approach that analyzes the magnitude and temporal occurrence of these biomarkers and estimates the extent to which exposure to these pollutants will result in subsequent deleterious consequences, not only at the individual level, but also at the population, community, and ecosystem levels and across species as well. Such information is invaluable because it helps establish the status of the overall health of the biota in the impacted environment, which is a necessary component for understanding the socioeconomic implications of pollution and its remediation.

ACKNOWLEDGMENTS

The author wishes to express his sincere appreciation to Dr. A. Renzoni of the University of Siena, Siena, Italy for the invitation to lecture at the "Summer School for Multidisciplinary Assessment of Environmental Risks for Human Health." This document represents a synopsis of my lectures at this school. Many of the ideas and concepts expressed herein were adapted and synthesized from previously published papers by the author on the subject of *Biomarkers*. The reader is referred in particular to those references containing the author's name for more detailed discussion of this topic.

The Oak Ridge National Laboratory is managed by Martin Marietta Energy Systems, Inc., under contract DE-AC05–84OR21400 for the U.S. Department of Energy. This is Environmental Sciences Division Publication No. 4266.

REFERENCES

1. **Shugart, L. R., Adams, S. M., Jimenez, B. D., Talmage, S. S., and McCarthy, J. F.,** Biological markers to study exposure in animals and bioavailability of environmental contaminants, in *ACS Symposium Series 382, Biological Monitoring for Pesticide Exposure: Measurement, Estimation, and Risk Reduction,* Wang, R. G. M., Franklin, C. A., Honeycutt, R. C., and Reinert, J. C., Eds., American Chemical Society, Washington, D.C., 1989, 86.
2. **McCarthy, J. F., and Shugart, L. R.,** *Biological Markers of Environmental Contamination,* Lewis Publishers, Boca Raton, FL, 1990.
3. **Shugart, L. R., McCarthy, J. F., and Halbrook, R. S.,** Biological markers of environmental and ecological contamination: An overview, *Risk Analy.,* 12, 353, 1992.
4. **Peakall, D. B., and Shugart, L. R.,** *Biomarkers: Research and Application in the Assessment of Environmental Health,* NATO ASI Series H, Cell Biology, Vol. 68, Springer-Verlag, Heidelberg, 1993.
5. **Adams, S. M.,** *Biological Indicators of Stress in Fish,* American Fisheries Society, Bethesda, MD, 1990.
6. **Peakall, D.,** *Animal Biomarkers as Pollution Indicators,* Ecotoxicology Series 1, Depledge, M. H., and Sanders, B., Eds., Chapman and Hall, London, 1992.
7. **Messer, J. J., Linthurst, R. A., and Overton, W. S.,** An EPA program for monitoring ecological status and trends, *Environ. Monitor, Assess.,* 17, 67, 1991.
8. **Lyne, T. B., Bickham, J. W., Lamb, T., and Gibbons, J. W.,** The application of bioassays in risk assessment of environmental pollution, *Risk Analy.,* 12, 361, 1992.
9. **Committee on Biological Markers of the National Research Council,** Biological markers in environmental health research, *Environ. Health Perspec.,* 74, 3, 1987.
10. **Committee on National Monitoring of Human Tissues,** *Monitoring Human Tissues for Toxic Substances,* National Academy Press, Washington, D.C., 1991.
11. **National Research Council/Committee on Environmental Epidemiology,** *Environmental Epidemiology: Public Health and Hazardous Wastes,* National Academy Press, Washington D.C., 1991.
12. **McBee, K., and Bickham, J. W.,** Mammals as bioindicators of environmental toxicity, in *Current Mammalogy,* Vol. 2, Genoways, H. H., Ed., Plenum, New York, 1990.
13. **Huggett, R. J., Kimerle, R. A., Mehrle, P. M., and Bergman, H. L.,** *Biomarkers: Biochemical, Physiological, and Histological Markers of Anthropogenic Stress,* Lewis Publishers, Boca Raton, FL, 1992.
14. **de Serres, F., Geldhill, B. L., and Sheridan, W.,** *DNA Adducts: Dosimeters to Monitor Human Exposure to Environmental Mutagens and Carcinogens,* Vol. 62, Environmental Health Perspective, U.S. Department of Health and Human Services Publication No. 85–218, Research Triangle Park, NC, 1985.
15. **Bartsch, H., Hemminki, K., and O'Neill, I. K.,** *IARC Scientific Publication No. 89, Methods for Detecting DNA Damaging Agents in Humans: Application in Cancer Epidemiology and Prevention,* IRAC, Lyon, France, 1988.
16. **Jenkins, K. D., and Sanders, B. M.,** Monitoring with Biomarkers: A multi-tiered framework for evaluating the ecological impact of contaminants, in *Ecological Indicators, Vol. 2,* McKenzie, D., Hyatt, E., and McDonald, J., Eds., Elsevier Applied Science Publishers, New York, 1992, 1279–1293.
17. **DiGiulio, R. T., Washburn, P. C., Wenning, R. J., Winston, G. W., and Jewell, C. C.,** Biochemical responses in aquatic animals: A review of determinants of oxidative stress, *Environ. Toxicol. Chem.,* 8, 1103, 1989.
18. **McCarthy, J. F., Halbrook, R. S., and Shugart, L. R.,** *Conceptual Strategy for Design, Implementation, and Validation of a Biomarker-Based Biomonitoring Capability,* Oak Ridge National Laboratory/TM-11783, Oak Ridge, TN, 1992.
19. **Loar, J. M.,** *First Report on the Oak Ridge Y-12 Plant Biological Monitoring and Abatement Program for East Fork Poplar Creek,* Y/TS-886, Oak Ridge National Laboratory, Oak Ridge, TN, 1992.
20. **Adams, S. M., Shepard, K. L., Greeley, M. S., Jimenez, B. D., Ryon, M. G., Shugart, L. R., and McCarthy, J. F.,** The use of bioindicators for assessing the effects of pollutant stress on fish, *Marine Environ. Res.,* 28, 459, 1989.
21. **Shugart, L. R.,** DNA damage as an indicator of pollutant-induced genotoxicity, in *13th Symposium on Aquatic Toxicology and Risk Assessment: Sublethal Indicators of Toxic Stress,* Landis, W. G., and van der Schalie, W. H., Eds., ASTM, Philadelphia, 1990, 348.

CHAPTER 4

Pollution Biomonitoring: A Summing Up

Henri M. André

INTRODUCTION

The terms "biomonitoring" and "bioindicators" and their equivalent (biological monitoring and biological indicators) have become an important part of the vocabulary of ecologists and other applied biologists and they are more and more frequently listed as key words in publications dealing with environmental problems.

As a result of this increasing success, some confusion revolves around those terms. For instance, some authors seem to restrict the biomonitoring approach to the observation and description of the effects of pollution. This is unsatisfactory for several reasons. Conceptually, surveying the effects of pollutants is not necessarily tantamount to monitoring the pollution itself, i.e., the global and complex process that causes those effects. For instance, the influence of industrial emissions on the health of plants was noted as early as 1661 (quoted by Cowling[1]). However, the idea of using plants for assessing the air pollution level was proposed much later in 1866 when the famous Finnish botanist, Nylander,[2] wrote: "Les lichens donnent à leur manière la mesure de salubrité de l'air et constituent une sorte d'hygiomètre très sensible."

In addition, even if biological assessment of pollution had depended historically on observing effects directly in natural conditions, direct observation of environmental impacts cannot be the sole basis for the development of a real biomonitoring program. Indeed, the cause–effect relationship between a pollution and its effects is rarely a one-to-one correspondence. On the one hand, there are a number of different responses to a single stimulus (pollutant) from the molecular to the biocenotic level and those responses are likely to interact. On the other hand, quite different stimuli can induce similar responses in natural systems.[3] Therefore, analytical and experimental approaches are also essential in biomonitoring programs if one wishes to understand and quantify any cause–effect relationship.

Lastly, as stated by Herricks and Cairns,[4] biomonitoring data must be differentiated from the information contained in the data. Data in the form of numbers or activity of organisms and physical and chemical parameter values only provide the opportunity of generating information. In other words, the symptoms and resulting syndrome must not be confused either with the disease itself (i.e., the pollution process) or with the health and well-being of a biological system. Collecting, describing, and analyzing data are just a prerequisite to making a diagnosis.

Another problem arising from the success of the biomonitoring approach lies in the large variety of biological techniques and methods involved. This variety of techniques superimposes onto the high diversity of species used as indicators. As a result, almost as many methods have been proposed or suggested as biologists working in the field and this situation has been partly responsible for less acceptance of biological methods than physicochemical ones.[5]

In addition, information does not always circulate as it should do within the scientific community. Barriers do exist, for instance between environmentalists involved in water pollution and others working on air pollution. More serious is the linguistic barrier, which still persists between biologists. Just an example: in the series of review papers devoted to water biological monitoring,[4,6-10] there is no reference to the work of Verneaux and Tufféry who are leaders in water pollution monitoring in France. Conversely, in the master book published in France by Pesson[11] and entitled "La pollution des eaux continentales," there is just one reference to Cairns who has written a considerable number of papers on the subject.

Consequently, the biomonitoring approach and philosophy appear to be heterogeneous, heteroclitous, and sometimes full of contradistinctions. A descriptive example is offered by the properties that scientists require for what they call a good indicator. Some scientists, as Cairns and Schalie,[6] require a time response of a few minutes. In contrast, botanists use as indicators lichens whose time response may exceed several months or even 1 year.

Beyond this apparent heterogeneity, it is thus necessary to draw up a biomonitoring philosophy that takes into account the numerous facets of environmental pollution and degradation, set up a conceptual frame likely to integrate the different techniques and methods, and thus to lead biologists to a fundamental unity through the current formal diversity. It is the purpose of the present chapter to present a general typology of the different approaches in biomonitoring and analyze, for each monitor type and irrespective of the substrate (soil, water, air), some commonly used methods. From the particular properties and requirements of each approach, it is hoped a fundamental paradigm may be extracted that reflects both the uniqueness of the biomonitoring philosophy and the diversity of methodologies.

GENERAL TYPOLOGY

Analytical Approach

Biological changes due to pollution occur at all levels of organization, from the molecular to the community level. Accordingly, biomonitoring techniques might be divided into several groups depending on the selected level. Roughly, three major groups could be distinguished: infraindividual (molecular to organic level), individual (population or single species studies), and biocenotic (multiple-species systems). The last group is of major importance as it is the only one that takes into account what Blanck et al.[12] call environmental realism, and emphasizes an holistic approach.

On the other hand, biological systems, whatever the organization level considered, are affected in their structure and functioning. As in classical system theory, state variables can be distinguished from activity variables. State variables comprise essentially morphological, structural, and/or taxonomic information while activity variables require functional studies. However, although functional information is a prerequisite for modeling systems and is vital to a clear understanding of system responses, the distinction between state and activity variables seems less essential for biomonitoring than the distinction between qualitative and quantitative variables. Indeed, the former is directly related to the notion of threshold, however precisely it is defined. In other words, a system is supposed to be in a normal state or function normally vs. the system departs from its normal state or activity. Quantitative variables should allow a more subtle monitoring strategy and forecasting.

Most variables used in biomonitoring express the impact of environmental conditions that affect any biological system more or less severely depending on its sensitivity. Such biological systems used for monitoring the environment are thus sensors. On the other hand, some biological systems are not necessarily affected by pollutants but have the property of accumulating them: they are accumulators. Despite the fact that most organisms or biological systems are both sensors and accumulators, a clear distinction must be made at least from a theoretical and methodological point of view. A sensor is affected positively or negatively by a toxicant and thus reacts. In contrast, accumulators may display a quite passive behavior toward pollutants: they just accumulate without necessarily suffering from the process.

A last dichotomy in the classification of monitors appears when biological systems observed *in situ* are opposed to those introduced in the study area. This distinction, initially proposed by Jenkins[13] and also stressed by Blandin[14] who opposes "allochthonous" and "autochthonous" species, results in fundamental differences little explored. For example, organisms observed *in situ* are relatively or entirely uncontrolled; conversely, the introduction of organisms allows the biologist to have at least some degree of control over the experimental parameters.

Synthetic Typology

The different points of view and dichotomies propounded in the previous section lead to the construction of a multiple-way table (Table 1). The first column of the table refers to the different levels of organization taken into account, from the molecular to the biocenotic levels. Vertically, the table is divided into two major parts depending on whether the monitoring systems are considered to be accumulators or sensors. The latter are again divided into two categories depending on the type of variables. Activity variables are related to the system functioning while state variables are related to the structure and composition of systems. State variables are either quantitative or qualitative. Lastly, each box of the table is again divided into two parts (black vs. blank areas) depending on whether systems are introduced or merely observed *in situ*.

This multiple-way table allows different types of monitors to be defined. Types are named after the oldest nomenclature ever proposed, that of Jenkins.[13] However, the initial typology has been somewhat reorganized to fit the general frame proposed in Table 1.

SENTINEL SPECIES AND MICROCOSMS

Definition

The term "sentinel" has been used in different ways. It may designate organisms that grow in polluted areas[15] and, possibly, accumulate pollutants in their body tissues.[16] According to Jenkin's terminology,[13] sentinels are "highly sensitive organisms introduced into the environment as early warning devices to give an alarm of an approaching or presently dangerous situation." The typical sentinel organism is the mine canary used as a warning where death indicated a dangerous concentration of invisible, odorless methane.

This illustrative example reveals the major features of sentinels: (1) they are sensors, i.e., they react to pollutants, (2) the species is (are) introduced, (3) the monitor(s) exhibit(s) either a normal or abnormal state or function; the critical response is thus qualitative, and most often binary (dead vs. alive; normal vs. abnormal), and (4) the critical response is directly related to a threshold level of pollutant and occurs whenever the threshold value is exceeded.

However, it must be pointed out that the threshold level of pollutants must not be necessarily known with great precision, as was the case with the mine canary. Furthermore, as an extension of the above definition (especially point 3), the response may be measured through a quantitative

40 CONTAMINANTS IN THE ENVIRONMENT

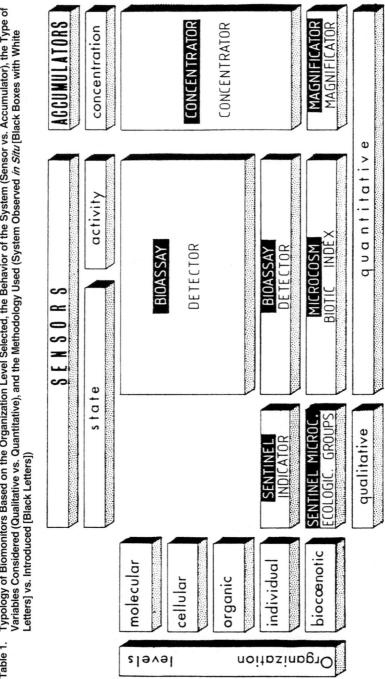

Table 1. Typology of Biomonitors Based on the Organization Level Selected, the Behavior of the System (Sensor vs. Accumulator), the Type of Variables Considered (Qualitative vs. Quantitative), and the Methodology Used (System Observed *in Situ* [Black Boxes with White Letters] vs. Introduced [Black Letters])

parameter (as a rate for instance) but it is judged critically only if a given threshold value is exceeded. If it is, the response is said to be abnormal and thus critical; if not, everything is supposed to be normal whatever the values recorded.

Examples

The most illustrative examples are the early warning systems, called in French "sonnettes d'alarmes,"[11] and reviewed by Cairns and Schalie.[6] A detailed description of such a system is offered by Cairns et al.[17] Briefly, a monitor unit is settled in the vicinity of a plant along a river. The monitoring system consists of several aquaria. Control fish are exposed to upstream water alone while test fish are exposed to downstream water alone or plant waste diluted with upstream water. Such a system, called an "in-plant" system in opposition to "in-stream" systems, measures changes in the swimming or breathing rates of fishes. Any significant change from previous behavior in "breathing" (ventilation) or movement in the test (exposed to plant waste) series, and/or compared to controls is considered to be a critical response and produces a warning or alert signal. Of course, different salt and freshwater fish species may be used as sensors[18] and different more or less sophisticated instrumentation may be involved.[19]

Another approach in early warning systems is offered by the automation of behavior analysis, especially the analysis of fish preference, and avoidance of certain physical and/or chemical conditions. This approach is most interesting in so far as behavior has long been recognized as an excellent way to assess the condition and well-being of a particular organism, even at low pollutant concentrations. Preference and avoidance studies have been reviewed by Cherry and Cairns.[9]

Of course, organisms other than fish could be used as sentinels. Experiments by Novak et al.,[20] for instance, suggest the use of small crustaceans as monitors, and different exposure systems have been designed to accommodate small or medium-sized aquatic organisms.[21,22] Bacteria might also be used as sentinels. Indeed, bacterial respiration, a widely tested material monitoring technique, is inhibited by different toxic materials. Accordingly, the presence of toxicants may be detected through a decrease of the bacterial oxygen consumption below a given threshold and produces a warning signal.[23–25]

Although most sentinels have been proposed in the context of water pollution, such monitors could also be used to detect air pollutants. For instance, Craker and Fillatti[26] have shown an increase in ethyle production from seedlings exposed to phytotoxic air pollutants during two hours, while Newman[27,28] has proposed the use of bees as sentinels of SO_2 and fluoride.

Discussion

A monitoring system using sentinel organisms or microcosms should have the following properties:

1. The organism(s) is (are) held in a laboratory situation or in the field under controlled conditions and are exposed on a frequent or continuous flow basis to the water or air being tested.
2. Physiological or behavioral parameters of the organism(s) are monitored by a recording device with the capability of responding to abnormal conditions indicated by the organism(s).
3. The function of the monitor is primarily for the detection of short-term changes in toxicity as opposed to chronic or cumulative effects of a toxicant.

The fundamental properties, proposed by Cairns and Schalie,[6] entail several practical requirements to select sentinel organism(s) on the one hand and to design the monitoring system on the other. Practical requirements concerning the organism(s) are as follows:

1. The sentinel selected should be sensitive to a large variety of toxic material; it should respond repeatedly to these materials and not lose its sensitivity to toxicants following long-term exposure to very low levels of toxic materials. This means that the number of false normal responses must be minimal. Practically speaking, a sentinel must not be so tolerant to a particular toxicant that it will pass undetected and harm members of the biota living in the receiving system.
2. In the same way, the sentinel should exhibit a minimum of false alarm responses to nonharmful variations in water or air quality.
3. Rapid detection of developing toxic conditions is essential. Delays of several hours between introduction of a pollutant and a critical response of the sentinel may be unacceptable.
4. The organism(s) used as sentinel should be fairly inexpensive and easy to acquire.
5. The physiological or behavioral parameter of the organism(s) selected for monitoring should be recordable and/or quantifiable through appropriate interfacing techniques (computer or other electronic recording equipment); this enables the monitoring process to be continuous and automatic. However, the recording system should not result in undue stress on the organism(s).

The last point makes the transition between requirement concerning the sensor itself and those pertaining to the monitoring system. Other desirable properties of the monitoring apparatus are as follows:

1. Appropriate methods for the analysis of data must be developed. For instance, if the parameters being monitored have a diurnal periodicity, it may be necessary to compute separately a normal range of values and corresponding threshold values for several different periods of the day.
2. Monitoring systems should be relatively easy to operate and should produce results easy to interpret.
3. The monitoring apparatus should be reliable and require as little maintenance as possible.

In conclusion, the major merit of the early warning systems and associated sentinel organisms is of course the quick response to a possible toxicant. Of course, the method is not a panacea. This methodology emphasizes the detection of short-term changes in toxicity irrespective of long-term effects. It is likely that low level of materials with cumulative toxicity may not be detected soon enough for the response to be useful.[29] In fact, this drawback just illustrates the major default of the early warning systems, their lack of ecological realism.

INDICATOR SPECIES AND ECOLOGICAL GROUPS

Definition

According to Jenkins,[13] indicators are specific species or organisms whose presence indicates probability of pollution of other dangerous organisms associated with pollution.

Indicators, also called thrivers by Newman,[28] are thus quite different from sentinels as the latter are introduced or transplanted organisms while the former are autochthonous species observed *in situ*. However, both sentinels and indicator species have in common the fact that a qualitative response is expected: the indicator is present or absent—that's all.

However, compelling reasons exist for abandoning the practice of placing sole reliance on a single species. The observation of an entire group of recurrent species provides considerably more

Table 2. Lichen Ecological Groups in Relation to Pollution

Zone	SO$_2$ level (μg/m^3)	Epiphytes[b]
0	>170	Epiphytes absent
1	≈170	(C) *Protococcus viridis*
2	≈150	(C) *Lecanora conizaeoides*, (C) *Protococcus viridis*
3	≈125	(C) *Lecanora conizaeoides*, (C) *L. expallens*, (C) *Lepraria incana*
4	≈70	(Fo) *Hypogymnia physodes*, (Fo) *Parmelia sulcata*
5	≈60	(Fo) *Parmelia saxatilis*, (Fo) *Physcia adscendens*, (Fo) *Ph. tenella*
6	≈50	(Fo) *Parmelia caperata*, (Fo) *P. acetabulum*, (C) *Pertusaria spp.*
7	≈40	(Fr) *Ramalina fastigiata*, (Fr) *Ramalina farinacea*, (Fr) *Anaptychia ciliaris*
8	≈35	(Fo) *Parmelia perlata*, (Fr) *Usnea ceratina*, (Fr) *Ramalina fraxinea*
9–10	<30	(Fr) *Lobaria pulmonata*

[a]Such a table may vary depending on the region, the climatic conditions, the tree species (basic vs. acid bark), etc.
[b]C, crustose epiphytes; Fo, foliose lichens; Fr, fruticose lichens.

assurance on the probability of pollution, hence the use of ecological groups as indicators of pollution.

Examples

The classic example of indicators is the saprobic system proposed by Kolkwitz and Marsson[30] as early as 1908. The system was modified and revised subsequently[31–34] and gave rise to many discussions.[14,35–40] Roughly, the saprobic system rests on the idea that when sewage is introduced into a river, this may be divided into a succession of zones downstream from the rejection point. The successive zones below the point of introduction were designated as polysaprobic, α- and β-mesosaprobic, and finally oligosaprobic, on the basis of their physical and chemical properties. Kolkwitz and Marsson[30,41] published long lists of the species they observed to be associated with each of the zones, the species being classified as polysaprobic (present in the zone of polysaprobia, close to the introduction point), α- or β-mesosaprobic, or oligosaprobic, and they used the presence and absence of those species to indicate the degree of organic pollution.

In the context of air pollution, the observation by Nylander[2] that lichens were "hygiometers" was the first step leading to the use of lichens as indicators of SO$_2$ and other air pollutants. Where the sulfur dioxide concentration is very high, no lichens are found, resulting in what has been called a "lichen desert." Practically, the absence of all lichens from an area indicates impure air. Next are found several lichens that tolerate quite high levels of pollution such as *Lecanora conizaeoides*. In contrast other lichens, especially foliose and fruticose lichens, are very sensitive to pollutants and are observed only in areas with pure air. Tables such as Table 2 are used to indicate the pollution intensity and estimate the sulfur dioxide concentration. Similar tables exist for water pollution.

Discussion

The idea that certain species can be used to indicate certain types of environmental conditions is an old one. A few years after Kolkwits and Marsson, Forbes and Richardson[42] also proposed to use specific organisms as indicators of pollution and stated that it was quite possible to arrange the plants and animals of a stream in the order of their preference for, or tolerance of, organic impurities, in such a way that a graded list of them may serve as an index to grades of contamination. However, the oldest systematic use of indicator species probably dates back to Clements[43] whose approach was masterly reviewed in 1928.[44]

The basic assumption is that the presence of a species indicates that the habitat is suitable to it. If the requirements of that species are known, it is thus possible to draw up conclusions about the nature of the environment in which it is found. Particularly, this means that the presence of a species furnishes assurance that certain minimal conditions have been met. Such conclusions are based on practical common sense observation but may also rely on laboratory experiments.

In the context of pollution biomonitoring, the presence of a species may indicate either clean conditions or a polluted environment. Trout are known to live in cold and well oxygened water, their presence implies the absence of organic or hot pollution: they are an indicator of salubrity. Conversely, other organisms are indicators of pollution or possible hazards to the public health. The classic example are indicators of feces, indicators of domestic sewage, and indicators of virus and other pathogens. To be acceptable, an indicator of pathogens must have two attributes: (1) it must be present when the pathogens are present and (2) it must be easy to detect and possibly quantify.

The practical meaning of both types of indicators is quite different. Indeed the presence of an indicator of salubrity furnishes assurance that certain minimal conditions of salubrity have been met. On the other hand, determining the significance of the absence of such a species is ludicrous and considerably more risky. For instance, the species may not be present because it has not had the opportunity to get into the area but might well survive if it was introduced. Thus the absence of such an indicator is less useful—let's say less meaningful—than its presence. Conversely, if an indicator of pollution must be always found when the pollution is present, it may be important, but not imperative, that it is absent when the pathogen is absent. In the latter type of indicator, the nonequivoqual information must be given by negative results, i.e., by the absence or at least a density below a given threshold.

However simple the indicator seems to be, it must also rely on an experimental approach and require a clear understanding of the processes involved. As already outlined by Clements,[45] the indicator value of any species can not be known until its functional responses have been measured and correlated with the structural. In the absence of such knowledge, the reliability of the indicator may be questioned by decidors. This is why it has been proposed to increase the safety margin of microbiological water quality evaluation because it was difficult to answer certain questions regarding the relation of coliform limits and the virological safety of drinking water. Similar objections have been made against the saprobic system since little is known about the response of most aquatic organisms to a variety of pollutants.[37,46]

Using a single species as an indicator of salubrity is, however, risky in the sense that a given organism may be tolerant to one type of pollution and sensitive to the other. In addition, if the absence of a species used as an indicator of salubrity is questionable, the absence of a group of species is much more reliable, hence the resort to ecological groups as indicators of environmental conditions.

Tuffery[38] commenting upon the use of indicators species stated that only experts specialized in systematics of aquatic organisms are able to use biological scales. The same remark has been made against the use of lichens. However, surveys made by school children in Great Britain are famous and obviously weaken this argument.[47,48]

DETECTORS

Definition

According to Jenkins,[13] detectors are individual species occurring in the environment that show a measurable response to pollutants or environmental changes such as pathology, mutation, death, change in physiology, reproduction, and so on. This definition may be conveniently extended to any part of organisms observed *in situ*.

Detectors are similar to indicator species in that both are observed *in situ*. However, detectors are fundamentally different from indicators as the former require the measurement of a quantitative response and not merely a presence/absence observation. On that point, detectors are similar to bioassays since both are based on the measurement of a quantitative response, the difference being that the former are observed *in situ* while the latter are transplanted organisms.

Examples

An illustrative example of a detector is the tar spot index (TSI) proposed by Bevan and Greenhalgh.[49] This index expresses the number of spots/100 cm^2 leaf area of the fungus *Rhytisma acerinum*, the cause of tar spot disease of sycamore. TSI values were shown to be negatively correlated with annual average SO_2 concentrations. A quite different application is given by the measure of the enzymatic activity. For instance, fish brain cholinesterase is inhibited *in vivo* by organic phosphorus insecticides at concentrations of 0.1 mg/L and lower. An application consists precisely in measuring the activity of the acetylcholinesterase. This method (applied to man or trout brain) is routinely used to assess the organophosphate concentrations present in the environment.

Discussion

The use of detectors is less successful than the recourse to bioassays. The reason is rather simple. Indeed the use of bioassays allows a high degree of standardization of the experimental design in laboratory experiments and in field conditions as well. This standardization makes it easier to interpret field results in light of experimental data obtained in the laboratory. In the absence of a real standardization, data based on detectors are much less easy to interpret and conclusions may often be questionable. For instance, the TSI values must be regauged whenever a new area is studied. Discrepancies between different areas or between different populations may be expected due to selective adaptations to pollution.

Furthermore, the drawbacks associated with single species methods and already mentioned about indicator species still hold. This is why biologists interested in observing phenomena *in situ*, rather than in playing with transplanted organisms, turn to multiple species methods and tend to use biotic indices based on biotic communities.

BIOASSAYS AND MICROCOSMS

Definition

Biassay monitors are organisms—or parts of them—introduced into the environment and used to estimate either a level of pollution (concentrations, doses) or a level of risk (probability). Bioassay monitors may be exposed to air, water, or soil in the field or test site, or, conversely, samples of air or water may be taken to the laboratory and bioassayed. When a mixture of species is used, microcosms are referred to.

The major features of bioassay species and microcosms—whatever the level of organization—are that (1) they are sensors, (2) the species or the organism(s) is (are) introduced, and (3) a quantitative response is used to estimate physicochemical parameters or assess a risk. The major difference from sentinels is thus the observation of a quantitative response.

The oldest example of bioassay monitoring is the phytometer method used by Clements and Weaver at Pilke's Peak in 1918 and 1919.[45,50] Different plants, such as sunflowers or beans, were grown in sealed containers and transferred into different stations. The responses measured (transpiration, growth, or photosynthesis) showed marked differences with reference to altitude, degree of shade, and seasonal factors. Clements concluded that his results left no question about the paramount importance of plants for the quantitative study of habitat and communities. The idea of using transplanted plants for monitoring pollution was a direct application of Clement's approach.

As already outlined by Clement about phytometers, bioassays imply a high level of standardization in the procedure (test conditions, dosage, route, timing). However, the more standardized the test or test animals, the less applicable the information obtained is likely to become.[51] Bioassays also require the use of a mathematical model likely to relate the quantitative response to physicochemical parameters. As emphasized by different authors,[51,52] the bioassays monitoring method must not be confused with toxicity tests. The latter allow the determination of the response of a biological system to a stimulus of known intensity while the former method, although based on the same dose–response relationship, aims at estimating the stimulus intensity from the response measured. Obviously, the establishment of a quantitative relationship requires preliminary laboratory studies.

Examples

There is a huge amount of toxicological tests reported in the literature. However, even if many organisms have been proposed as potential bioassays, few of them have been subjected to an integrated approach and are likely to be used as real bioassays. The case of *Humerobates rostrolamellatus* proposed as a bioassay monitor of air pollution is an illustrative example of the problem encountered to set up a good monitor.

H. rostrolamellatus is a corticolous oribatid mite usually abundant in orchards. Following several studies of corticolous mite communities affected by air pollution,[53–55] this species was selected and its sensitivity to SO_2 was tested in the laboratory.[56,57] Using the probit model,[58] the LC_{50} was estimated to be 0.48 ppm after a 96-h exposure. The second step was the validation of the model in the field conditions. It turned out that *H. rostrolamellatus* was also sensitive to SO_2 in the field and its mortality after a 1-week exposure was used for mapping air pollution in Brussels.[59] However the model was not satisfactorily validated in the sense that LC_{50} estimated in the field was too different from that observed in the laboratory. New experiments under controlled conditions revealed that the LC_{50} was 3180 µg/m SO_2 (fiducial limits: 1052–1179) while the LC_{50} estimated in the field was 10 times lower (311µg/m SO_2; fiducial limits: 289–332). A new series of laboratory experiments was necessary to highlight synergic interactions between SO_2 and NO_2 and explain the discrepancies between laboratory and field data.[60] Needless to say, this kind of experiment requires both sophisticated devices to generate and monitor gases under controlled conditions and special test chambers suited to test animals.[61]

Apart from *H. rostrolamellatus,* few animal species have been proposed as bioassays to assess air pollution. In contrasts, plants, especially lichens, are extensively used as phytometers. In the context of water pollution, there are many recommended "standard" bioassays using fish,[62–64] daphnia,[65] and other organisms.

Another type of bioassay must be mentioned as it does not rely on a dose–response relationship but is aimed at estimating prevalence from the results of screening tests. Very often, those bioassays are used to estimate the prevalence of genetic toxicants, i.e., mutagenetic, teratogenetic, or carcinogenetic substances. Such tests are characterized by their sensitivity, specificity and predictive values (Table 3). As outlined in Table 3, the positive predictive value of a test, p, varies with the prevalence even when sensitivity and specificity are constant. Mutagenetic testing in environmental toxicology has become a major field of research and the most widely used test is probably the famous Ames' *Salmonella*/mammalian microsomal assay system.[66] As with dose–response models, the relationship may be used either way, either as a toxicological test, or as a bioassay monitor. In the latter context, the Ames assay was used to investigate the mutagenetic activity of fish and sediments in rivers.[67] As with a dose–response curve, plants may be transplanted very easily in the field. An illustrative example is offered by the *Tradescentia* stamen hair test. This test detects somatic gene mutations that cause a change in the pigmentation of stamen hair cells from blue to pink. It has been used to monitor mutation frequencies in the vicinity of nuclear power plants in the United States and Japan.[68]

Table 3. Relationships among Sensitivity, Specificity, Predictive Values, and Prevalence of Bioassays Applied to Mutagenetic Substances[a,b]

	Observed frequencies	"True" frequencies	
		Mutagens	Nonmutagens
	Mutagens	a	b
	Nonmutagens	c	d

[a] Same relationships apply to teratogenetic and carcinogenetic substances.
[b] Sensitivity: $Se = a/(a + c)$
Specificity: $Sp = d/(d + b)$
True prevalence of mutagens: $P = (a + c)/(a + b + c + d)$
Observed prevalence of mutagens: $p = (a + d)/(a + b + c + d)$
Positive predictive value: $V^+ = a/(a + b)$ and $V^+ = 1/(1 + RK)$ where $R = (1-Sp)$ and $K = (1-P)/P$, $p = Se \cdot P + (1-Sp)(1-P)$.

Discussion

A monitoring system using bioassay organisms or microcosms needs much more detailed studies than sentinels due to the necessity of determining a quantitative response. Roughly, three major steps may be distinguished during the search for a good bioassay.

1. The choice of a biological model, bacteria, plant, or animal, which proves to be sensitive to the toxicant
2. The choice of a mathematical model likely to describe the response of the bioassay under controlled conditions, i.e., in the laboratory
3. The validation of the model in the field

To each step are associated different requirements that must be respected if the bioassay has to be used in biomonitoring. The choice of an animal or a plant generally results from observations in the field and preliminary screening tests carried out in the laboratory. Laboratory conditions imply that the plants or animals are easy to handle and their requirements must be easily satisfied (food supply, etc.). If a specific bioassay is searched for, the potential bioassay must be sensitive to a single toxicant. In such a simple case, the dose–response curve is easy to establish. The dose–response curve must, however, be stable in the sense that it cannot be overly affected by meteorological factors (such as temperature). Otherwise, it is necessary to determine several curves for different values of the modifying factor. Very often, plants or animals are sensitive to a large range of toxicants. In the case of mutagenicity testing, this wide sensitivity is even expected to be as high as possible to correctly identify all possible mutagens. With bioassay based on dose–response relationships, multiple sensitivity may seem to be an interesting property but it makes the next step very complex as explained further. A last point important in the choice of a bioassay concerns the time response that may vary to a great extent, from a few days in the case of *H. rostrolamellatus* to several weeks in transplanted lichens.

The choice of a mathematical model is also essential even if this step is sometimes disregarded in some applications. Numerous problems may arise. For instance, it is well known that the predictive value of prevalence models varies with the actual prevalence even if sensitivity and specificity are constant (Table 3). Dose–response curves based on the probit model also pose some problems especially when confidence intervals are considered.[69] Such problems become impossible to overcome once more than two pollutants interact synergistically.

The validation of the model is the last step in the selection process of a good bioassay. It is performed in a few cases. The example of *H. rostrolamellatus* illustrates the problems encountered at this step. Indeed, the dose–response curve set up from the laboratory data may not apply directly to observations made in the field because of synergistic interactions. Furthermore, drawbacks unnoticed in the laboratory conditions may arise in the field due to different environmental conditions. So, the mite was known to be able to survive at very low temperatures, but in the field it proved to lose its sensitivity to air pollution when exposed at temperatures below $-7°C$, temperatures that, of course, had not been used in the laboratory conditions. Consequently, any model built from laboratory results is useless until it is validated in the field. Similarly, models built directly from field data, and not supported by experiments carried out in controlled conditions, are not reliable. Furthermore, special attention must be given to confidence intervals if the bioassay is to be used predictively. Just recall that a correlation of 0.7—a high value in field conditions—means that 50% of the response variations are not explained by the stimulus under study, which implies a very poor predictive value.

BIOTIC INDICES

Definition

The fact that indicator species might be used to assess the pollution level has given rise to the idea that assemblages of organisms—or ecological groups—could be used collectively to assess the pollution. A further step has been to use not only the presence–absence of groups, but properties of biotic communities such as their structure or their diversity. Such properties are complex since communities form interlocking independent systems comprising numerous species. Therefore, empirical formulas and subsequently mathematical equations have been developed to summarize complex biological data into a single value. In this context, an index is considered to be any mathematical combination of two or more parameters that has utility in an interpretive sense.

Apart from biotic indices, others are based only on physical and chemical parameters and are used to estimate pollution over a long period. This is the case of the Oak Ridge Air Quality Index (ORAQUI).[70,71] Other indices—called mixed indices—combine physical and chemical characters with biological parameters. An illustrative example is offered by the Water Quality Index (WQI) developed at the National Sanitation foundation.[72] In addition to the temperature, nitrate concentration, etc., the WQI takes into account the fecal coliforms. Hereafter, only biotic indices will be treated.

Biotic indices are so numerous that they have been classified into different categories. First, there are structure and composition indices *per se* vs. comparison indices through which any station is compared to a reference station. Wilhm[5] defined community structure indices in opposition to activity indices based on biomass or metabolism expressions. Thomas et al.[73] distinguished biotic vs diversity vs saprobic indices.

Fundamentally, however, most biotic indices are based on a few elementary concepts and associated hypotheses. The first concept is richness, i.e., the number of species. The associated hypothesis is that the higher the pollution is, the fewer the species that will survive (Figure 1A). Possibly, the number of species will fall to zero as in lichen deserts. In other words, the richness varies negatively with pollution. A second concept is related to the coverage (when lichens are considered) or abundance. Indeed, if a species is sensitive to pollutants, its coverage should decrease with pollution (Figure 1B). A third concept is evenness. For instance, two trees may be covered by two species of lichens as in Figure 1C but correspond to quite different situations. Left, one of the lichens covers all the trunk while the other is represented by only one small thallus and is vanishing due to the pollution. In Figure 1C (right), both lichens are flourishing and share the trunk evenly. To an even distribution of lichens is thus associated a low level of pollution. Lastly, a fourth approach takes into account the sensitivity of organisms. In Figure 1D, trees shelter two species of lichens that are evenly distributed; however, they may reflect quite different situations depending on the species recorded and their respective sensitivity. Actually,

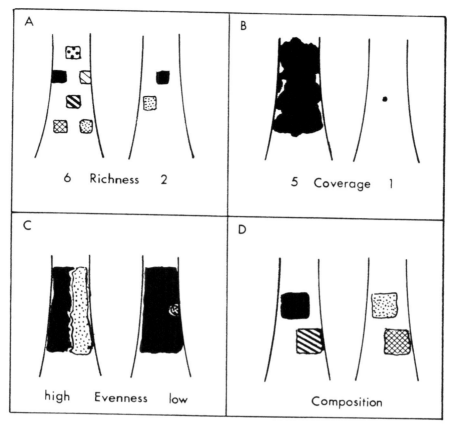

Figure 1. The four basic concepts (A–D) pertaining to community analysis as they apply to biomonitoring. In the four cases, the pollution gradient is descending from left to right. Patches with different hatching patterns represent various epiphytic lichens growing on trunks. A fifth concept—not illustrated—is the diversity (or heterogeneity depending on the authors), which is a combination of the richness and evenness.

all indices combine to some extent two or more basic concepts. For instance, when evenness is combined with richness, ecologists refer to diversity indices. But even when only two basic concepts are considered, the number of indices used may be high because the component concepts may be weighted differently.

Depending on the underlying concepts, biotic indices may be classified as in Table 4. Indices in the upper part emphasize the tolerance of component species and are close to the ecological group methods inasmuch as both approaches are based on the use of synoptic tables. Indices grouped in the lower part of Table 4 are based on the structure of communities, expressed in terms of diversity or richness, and do not take into account the sensitivity of the component species. For each type of index, different formulas or mathematical approaches have been proposed; the simplest are empirical (left part of Table 4) while others rely on more sophisticated methods such as multivariate analyses (on the right of Table 4). Nevertheless, formulas belonging to the same type of index are isomorphic and may be described by a type formula given in the right column of Table 4.

Examples

A classic example of biotic index used to evaluate pollution is the Index of atmospheric purity devised by Leblanc and De Sloover.[74] This index has been widely used all over the world and

Table 4. Typology of Biotic Indices

		Underlying basic concepts	Data analysis procedure from simple to complex			Formula isomorphism
			Empirical approach	Mathematical approach	Multivariate analysis	
COMMUNITY ANALYSIS	**COMPOSITION**	Tolerance	Verneaux and Tuffery biotic index[110]		Coste[111]	Synoptic tables
			Bellan[15]			$I = \dfrac{\sum \dfrac{A_p}{N}}{\sum \dfrac{A_s}{N}}$
		Richness and tolerance	Beck's Index[112] De Sloover's IAP[76] Chandler's score[113]	Leblanc and De Sloover's IAP[74] Empain's index[114]		$I = \sum_{i=1}^{s} c_i A_i$
		Diversity and tolerance	Pantle and Buck's saprobic index[115]		Descy's index[116]	$I = \dfrac{\sum_{i=1}^{s} c_i A_i}{\sum_{i=1}^{s} A_i}$
						$= \sum_{i=1}^{s} \dfrac{c_i A_i}{N}$
	STRUCTURE	Richness	Cairns et al. SCI[117] Cairns and Dickinson[118] Lafont[119]			$I = 1 - \sum_{i=1}^{s} \dfrac{A_i}{N}$
		Diversity		Simpson's index[120] Shannon's index[121]		$I = N_a = \left[\sum_{i=1}^{s} \dfrac{A_i}{N} \right]^{1/(1-a)}$ (Hill's formula[122])

Note: A_i, number of individuals of species i, i varying from 1 to s; s, number of species; N, total number of individuals; A_p, number of individuals of species indicator of pollution; A_s, number of individuals of species indicator of salubrity; c_i, coefficient of toxiphoby — or toxitolerance — of species i; N_a, Hill's diversity of order a.

its history deserves some comments. As already noted in the section devoted to ecological groups, lichens are vanishing in highly polluted areas and lichen desert may even be observed. Hence the idea of using the richness in epiphytic lichens to estimate a pollution level. A first index was proposed by Barkmann.[75]

However, the richness in lichens is a rough parameter. As already noted above, the same richness does not necessarily indicate the same level of pollution because the component species are not necessarily evenly distributed and also because their sensitivity may vary depending on the species. It is thus necessary to consider the frequency-coverage of lichens and also to weigh the frequency-coverage score of each component species by a specific index of toxiphoby. This lead to the construction of an index called I.A.P. used for the first time by De Sloover in 1963:[76,77]

$$\text{I.A.P.} = n/100(Q_i \times f_i) \qquad (1)$$

where n is the number of species at a station, f_i is the frequency-coverage value of species i, and Q_i is its index of toxiphoby. The value of Q_i was obtained by ranking the species more or less subjectively in a so-called increasing order of toxitolerance and giving each species a level ranging from 1 to 12.

In 1970, the I.A.P. was improved and took its final form:[74]

$$\text{I.A.P.} = 1/10 \, (Q_i \times f_i) \qquad (2)$$

where Q_i, renamed ecological index, was defined as the mean number of "companion" species of species i, i.e., the average number of other epiphytes occurring with it at all the investigated sites.

Numerous other examples are cited in Table 4.

Discussion

The dilemma faced by ecologists playing with biotic indices is well apparent in Patrick[78] who concluded: "To try to describe these communities by any one single index, such as a diversity index, is rather futile" but wrote on "However, such indices are of value." The same perplexity is found in Warren[79] who commented "A biological index is but a tool, albeit a useful tool" but also concluded "the biological complexity of pollution problems cannot be cut by a single tool, no matter how useful." Perplexity still increases if one considers the inconsistencies that can occur when various indices are used to demonstrate pollution.[80]

Some indices such as Leblanc and De Sloover's I.A.P. have been extensively used all over the world and, obviously, have some diagnostic value in mapping pollution. The hypotheses underlying the use of such indices are relatively simple and are supported, to some extent, by laboratory experiments.

However, at the community level, it appears that species abundance heterogeneity does not always conform to the conceptual model relating a decreasing diversity to increasing pollution. If in the case of epiphytic lichens, the model holds, it does not apply to the fauna sheltered by those lichens as observed by André and Lebrun.[55] To explain their observation, these two authors distinguished two phenomena that play a role at the community level. If the microhabitats maintain, which is the case of algae that are very resistant to pollution, dominant arthropod species succeed one another and there is an apparent alteration of the community correlated with increasing pollution. Each peak of dominance corresponds to a low diversity; on the other hand, each time a dominant species is vanishing and progressively being replaced by another, heterogeneity increases as illustrated in Figure 2A. This entails oscillatory variations of the α (or within-habitat) diversity. If the microhabitats are vanishing—which is the case of more sensitive fruticose lichens—there is a depression in the habitat diversity that entails the disappearance of the associated species and causes a depression in β (or between-habitat) diversity along the coenocline (Figure 2B). This deformation sensu Walter[81] of the original community thus consists of a real simplification and recalls to some extent the retrogression concept proposed by Whittaker and Woodwell.[82] However, at the limit, even a deformation may result in an increase in diversity. Owing to the disappearance of associated and dominant populations, the "noise" due to rare and incidental species becomes more and more important (Figure 2A). Both phenomena, alteration and deformation, plus the "noise" due to rare or incidental species interfere and makes more complex the interpretation of diversity values in relation to pollution. Roughly, at a low level of pollution, deformation is important and results in a decrease in γ (i.e., total) diversity of the fauna; at a high level, most microhabitats offered by lichens disappear and alteration becomes the predominant phenomenon.

A similar situation was observed in the study of phytoplankton community in Great Lakes by Dennis et al.[83] who observed an unexpected positive correlation between the community richness, evenness, and diversity on one hand and the pollution level on the other. Obviously, pollution alters the composition and structure of communities, but the way pollution modifies them may be more complex than is usually expected.

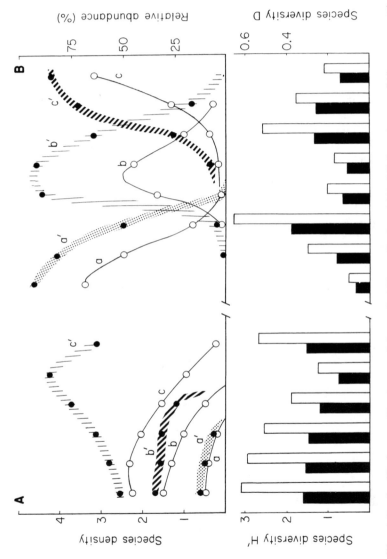

Figure 2. Deformation (A) and alteration (B) of communities along a pollution gradient (descending from right to left). In each case, solid lines represent the density of three dominant species (a, b, c). Corresponding dotted or hatched curves (a', b', c') indicate their respective relative abundance. Histograms, at the bottom, show the diversity estimated in some points scattered along the gradient [H' (Shannon index) = black bar; D (Simpson index) = white bar] (from André and Lebrun,[55] courtesy of Blackwell Scientific Publications).

Nevertheless, biotic indices are probably the indicators that offer the greatest degree of ecological realism. Any pollution, however subtle it is, affects the biotic communities in one way or another. The integration of multiple pollutants over space and time is probably optimal whenever biotic indices are considered. From this point of view, any departure of an index value from its "normal" range is likely to reflect a modification in the environment. The major problem, however, remains the interpretation of the variations observed in the values of indices. Indeed, if any modification in the value of a biotic index reflects variations of the environment, it is not necessarily related to pollution. This clearly highlights the difficulty in establishing a cause–effect relationship at the community level, a difficulty related to the rarity or even the lack of experimental data. Inasmuch as biotic indices really integrate over time all effects that may be deleterious to the growth and reproduction of the species involved, they do not allow a proper identification of pollutants.

Practically speaking, the use of biotic indices is not necessarily simple. In some cases, such indices require little taxonomic expertise. The use of diversity indices does not require the identification of species and ecologists have just to determine the right number of species present in their samples. In contrast, Leblanc and De Sloover's I.A.P.[74] and other similar indices based on species sensitivity require precise identifications and are thus less easy to use. Lastly, the use of biotic indices requires careful sampling protocol and standardization of sampling methods (surface area, number of samples, etc.). These requirements are essential to make possible the comparison of results.

A last remark bears on the functioning of ecosystems. Indeed, the eradication of a single species may be without significance, at least less important than the disturbance of the functions and the self-regulation of the system.[84] On that point, we are far short of having the necessary background data to properly assess what Hapke[84] called ecotoxicity.

ACCUMULATORS

Definition

Accumulators are organisms that absorb and accumulate a pollutant in measurable, usually large quantities.[13] Such species may be used as monitors to measure the concentration of a pollutant that occurred or is present in the environment.

Accumulators are basically different from other monitors as they are not supposed to suffer from the pollution. They just accumulate and concentrate pollutants to such a degree that measurements of the accumulation of pollutants by those organisms provide information on concentrations present in the environment. Inasmuch as the concentrations in such organisms may be orders of magnitude greater than the concentrations in the medium, measurements are made easier.

Two major pathways are involved in bioaccumulation. Either, high levels of pollutant are concentrated in organisms from very low concentrations in the environment, this process is referred to as bioconcentration. Or pollutants are concentrated along a food chain and the process is called biomagnification. Monitor organisms are respectively called concentrators and magnificators.

Examples

Numerous applications of this approach have been published. A famous example is the mussel watch concept proposed as early as 1975 by Goldberg.[85] This project developed by Goldberg et al.[86,87] and taken up by numerous other workers aims at monitoring marine pollutions on a large scale through the systematic use of mussels. Indeed, mussels are widely distributed in oceans and accumulate a great number of pollutants such as heavy metals, hydrocarbons, and various pesticides. Numerous applications of this method even lead to a standardization of the procedures and resulting classifications at an international level.[88] Oysters are also frequently used in the same way. Other organisms involved in this type of biomonitoring are fish, crustaceans, insects, algae, and mosses.

Terrestrial environments are also monitored through the use of accumulator organisms. Lichens, bryophytes, tree foliage, and mammal hairs have been used to assess the concentrations of heavy metals and many other pollutants.

Not only are concentrations in different pollutants measured in organisms found *in situ*, but some organisms are transplanted in the field and subjected to chemical analyses after a predetermined exposure time. For instance, Goodman et al.[89] have used mosses put into plastic bags and deposited along transects from a pollution source to assess the heavy metal concentrations. The "moss-bags" technique has been subsequently used by many other authors.[90-92] Transplanted lichens and plants grown in flowerpots have also been used in the same way.

Discussion

Direct measurements of pollutant concentrations, (metals, pesticides, etc.) for pollution monitoring may require extensive and time consuming procedures. In addition, they may require operation of the analytical apparatus at its limits of detection.

Another advantage of concentrators is that most direct measurements provide only estimates of an instantaneous concentration that can vary significantly and quickly over time. Conversely, biological monitors accumulate pollutants and integrate them over time in proportion to the concentrations observed in the environment.

A major problem is the choice of a model for estimating pollutant concentrations in the environment knowing pollutant concentrations in the monitor organisms. Developing meaningful relationships between a widely varying instantaneous variable (pollutant concentration in the environment) and a time-averaged variable (concentration in the monitor) in a field study with many varying factors is not always easy. For instance, Schultz-Baldes[93] derived a simple model for estimating lead concentrations in seawater from those measured in mussels. However, the applicability of the model in multiparameter situation, i.e., in the field, was questioned by Waldichuk.[94] Hence the necessity of developing other relationships based on multiple regression models such as those proposed by Popham and D'Auria.[95]

An alternative approach has been to develop models based on the dynamics of uptake and depuration of pollutants in monitor organisms. The bioconcentration process is then viewed as a balance between the two kinetic processes. A classic example of this approach is the fugacity model[96] (see this volume).

However, many practical problems still arise with accumulators. For instance, Flannagan et al.[97] monitored the methoxychlor present in aquatic invertebrates; surprisingly, they found that caged animals had significantly different residues than the natural populations submitted to the same treatment and concluded that the use of caged animals was therefore questionable. However, Bradshaw[98] reached a quite different conclusion. Indeed, pollution is also a selection factor, which means that bioaccumulation mechanisms are likely to change with time. Hence the advantage of using introduced organisms, selected and genetically standardized.[98] Discrepancies between laboratory and field conditions are also observed. For instance, uptake of DDT residues by slugs was found to be different under glass house conditions than in outdoor enclosures due to differences in climatic conditions.[99] Effects of temperature on the trophic transfer was also highlighted by Boudou and Ribeyre.[100] Interactions among metals may also affect metal availability to organisms, which means that presence of one metal may inhibit or enhance the uptake of another.[101] Animals may accumulate metals from both food and solution, with the relative importance of the source being dependent upon the available concentrations.[102] Lastly, sampling strategy is also of vital importance as emphasized by Strong and Luoma.[103] Surprising results—such as a negative relationship between the concentrations in the animal and the environment—may occur because of a sampling strategy neglecting the effects of body size, seasonal variations, and sex.

However attractive is the accumulator concept, its use seems to be restricted to the mapping of pollution zones. Its use as an "early warning system" has also been stressed by different authors. Nevertheless in most cases, it seems still illusive to precisely assess the concentrations present in the environment just by a simple dosage of pollutants in living beings. Bioaccumulation determination remains, however, an important complement to chronic toxicity test as chemicals with a pronounced tendency to bioaccumulate are likely to have at least predictable biological effects in long-term, low-level exposures.[8]

CONCLUSIONS

This overview of the different types of biomonitors has revealed the diversity of techniques and organisms involved, the heterogeneity of approach, and, in some occasions, contradistinctions between authors. Even then, is it possible to advance a definition of what is a biomonitor or to propose a paradigm including all the aspects of biomonitoring?

Numerous definitions have been proposed, but most of them refer to a special type of biomonitor alone, for instance to indicators. Some combine two or more different aspects of biomonitoring. For instance, Bellan[106] defined "biological indicators as detectors revealing the existence of. . . ."

Cairns[46] defines biological monitoring as the "regular, systematic use of organisms to determine environmental quality." Cairns[105] stresses that monitoring implies regular and continuous assessment of one or more parameters. This conception may sound too restrictive in the sense that numerous studies based on irregular sampling or spot checks are implicitly rejected. Matthews et al.[7] opinion is more shaded: "Biological monitoring can be defined as the systematic use of biological responses to evaluate changes in the environment with the intent to use this information in a quality control program." Another comprehensive definition is given by Iserentant and De Sloover[52] who designate as indicator "any species or biological system used to assess a modification—generally a deterioration—of the environment quality, whatever its organization level and the use made of it." Lebrun[106] also emphasizes the richness of methods: "A biomonitor is any biological parameter, qualitative or quantitative, measured on an individual, a population, a guild or a community, likely to indicate special conditions corresponding to a given state, or natural variations or perturbations of the environment." All these definitions have in common that they embrace all the aspects of biological monitoring, refer explicitly to biological systems, and designate the goal to achieve.

A fundamental question remains, however. Why to resort to biological monitoring? In other words, wouldn't chemical and physical monitorings be sufficient? The answer is both simple and complex. Chemicophysical monitoring determines the concentration or level (e.g., temperature) of various components, that's all. Although this information is objective and necessary, it is, however, incomplete if not irrelevant. This information becomes irrelevant because the ultimate goal of monitoring is not the scientific knowledge per se, but the protection of human health and well-being and chemicophysical monitoring alone does not allow the prosecution of such a goal for several reasons. First, any chemical compound may affect biological systems in many different ways, depending on the nature on the system (species, organization level) and environmental conditions. Second, biological systems are exposed, at the same time, to an aggregate of stresses, the sum of which is unpredictable even if an exhaustive list of stresses was available. Living beings do not react to stresses separately but integrate them. Such synergistic effects are intrinsic properties of biological systems and are not predictable from chemicophysical parameters alone. Third, living beings integrate stresses over time. Integration over time corresponds to several cases: (1) living beings are *continuously* exposed during all their life to small amounts of the

Table 5. Objectives Prosecuted in Biological Monitoring

1. Early warning systems
2. Diagnosis
3. Pollution zones mapping
4. Pollution source identification
5. Risk assessment
6. Forecasting

same pollutant (chronic effects), (2) they are submitted *at regular intervals* to the *same* pollutant, and (3) they are submitted to a succession of *different* stresses. Chronic effects are often related to concentrations almost undetectable. Intermittent pollution (e.g., discharges into a river) may be missed in a traditional chemical sampling. Lastly, successive exposures imply that an organism already weakened by an exposure to a first pollutant will not necessarily react to a new toxicant as a healthy organism. In all three cases, living beings integrate pollution over time and give useful information even if they are observed irregularly. Fourth, some effects are delayed and amplified in time. Some toxicants, first safe, may become dangerous because of biological reactions within an organism or along the food chain. This relates also to the biomagnification processes: first, the concentration is low and little toxic to most organisms and, later, concentrations increase and reach lethal thresholds. Other effects become apparent only when they affect the ecosystem equilibrium, this is part of the ecological realism already referred to in the introduction.

Biological monitoring is thus an essential part of environment management since living beings are amplifiers, integrators in space and time, and even premonitors (they allow the forecasting). Biomonitoring is fundamental since it is the only way to get information meaningful in biological terms, information relevant to this special biological system called man, information useful to the protection of his health and well-being.

Apart from these fundamental properties, biological monitoring may present other advantages in comparison to physicochemical monitoring. For instance, easiness may be a key factor: it is much easier to repeatedly measure the acetylcholinesterase activity in fish brain than to measure the parathion concentrations in a river. Another advantage of biological methods often cited in the literature is their low cost.

Other attributes of a good biomonitor have been extensively discussed in the literature. Some of them are general and apply to any type of monitors. For instance, Wielgolasky[107] requires that physiology and ecology of an organism should be well known. In contrast, the requirement that a monitor should be preferably stenovalent to a key factor will not be accepted in all biomonitoring applications. The discrepancies between different authors concerning the time response of a good monitor—from a few minutes to several months—was already stressed in the introduction. This clearly shows that attributes of a good biomonitor depends to a great extent on the types of monitoring as detailed in the sections above.

Biological monitoring may be used to achieve different objectives (Table 5). Depending on the objectives, the geographic scale, the precision searched for, different advantages may be essential. A comparison between a bioassay method and a biotic index applied to lichens to assess air pollution is presented in Table 6. Another comparative study of indicator species vs. biotic indices is given by Cairns.[108]

From the mine canary, passing through Nylander's hygiometers,[2] Clements' indicators,[43] Clements and Weaver's phytometers,[50] and Axt's toximeter,[25] a large variety of biological techniques and methods have been proposed. All types of organisms—from bacteria to animals passing through higher plants—have been used to some extent as biological monitors of pollution in air, water, and soil. However, all those methods must not be perceived as antagonistic. Not only do they present peculiar advantages and drawbacks, but they supplement each other. Each of them is likely to give us relevant information and highlight a special point of view. Beyond this diversity, the uniqueness of the biological monitoring must be emphasized. Each method implies

Table 6. Comparison between the Use of Biotic Indices Based on Lichens and a Corticolous Mite as a Bioassay Monitor of Air Pollution (Adapted from André et al.[60])

Characteristics	Dependence [a]	Biotic index[b]	Bioassay
Type of test	M	Multiple species	Single species
Species	M	Observed *in situ*	Transferred
Ecological realism	M	+++	+
Reliability	M	+++	+
Accuracy	M	+	+++
Location of stations	M	Not anywhere	Anywhere
Replication	M	Stochastic	Possible
Taxon identification	S-M	Easy to tedious	Easy
Time required for handling and yielding data	S-M	Short to long	Short
Time response	S	Long	Short
Season effect	S	Anytime	Only in winter

[a]Characteristic are either method dependent (M) or species dependent (S), or both (S-M).
[b]+++, in a given respect, a method is better than the other indicated by a single +.

the choice of a biological model (plant, animal, ecosystem), the choice of a mathematical model (however simple it may be), the test of the latter in experimental (controlled) conditions, and its validation in field conditions. Depending on the method, these different steps are more or less difficult to overcome. The selection of a biological model most often follows a screening procedure and, possibly, preliminary observations in the field, this step may be time-consuming but is essential as it determines the success of the biomonitoring method. Schafer and Brunton[109] state that even if a considerable amount of toxicological information is available on the relative susceptibility of different mammal species, it is generally not sufficient to extrapolate a valid relationship between the indicator species and the target species. Testing is another delicate step, especially with complex systems. Among the numerous biological models, ecological ones are probably the most difficult to test in experimental conditions. However, experimental data are a prerequisite to the use of monitors because direct observations of an environmental impact cannot be the sole basis for the development of a predictive capability. This point was made clear as early as 1918 by Clements.[45] Mathematical models, even simple, are also necessary because they do provide a link between observations and predictions. Lastly, the validation of the model in field conditions is also essential. Indeed laboratory results cannot be extrapolated directly to the field.[60] If the model is not validated in the field, it is necessary to go back to experimental procedures or to revise the mathematical model.

In spite of numerous drawbacks and seeming contradistinctions, biological monitoring remains the only way of collecting pertinent information on our environment. This is why much has to be done to improve the methodologies already in use and possibly propose new techniques if biomonitoring has to be accepted by the academic community, potential industrial users, regulatory agencies, and other decision-makers.

ACKNOWLEDGMENTS

This contribution is a summing up of works carried out during several years in the department of ecology and biogeography of the "Université Catholique de Louvain" at Louvain-la-Neuve (Belgium), until I turned from studying biomonitoring to evolutionary biology. It results also from the many discussions I had with my colleagues who worked on bioindicators from lichens to mites, from terrestrial organisms such as beetles to diatoms living in water. I wish to express my gratitude to all of them, especially Ph. Lebrun, J. De Sloover, and R. Iserentant. I thank R. Kime for reviewing the language. I thank also Blackwell Scientific Publications who have given permission for Figure 2 to be included here.

REFERENCES

1. **Cowling, E. B.**, Acid precipitation in historical perspective, *Environ. Sci. Technol.,* 16, 110A, 1982.
2. **Nylander, W.**, Les lichens du Jardin du Luxembourg, *Bull. Soc. Bot. Fr.,* 13, 364, 1866.
3. **Woodwell, G. M.**, Effects of pollution on the structure and physiology of ecosystems, *Science,* 168, 429, 1970.
4. **Herricks, E. E., and Cairns, J., Jr.**, Biological Monitoring. III. Receiving system methodology based on community structure, *Water Res.,* 16, 141, 1982.
5. **Wilhm, J. L.**, Biological indicators of pollution, in *River Ecology,* Whitton, B. A., Ed., Blackwell Scientific Publications, London, 1975, 375.
6. **Cairns, J., Jr., and van der Schalie, W. H.**, Biological monitoring. I. Early warning systems, *Water Res.,* 14, 1179, 1980.
7. **Matthews, R. A., Buikema, A. L., Jr., Cairns, J., Jr., and Rodgers, J. H., Jr.**, Biological monitoring. IIA. Receiving system functional methods, relationships and indices, *Water Res.,* 16, 129, 1982.
8. **Buikema, A. L., Jr., Niederlehner, B. R., and Cairns, J., Jr.**, Biological monitoring. Part IV. Toxicity testing, *Water Res.,* 16, 239, 1982.
9. **Cherry, D. S., and Cairns, J., Jr.**, Biological monitoring. Part V. Preference and avoidance studies, *Water Res.,* 16, 263, 1982.
10. **Cairns, J., Jr.**, Biological monitoring. VI. Future needs, *Water Res.,* 15, 941, 1981.
11. **Pesson, P.**, Ed., *La pollution des eaux continentales. Incidences sur les biocénoses aquatiques,* Gauthier-Villars, Paris, 1980.
12. **Blanck, H. G. D., and Gustafsson, K.**, An annotated literature survey of methods for determination of effects and fate of pollutants in aquatic environments, Report from the National Swedish Environmental Protection Board, 1978.
13. **Jenkins, D. W.**, Global biological monitoring, in *Man's Effects on Terrestrial and Oceanic Ecosystems,* Matthews, W. D., Smith, F. E., and Goldberg, E. D., Eds., M.I.T. Press, Cambridge, 1971, 351.
14. **Blandin, P.**, Bioindicateurs et diagnostic des systèmes écologiques, *Bull. Ecol.,* 17, 215, 1986.
15. **Bellan, G.**, Annélidés Polychètes des substrats solides de trois milieux pollués sur les côtes de Provence (France): Cortiou, Golfe de Fos, Vieux Port de Marseille, *Téthys,* 9, 260, 1980.
16. **Burns, K. A., and Smith, J. L.**, Biological monitoring of amibent water quality: The case for using bivalves as sentinel organisms for monitoring petroleum pollution in coastal waters, *Estuarine Coastal. Shelf Sci.,* 13, 433, 1981.
17. **Cairns, J., Jr., Hall, J. W., Morgan, E. L., Sparks, R. E., Waller, W. T., and Westlake, G. F.**, The development of an automated biological monitoring system for water quality, *Virginia Polytechnic Inst. Water Res. Res. Center Bull.,* 59, 1, 1973.
18. **Cairns, J., Jr., Thompson, K. W., Landers, J. D., Jr., McKee, M. J., and Hendricks, A. C.**, Suitability of some fresh and marine fishes for use with a minicomputer interfaced biological monitoring system, *Water Res. Bull.,* 16, 421, 1980.
19. **Cairns, J. Jr., and Gruber, D.**, A comparison of methods and instrumentation of biological early warning systems, *Water Res. Bull.,* 16, 261, 1980.
20. **Novak, A. J., Berry, D. F., Walters, B. B., and May Passino, D. R.**, New continuous-flow bioassay technique using small crustaceans, *Bull. Environm. Contam. Toxicol.,* 29, 253, 1982.
21. **Arnold, D. J., and Keith, D. E.**, A simple continuous-flow respirometer for comparative respiratory changes in medium-sized aquatic organisms, *Water Res.,* 19, 261, 1976.
22. **Benoit, D. A., Mattson, V. R., and Olson, D. L.**, A continuous mini-diluter system for toxicity testing, *Water Res.,* 457, 1982.
23. **Axt, G.**, Bewertungsgrundlagen für schudstoffgerechte Abwassergebuhren, *Vom Wass.,* 39, 299, 1972.
24. **Axt, G.**, Ergebnisse Kontinueierlicher Toxizitatsmessungen mit Bakterien, *Vom Wass.,* 41, 1973.
25. **Axt, G.**, Kontinuierliche Toxizitatsmessungen mit Baktieren (Toximeter), *Gewasserschuts Wass. Abwass.,* 10, 297, 1973.
26. **Craker, L. E., and Fillatti, J. J.**, Development of a test-tube stress-ethylene bioassay for detecting phytotoxic gases, *Environ. Pollution* (ser. A), 28, 265, 1982.
27. **Newman, J. R.**, Effects of air emissions on wildlife resources, U.S. Fish and Wildlife Service, Biological Program, National Power Plant Team, FWS/OBS-80/40.1, 1980, 1.

28. **Newman, J. R.,** The effects of air pollution on wildlife and their use as biological indicators, in *Animals as Monitors of Environmental Pollutants,* National Academy of Sciences, Washington D.C., 1979, 223.
29. **Brown, V. M.,** Advances in testing the toxicity of substances to fish, *Chemy Ind.,* 4, 143, 1976.
30. **Kolkwitz, R., and Marsson, M.,** Ökologie der pflanzlichen Saprobien, *Ber. Dt. Bot. Ges.,* 26, 505, 1908.
31. **Kolkwitz, R.,** *Pflanzenphysiologie,* 3, Aufl., Jena, 1935.
32. **Kolkwitz, R.,** Ökologie der Saprobien, *Schriftenreihe Veriens Wasser-, Boden- und Lufthygiene,* 4, 1, 1950.
33. **Liebmann, H.,** *Handbuch der Firschwasser- un Abwasserbiologie,* Bd 1, 1 ed., Munchen, 1951.
34. **Liebmann, H.,** *Handbuch der Firschwasser- un Abwasserbiologie,* Bd 1, 2 ed., G. Fisher, Jena, 1962.
35. **Sládecek, V.,** The future of the saprobity system, *Hydrobiologia,* 25, 518, 1965.
36. **Sládecek, V.,** The reality of three British biotic indices, *Water Res.,* 7, 995, 1973.
37. **Cairns, J. Jr.,** Biological monitoring—Concept and scope, in *Environmental Biomonitoring, Assessment, Prediction, and Management,* Cairns, J. Jr., Patil, G. P., and Waters, W. E., Eds., International Co-operative Publishing House, Fairland, MD, 1981, 3.
38. **Tuffery, G.,** Incidences écologiques de la pollution des eaux courantes. Révélateurs biologiques de la pollution, in *La pollution des eaux continentales. Incidences sur les biocénoses aquatiques,* Pesson, P. Ed., Gauthier-Villars, Paris, 1980, 243.
39. **Verneaux, J.,** Cours d'eau de Franche-Comté (massif du Jura). Recherches écologiques sur le réseau hydrographique du doubs—essai de biotypologie, *An. Sci. Univ. Besanon,* 3e sér., 9, 1, 1983.
40. **Verneaux, J.,** Méthodes biologiques et problèmes de la détermination des qualités des eaux courantes. *Bull. Ecol.,* 15, 47, 1984.
41. **Kolkwitz, R., and Marsson, M.,** Ökologie der tierischen Saprobien, *Int. Rev. ges. Hydrobiol. Hydrogr.,* 2, 126, 1909.
42. **Forbes, S. A., and Richardson, R. E.,** Studies on the biology of the upper Illinois river, *Ill. St. Nat. Hist. Surv. Bull.,* 9, 481, 1913.
43. **Clements, F. E.,** *Research Methods in Ecology,* University Publishing Company, Lincoln, NE, 1905.
44. **Clements, F. E.,** *Plant Succession and Indicators,* The H. W. Wilson Co., New York, 1928.
45. **Clements, F. E.,** Scope and significance of paleo-ecology, *Bull. Geol. Soc. Am.,* 29, 369, 1918.
46. **Cairns, J. Jr.,** Hazard evaluation with microcosms, *Int. J. Environ. Studies,* 13, 95, 1979.
47. **Gilbert, O. L.,** An air pollution survey by school children, *Environ. Pollut.,* 6, 175, 1974.
48. **Mellanby, K.,** A water pollution survey, mainly by British school children, *Environ. Pollut.,* 6, 162, 1974.
49. **Bevan, R. J., and Greenhalgh, G. N.,** *Rhytisma acerinum* as a biological indicator of pollution, *Environ. Poll.,* 10, 271, 1976.
50. **Clements, F. E.,** Plant Indicators, *Carnegie Inst. Wash. Pub.,* 290, 1, 1920.
51. **Brown, V. M.,** Concepts and outlook in testing the toxicity of substances to fish, in *Bioassay Techniques and Environmental Chemistry,* Glass, G. E., Ed., Ann Arbor Science Publishing, Ann Arbor, MI, 1973, 73.
52. **Iserentant, R., and De Sloover, J.,** Le concept de bioindicateur, *Mém. Soc. Roy. Bot. Belg.,* 7, 15, 1976.
53. **Lebrun, P.** Effets écologiques de la pollution atmosphérique sur les populations et communautés de microarthropodes corticoles (Acariens, Collemboles et Ptérygotes). *Bull. Ecol.,* 7, 417, 1976.
54. **André, H. M.,** Introduction à l'étude écologique des communautés de Microarthropodes corticoles soumises à la pollution atmosphérique. II. Recherche de bioindicateurs et d'indices biologiques de pollution, *Ann. Soc. R. Zool. Belg.,* 106, 211, 1977.
55. **André, H. M., and Lebrun, P.** Effects of air pollution on corticolous microarthropods in the urban district of Charleroi (Belgium), in *Urban Ecology,* Bornkamm, R., Lee, J., and Seaward, M. R. D., Eds., Blackwell Scientific Publishers, London, 1982, 191.
56. **Lebrun, P., Wauthy, G., Leblanc, C., and Goossens, M.,** Tests écologiques de toxitolérance au SO_2 sur l'oribate corticole *Humerobates rostrolamellatus. Annls Soc. R. Zool. Belg.,* 106, 193, 1977.
57. **Lebrun, P., Jacques, J. M., Goossens, M., and Wauthy, G.,** The effects of interaction between the concentration in SO_2 and the relative humidity of air in the survival of the bark-living bioindicator mite *Humerobates rostrolamellatus, Water, Air Soil Pollut,* 10, 269, 1978.

58. **Finney, D. J.,** *Probit analysis,* 3rd ed., Cambridge University Press, London, 1971, 333.
59. **André, H. M., Bolly, C., and Lebrun, P.,** Monitoring and mapping air pollution: A new and quick method. *J. Appl. Ecol.,* 19, 107, 1982.
60. **André, H. M., De Sloover, J., Iserentant, R., and Lebrun, P.,** The use of biological monitors of acide pollution, in *Proc. "Acide Deposition and Sulphur Cycle,"* SCOPE Belgium, Brussels, 1984, 109.
61. **André, H. M.,** Design and use of an exposure chamber for air pollution studies on microarthropodes, *Environ. Entomol.,* 11, 1123, 1982.
62. **Sprague, J. B.,** Measurement of pollutant toxicity to fish. I. Bioassay methods for acute toxicity, *Water Res.,* 3, 793, 1969.
63. **Sprague, J. B.,** Measurement of pollutant toxicity to fish. II. Utilizing and applying bioassay results, *Water Res.,* 4, 3–32, 1970.
64. **Sprague, J. B.,** Measurement of pollutant toxicity to fish. III. Sublethal effects and "safe" concentrations, *Water Res.,* 5, 245–266, 1971.
65. **Anderson, B. G.,** The toxicity treshold of various substances found in industrial wastes as determined by the use of *Daphnia magna, Sew. Works J.,* 16, 1156, 1944.
66. **Ames, B., McCann, J., and Yamasaki, E.,** Methods for detecting carcinogens and mutagens with the *Salmonella*/mammalian-microsome mutagenicity test, *Mut. Res.,* 31, 347, 1975.
67. **Osborne, L. L., Davies, R. W., Dixon, K. R., and Moore, R. L.,** Mutagenic activity of fish and sediments in the Sheep River, Alberta, *Water Res.,* 16, 899, 1982.
68. **Hoffman, G. R.,** Mutagenicity testing in environmental toxicology, *Environ. Sci. Technol.,* 16, 560A, 1982.
69. **Kooijman, S. A. L.,** Parametric analyses of mortality rates in bioassays, *Water Res., 15, 107, 1981.*
70. **Thomas, W. A., Babcok, L. R., and Shultz, W. D.,** Oak Ridge Air Quality Index Report (ORNL-NSF-EP-8), Oak Ridge National Laboratory, Oak Ridge, TN, 1971.
71. **Babcok, L. R. Jr., and Nagda, N. L.,** Indices of air quality, in *Indicators of Environmental Quality,* Thomas, W. A., Ed., Plenum Press, New York, 1972, 183.
72. **Brown, R. M., McClelland, N. I., Deininger, R. A., and O'Connor, M. F.,** A water quality index — crashing the psychological barrier, in *Indicators of Environmental Quality,* Thomas, W. A., Ed., Plenum Press, New York, 1972, 173.
73. **Thomas, W. A., Goldstein, G., and Wilcox, W. H.,** *Biological Indicators of Environmental Quality,* Ann Arbor Science, Ann Arbor, MI, 1973, 254.
74. **LeBlanc, F., and De Sloover, J.,** Relation between industrialization and the distribution and growth of epiphytic lichens and mosses in Montreal, *Can. J. Bot.,* 48, 1485, 1970.
75. **Barkmann, J. J.,** De epiphyten-flora en vegetatie van Midden-Limburg (België), *Verh. Kon. Nederl. Akad. Wetensch. (2de reeks),* 54, 1, 1963.
76. **De Sloover, J.,** Végétaux épiphytes et pollution de l'air, *Rev. Q. Sci.,* 25, 531, 1964.
77. **Anselin, N.,** Ricthplan voor de ruimtelijke ordering en ontwikkeling van de Denderstreek, Ministerie van Openbare Werken, Brussels, 1963.
78. **Patrick, R.,** Aquatic communities as indices of pollution, in *Indicators of Environmental Quality,* Thomas, W. A., Ed., Plenum Press, New York, 1972, 93.
79. **Warren, C. E.,** *Biology and Water Pollution Control,* Saunders, Philadelphia, 1971.
80. **Cook, S. E. K.,** Quest for an index of community structure sensitive to water pollution, *Environ. Pollut.,* 11, 269, 1976.
81. **Walter, H.,** Impact of human activity on wildlife, in *Sourcebook on the Environment,* Hammond, K. A., Macinko, G., and Fairchild, W., Eds., The University of Chicago Press, Chicago, 1978, 241.
82. **Whittaker, R. H., and Woodwell, G. M.,** Retrogression and coenocline distance, in *Ordination and Classification of Communities,* Whittaker, R. H., Ed., Junk, The Hague, 1973, 53.
83. **Dennis, B., Patil, G. P., and Rossi, O.,** The sensitivity of ecological diversity indices to the presence of pollutants in aquatic communities, in *Environmental Biomonitoring, Assessment, Prediction, and Management,* Cairns, J., Jr., Patil, G. P., and Waters, W. E., Eds., Internationsal Co-operative Publishing House, Fairland, MD, 1981, 379.
84. **Hapke, H.-J.,** Possibilities and limitations of ecotoxicological testing of chemicals, *Toxicol. Environ. Chem.,* 3, 227, 1981.
85. **Goldberg, E. D.,** The mussel watch, *Mar. Pollut. Bull.,* 6, 111, 1975.

86. **Goldberg, E. D., Bowen, V. T., Farrington, J. W., Harvey, G., Martin, J. H., Parker, P. L., Riseborough, R. W., Robertson, W., Schneider, E., and Gamble, E.,** The mussel watch, *Environ. Conserv.,* 5, 101, 1978.
87. **Goldberg, E. D., Koide, M., Hodge, V., Flegal, A. R., and Martin, J.,** U.S. mussel watch: 1977–1978 results on trace metals and radionucleides, *Estuarine, Coastal Shelf Sci.,* 14, 69, 1983.
88. **National Academy of Sciences,** *The International Mussel Watch,* National Academy of Sciences, Washington, D.C., 1980.
89. **Goodman, G. T., Smith, S., Parry, G. D. R., and Inskip, M. J.,** The use of moss-bags as deposition gauges for air-borne metals, *Proc. Conf. Natl. Soc. Clean Air,* 1974.
90. **Mäkinen, A.,** Moss- and peat-bags in air pollution monitoring, *Suo,* 28, 79, 1977.
91. **Temple, P. J., McLaughlin, D. L., Linzon, S. N., and Wills, R.,** Moss bags as monitors of atmospheric deposition, *J. Air Pollut. Control Assoc.,* 31, 668, 1981.
92. **Davies, I. M., and White, H. M.,** Environmental pollution by wind blown lead mine waste. A case study in Wales, U.K., *Sci. Total Environ.,* 20, 57, 1981.
93. **Schultz-Baldes, M.,** Die Miesmuschel *Mytilus edulis* als Indikator für die Bleikonzentration in Weserastuer und in der Deutzchen Butch, *Mar. Biol.,* 21, 98, 1974.
94. **Waldichuk, M.,** The assessment of sublethal effects of pollutants in the sea. Review of the problems, *Phil. Trans. R. Soc. London,* 286, 399, 1979.
95. **Popham, J. D., and D'Auria, J. M.,** Statistical models for estimating seawater metal concentrations from metal concentrations in mussels (*Mytilus edulis*), *Bull. Environ. Contam. Toxicol.,* 27, 660, 1981.
96. **Mackay, D.,** Finding fugacity feasible, *Environ. Sci. Technol.,* 13, 1218, 1979.
97. **Flannagan, J. F., Townsend, B. E., de March, B. G. E., Friesen, M. K., and Leonhard, S. L.,** The effect of an experimental injection of methoxychlor on aquatic invertebrates: accumulation, standing crop, and drift, *Can. Ent.,* 111, 73, 1979.
98. **Bradshaw, A. D.,** Pollution and evolution, in *Experimental Studies on the Biological Effects of Environmental Pollutants,* Mansfield, D. A., Ed., Soc. Exp. Biol., (Sem. Ser.), 1, 135, 1976.
99. **Forsyth, D. J., and Peterle, T. J.,** Uptake of ^{36}Cl-DDt residues by slugs and isopods in the laboratory and field, *Environ. Pollut.,* (Ser. A), 29, 135, 1982.
100. **Boudou, A., and Ribeyre, F.,** Comparative study of the trophic transfer of two mercury compounds—$HgCl_2$ and CH_3HgCl—between *Chlorella vulgaris* and *Daphnia magna*. Influence of temperature, *Bull. Environ. Contam. Toxicol.,* 27, 624, 1981.
101. **Heisinger, J. F., Hansen, C. D., and Kim, J. H.,** Effect of selenium dioxide on the accumulation and acute toxicity of mercuric chloride in goldfish, *Arch. Environ. Contam. Toxicol.,* 8, 279, 1979.
102. **Luoma, S. N., and Bryan, G. W.,** A statistical study of environmental factors controlling concentrations of heavy metals in the burrowing bivalve *Scrobicularia plana* and the polychaete *Nereis diversicolor, Estuarine Coastal, Shelf Sci.,* 15, 95, 1982.
103. **Strong, C. R., and Luoma, S. N.,** Variations in the correlation of body size with concentrations of Cu and Ag in the bivalve *Macoma balthica, Can. J. Fish. Aquat. Sci.,* 38, 1059, 1981.
104. **Bellan, G.,** Indicateurs et indices biologiques dans le domaine marin, *Bull. Ecol.,* 15, 13, 1984.
105. **Cairns, J., Jr,** Introduction to biological monitoring, in *Water Quality Measurements,* Mark, H. B., Jr., and Mattson, J. S., Eds., Marcel Dekker, New York, 1981, 375.
106. **Lebrun, P.,** L'usage des bioindicateurs dans le diagnostic du milieu de vie, in *Ecologie appliquée: Indicateurs biologiques et techniques d'études,* Assoc. Fran. Ingénieurs Ecologues, Mainvilliers, 1981, 175.
107. **Wielgolasky, F. E.,** Biological indicators on pollution, *Urban Ecol.,* 1, 63, 1975.
108. **Cairns, J., Jr.,** Indicator species vs. the concept of community structure as an index of pollution, *Water Res. Bull.,* 10, 338, 1974.
109. **Schafer, E. W., Jr., and Brunton, R. B.,** Indicator bird species for toxicity determinations: Is the technique usable in test method development?, in *Vertebrate Pest Control and Managemement Materials,* Beck, J. R., Ed., American Society for Testing and Materials, Philadelphia, 1979, 157.
110. **Verneaux, J., and Tuffery, G.,** Une méthode zoologique pratique de détermination de la qualité des daux courantes, *Ann. Sci. Univ. Besançon,* 3, 79, 1967.
111. **Coste, M.,** Etude sur la mise au point d'une méthode biologique de détermination de la qualité des eaux en milieu fluvial, Etude Lab. Hydroécologie, Div. Qualité Eaux, Pêche, Pisciculture, C.T.G.R.E.F., Paris, 1974, 80.

112. **Beck, W. M.,** Suggested method for reporting biotic data, *Sew. Ind. Wastes,* 27, 1193, 1955.
113. **Chandler, J. R.,** A biological approach to water quality management, *J. Water Pollut. Control,* 69, 415, 1970.
114. **Empain, A.,** Relations quantitatives entre les populations de bryophytes aquatiques et la pollution des eaux courantes. Définition d'un indice de qualité des eaux, *Hydrobiologia,* 60, 49, 1978.
115. **Pantle, R., and Buck, H.,** Die biologische Uberwachung der Gewässer und die Darstellung der Ergebnisse, *Gas- Wasserfach,* 96, 604, 1955.
116. **Descy, J. P.,** Utilisation des algues benthiques comme indicateurs biologiques de la qualité des eaux courantes, in *La Pollution des Eaux continentales. Incidences sur les biocénoses aquatiques,* Pesson, P., Ed., Gauthiers-Villars, Paris, 1976, 149.
117. **Cairns, J., Jr., Algaugh, D. W., Busey, F., and Chanay, M. D.,** The sequential comparison index—a simplified method for nonbiologists to estimate relative differences in biological diversity on aquatic bottom-dwelling organisms, *J. Water Pollut. Control Fed.,* 40, 1607, 1968.
118. **Cairns, J., Jr., and Dickson, K. L.,** A simple method for the biological assessment of the effects of waste discharges on aquatic bottom-dwelling organisms, *J. Water Pollut. Control Fed.,* 43, 755, 1971.
119. **Lafont, M.,** Oligochaete communities as biological descriptors of pollution in the fine sediments of rivers, *Hydrobiologia,* 115, 127, 1984.
120. **Simpson, E. H.,** Measurement of diversity, *Nature (London),* 163, 688, 1949.
121. **Shannon, C. E., and Weaver, W.,** *The Mathematical Theory of Communication,* University of Illinois Press, Urbana, 1949, 117.
122. **Hill, M. O.,** Diversity and evenness: Unifying notation and its consequence, *Ecology,* 54, 427, 1973.

SECTION II

Metabolism and Toxicity

CHAPTER 5

Comparative Metabolism and Selective Toxicity

Colin H. Walker

Introduction

In terrestrial animals, the efficient elimination of liposoluble xenobiotics depends upon their metabolism to water soluble, readily-excretable metabolites and conjugates. Although enzymic metabolism usually leads to detoxication, in some cases (e.g., polycyclic aromatic carcinogens, some organophosphorus insecticides) it can lead to activation. Differences in regard to these enzymes between species, sexes, strains and age groups may give rise to corresponding differences in susceptibility to xenobiotics.[10,11,12]

In the following account, which will be restricted to Phase 1 enzymes known to have an important role in the detoxication of xenobiotics, emphasis will be given to species differences in monooxygenases, esterases, and epoxide hydrolases, with some discussion of the toxicological significance of trends. All of these types of enzymes are represented in vertebrates and insects although there is considerable variation between species and groups in regard to the forms.

MONOOXYGENASES

Of particular interest are the hepatic microsomal monooxygenases (HMOs) of vertebrates. Similar enzymes are found in the gut and fat body of insects. These enzymes have a heme protein, cytochrome P-450, which provides their catalytic site. Cytochrome P-450, which exists in many different forms, binds and activates molecular oxygen thereby facilitating oxidative attack upon lipophilic substrates that are also bound to the heme protein. Many of these forms have recognized roles in detoxication, but two forms (cytochromes P-450 1A1 and 1A2) are known to activate certain polycyclic aromatic hydrocarbons that are carcinogens.[2]

There are marked differences between the phyla regarding the cytochrome P-450 content of hepatic microsomes. In vertebrates these follow the general trend mammals > birds > fish, although there is considerable overlap between the groups.[8] The activity of the HMO system toward diverse lipophilic substrates varies markedly between groups. In mammals there is a strong inverse correlation between HMO activity, (expressed on a body weight basis), and body size.

HMO activity in fish tends to be very much lower (10 to 15-fold on average) than in mammals of similar body size. Most birds fall between these two extremes but eight species of fish-eating birds had activities comparable to those of fish of the same body size. The sparrowhawk (*Accipiter nisus*) had similarly low activity (this is a specialized predator feeding almost exclusively on birds).[8] On limited evidence reptiles and amphibia also have low HMO activities.[4]

Recent work on the molecular biology of cytochrome P-450 forms has led to the proposal of a phylogenetic classification for them that identifies a number of different families of enzymes.[4,5] One of these families, contains several cytochrome P-450s of wide substrate specificity and inducible by phenobarbitone, which appear to have evolved in response to "plant-animal warfare."

Over a range of species there is an inverse relationship between the biological half lives of certain lipophilic compounds and HMO activities (these are all cases where HMO has a major role in detoxication).[7] This relationship has been shown not only between individuals of one species, but also between different species. The very low HMO activities of fish-eating birds are related to a strong tendency to bioaccumulate lipophilic pollutants such as dieldrin, DDE, and PCBs. It is noteworthy that birds, which have relatively low HMO activities, are more susceptible to carbamate insecticides than mammals.[9] Since HMOs have a major role in the detoxication of carbamates, this may be a causal relationship.

Monooxygenases have been implicated in a number of cases of resistance of insects to insecticides. One example is the marked resistance of one strain of the tobacco budworm (*Heliothis virescens*) to cypermethrin.[3] The resistant strain excretes conjugates of oxidative metabolites much more rapidly than a susceptible strain, and the generation of these metabolites is strongly inhibited by piperonyl butoxide, which acts against the monooxygenase system. Piperonyl butoxide is a powerful synergist for cypermethrin in the resistant strain, but much less so in the susceptible strain, pointing to the importance of the monooxygenase as a resistance factor.

ESTERASES

According to the classification of Aldridge, "A" esterases metabolize organophosphates whereas "B" esterases are inhibited by them.[6] "A" esterases are either absent or at very low levels in the plasma or serum of birds. "A" esterases of mammals can carry out the rapid detoxication of the active (oxon) forms of certain organophosphorus insecticides, but birds do not have this capacity. This appears to be the main reason for the greater toxicity of organophosphorus compounds such as diazinon, pirimiphos-methyl, and pirimiphos-ethyl to birds as compared to mammals.[9]

Differences in "B" esterases have also been related to selectivity. The organophosphorus insecticide, malathion, contains carboxylester bonds that can be hydrolyzed by "B" esterases, with consequent loss of toxicity.[10] The products of hydrolysis, unlike malathion, cannot be activated by HMO. Mammals are less susceptible to malathion than are insects because they have "B" esterases that readily hydrolyze the insecticide; the "B" esterases of insects are ineffective against this insecticide. However, it is interesting that some resistant strains of insects have enhanced levels of "B" esterases that can detoxify malathion. Some strains of aphid (*Myzus persicae*), which have marked resistance to a range of organophosphorus insecticides, have large quantities of a "B" type esterase that can act as a "sink" for organophosphates.[1]

EPOXIDE HYDROLASES

The epoxide hydrolase activity of liver microsomes varies markedly between species when assayed with certain substrates, e.g., the dieldrin analogue HEOM and benzo[*a*]pyrene 4,5-oxide.[8]

Comparing different groups, the activity towards these substrates follows the descending order mammals > birds > fish. Particularly high levels of activity are found in rabbits, humans, and pigs.

Epoxide hydrolases have a role in the detoxication of carcinogens (e.g., the epoxides of benzo[a]pyrene) and certain chlorinated insecticides (e.g., dieldrin). HMOs can generate reactive epoxides by attacking benzene rings and aliphatic double bonds. Species differences in epoxide hydrolases have been shown to be related to different rates of production of mutagenic epoxides in *in vitro* systems.[10] Thus mice have a substantially lower level of epoxide hydrolase activity in relation to HMO activity than do rats; the hepatic microsomes of mice are much more effective than those of rats in activating benzo[a]pyrene in mutagenicity tests. However, the influence of species differences in epoxide hydrolase activity upon the response of animals to potential carcinogens and mutagens still awaits detailed investigation.

CONCLUSIONS

An understanding of the phylogenetic distribution of enzymes that metabolize xenobiotics provides a basis for understanding and predicting risks associated with environmental chemicals. It enables the identification of species or groups that may be especially at risk from particular pollutants, because they are poor in detoxifying ability or strong in activating ability. It may also enable the prediction of points in ecosystems where marked bioaccumulation of certain pollutants may occur, because species lack enzymes that can metabolize them effectively. Background knowledge of this kind should be considered when selecting species for laboratory tests designed to predict environmental hazards of chemicals.

REFERENCES

1. **Devonshire, A. L.,** Role of esterases in resistance of insects to insecticides, *Biochem. Soc. Transact.,* 19(3), 755–758, 1991.
2. **Ioannides, C., Lum, P. Y., and Parke, D. V.,** Cytochrome P-448 and the activation of toxic chemicals and carcinogens, *Xenobiotica,* 14, 119, 1984.
3. **Little, E. J., McCaffery, A. R., Walker, C. H., and Parker, T.,** Evidence for an enhanced metabolism of cypermethrin by a monooxygenase in a pyrethroid resistant strain of *H. virescens, Pest. Biochem. Physiol.,* 34, 58–68, 1989.
4. **Livingstone, D. R., and Stegeman, J. R., Eds.,** Cytochrome P-450 forms and functions in mammals, *Xenobiotica,* 19(10), 1989.
5. **Nebert, D. W., and Gonzalez, F.,** P_{450} genes: Structure, evolution and regulation, *Annu. Rev. Biochem.,* 56, 945–993, 1987.
6. **Reiner, E., Aldridge, W. H., and Hoskin, P. C. G.,** *Proceedings of International Meeting on Enzymes Hydrolyzing Organophosphorus Compounds,* Ellis Horwood, 1989.
7. **Walker, C. H.,** Species differences in microsomal monooxygenase activity, and their relationship to biological half lives, *Drug Metab. Rev.,* 7(2), 295–323, 1978.
8. **Walker, C. H.,** Species variations in some hepatic microsomal enzymes that metabolize xenobiotics, *Prog. Drug. Metab.,* 5, 113–164, 1980.
9. **Walker, C. H.,** Pesticides and birds—mechanisms of selective toxicity, *Agric. Ecosyst. Environ.,* 9, 211–226, 1983.
10. **Walker, C. H., and Oesch, F.,** *Enzymes in Selective Toxicity. Biological Basis of Detoxication,* Academic Press, New York, 1983, 349–368.
11. **Walker, C. H., Ed.,** The role of enzymes in regulating the toxicity of xenobiotics, *Biochem. Soc. Trans.,* 19, 731–767, 1991.
12. **Walker, C. H.,** Comparative toxicology, in *Introduction to Biochemical Toxicology,* 2nd ed., Hodgson, E., and Levi, P. E., Eds., Appleton and Lange, 1994, 193–218.

CHAPTER **6**

Application of Metabolic Studies to the Evaluation of the Toxicity of Chemicals to Man

Luciano Vittozzi

The interaction of a substance with a living organism can be schematically represented as in Figure 1. Each one of the steps indicated in this scheme is relevant in determinig the final effect, and its quantitative description is particular to the substance and organisms considered. Usually, polar hydrophylic chemicals are rapidly metabolized and/or excreted; ionized bulky molecules are absorbed poorly or not at all; lipophylic nonpolar substances tend to bioaccumulate in fatty tissues, and metabolism is often a prerequisite for their excretion.

Metabolic studies play two different roles in the safety evaluation of chemicals. Preliminary information on basic toxicokinetic parameters, such as the fraction of the dose that is absorbed, and the distribution and excretion kinetics, is very useful to identify the appropriate animal for testing and to attain and maintain the concentration of a chemical within the desired range of values in the organs of the experimental animal. However, a deeper understanding of the disposition and elimination of a chemical from a body, as well as several features of its mechanism of toxicity can be achieved only with the knowledge of its transformation in the body, which are supported often by enzymes located in the liver. As a result of these biotransformations a chemical can be converted (1) to a normal body constituent, (2) to products that are quickly excreted and usually, but not always, endowed of lower toxicity than the parent compound, and (3) to more toxic metabolites that can even react chemically and irreversibly with cellular molecules, such as glutathione, membrane lipids, proteins, and DNA.

On the other hand, knowledge of the metabolic processes involved especially in the exposure and toxicokinetic phases (Figure 1) of the interaction between a chemical agent and a living organism is of fundamental importance in any scientifically consistent extrapolation of experimental toxicity data to man. The extrapolation should indeed rely as much as possible on the assessment of the critical concentration of the toxicologically active substance at the target site(s).

A relatively simple example can be represented by cadmium toxicity. Availability of cadmium salts for absorption is related to their solubility. The absorbed fraction of an ingested dose of cadmium is dependent on the species considered: it is 2% in rats, 5% in swines and lambs, 16% in cows, and about 3–8% in humans.[1] Only a very minor part of the absorbed dose (10%) is excreted in a few days after absorption; the rest is excreted at a daily rate of 0.005%. The body

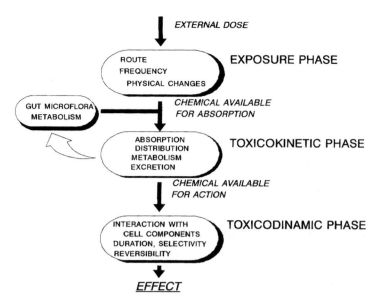

Figure 1. Schematic description of the interaction of a toxicant with a living organism.

burden is essentially located in the kidneys, liver, and lungs, and increases steadily because of the continuous exposure to cadmium, especially through foodstuffs. The toxic effect subsequent to chronic exposure of humans to cadmium is renal damage and the early symptom is proteinuria (appearance of retinol binding proteins) due to inefficient absorption of protein in the glomerular filtrate. Animals showed the same critical damage, and it was possible to determine in the experimental animals that the critical concentration of Cd in the kidney cortex was 200 ppm. This value agreed with theoretical estimations and necropsy measurements of tissue Cd concentrations relatable with the threshold of symptoms of workers exposed for long time periods to Cd. From this value and the other quantitative information on the distribution and kinetics of Cd, it was possible to set the no observed effect level (NOEL) for ingestion of Cd in foodstuffs at 250 μg/person/day.

Usually, there are interspecies differences in susceptibility to foreign compounds, which make the extrapolation of animal data to humans difficult. Although toxicity extrapolation has been based usually on the effects on the most sensitive animal species (conservative approach), the most scientifically correct procedure should use toxicity data on species that are very similar to man as far as the metabolism of the substance under study is considered, and possibly include relevant cases of interindividual variability. This is especially useful in cases of drugs with polymorphic metabolism in humans. Often, however, the metabolic similarity of experimental species to man is not sufficient and therefore knowledge of some aspects of the mechanism of toxicity is needed in order to make appropriate extrapolations. These aspects are mainly related to the metabolic profile of a chemical and do not necessarily involve a complete understanding of the mechanism of action. Within this frame a rather important aspect is the dose dependence of the metabolic profile. Indeed, the very high doses that are often required to observe experimentally overt toxicity in a practicable time frame may give metabolic patterns that are not at all representative of the situation related to the real exposure levels. This happens usually because of saturation of excretory processes and/or metabolic pathways.

Dichloromethane (DCM) has been under assessment for its carcinogenic effect in humans. This compound, when administered at high doses, produces lung and liver tumors in mice but not in rats, and undergoes in the liver two types of biotransformation, one mediated by glutathione-S-transferase and the other catalyzed by cytochrome P-450 (Figure 2). Both pathways give rise to chemically reactive and potentially toxic intermediates, a glutathionyl adduct and formyl chloride,

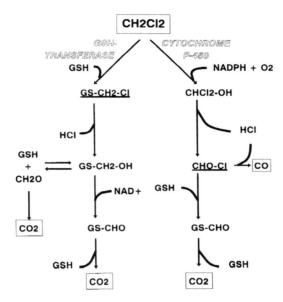

Figure 2. The pathways of dichloromethane metabolism. The reactive intermediates produced along the two pathways are underlined. Metabolic products measured in the expirate are reported in a box. GSH, reduced glutathione.

respectively.[2] Many studies, however, indicate that the glutathione adduct formation is a key step in the DNA damaging action of DCM, whereas formyl chloride is of minor importance to DCM carcinogenicity. The identification of the relevant biotransformation pathway allowed the application of a physiological model for the extrapolation of the carcinogenicity data from experimental animals to humans, based on the metabolic parameters of DCM measured in experimental animals and human tissues. With this model the extrapolation of the carcinogenesis data from the experimental doses to the real levels of human exposure was based on the relationship observed between formation rate of the end product of the GSH-transferase pathway and DCM environmental concentration (Figure 3). Would the oxidative metabolism have been relevant, the extrapolation procedure would have been quite different, since the latter pathway shows high affinity and saturation features (Figure 3) and its activities show a relationship across the species different from those of the glutathione-S-transferase pathway.

When the ingested substance is converted to normal body constituents, the toxicity assessment is simplified. However, this information alone gives no guarantee that the compound is safe: indeed many compounds of the intermediate metabolism are toxic at concentrations higher than the physiological ones and the body may have a limited ability to metabolize them. The possibility exists that the processes of catabolism and/or excretion of the natural component become saturated. This possibility should be kept in mind when assessing the acceptable intake of a food additive or contaminant. Usually, however, there is no need to perform extensive toxicological studies when the additive makes only a small contribution to the normally existing pools. A typical example that can be cited with reference to this problem is methanol. This compound produces lethal effects in man at doses of 60–240 ml per person. Sublethal doses of methanol give rise to several symptoms, the most relevant of which are acidosis and blindness. They are associated with the metabolism of methanol to formaldehyde and formic acid, initiated by alcohol dehydrogenase.[3] Both these compounds, in a bound form, are normally present in the body, taking part to the one-carbon pool associated with tetrahydrofolate coenzymes; most formaldehyde is, however, associated with glutathione, to give a complex that is very effectively metabolized by aldehyde dehydrogenase to formic acid. However, when glutathione is not available, free formaldehyde, which is not effectively metabolized by the dehydrogenase, accumulates in the

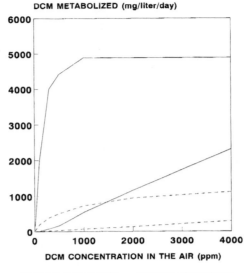

Figure 3. The hepatic metabolism of dichloromethane along the glutathione-S-transferase (GST) and monooxygenase (MFO) pathways, based on a computer simulation of inhalatory exposure of B6C3F1 mice and humans. The simulation is based on a physiological model incorporating toxicokinetic data of DCM on experimental animals as well as metabolic data obtained with preparations of different animal tissues and human autoptic samples.[2]

body reacting with cellular structures. On the other hand when formic acid saturates the available THF coenzymes and the excretory processes in the kidney, intracellular pH falls to low values and a generalized derangement of cellular functions occurs.

Methanol also gives the opportunity to exemplify another interesting situation: the interaction between chemicals. When interactions are known to occur, the possible toxicological consequences should be kept in consideration during the risk assessment procedure. In the case of methanol, an interaction can take place with ethanol at the metabolic level. This is one of the ways in which chemicals, either food components or not, may interact with each other. Methanol is present in nonnegligible amounts in many strong alcoholics and distillates. However, the presence of ethanol, which is also metabolized by alcohol dehydrogenase, inhibits the metabolism of methanol to the aldehyde, favoring the excretion of methanol through the lungs, and allowing the "safe" consumption of these drinks (in fact, the most important toxic effects are due to ethanol).

In addition to the metabolism taking place in the tissues of the exposed organism, it must be kept in mind, when assessing the toxicity of a chemical, that gut microflora may have a pronounced effect on the absorbed form of the ingested compound and on its excretion. The action of microflora is particularly likely on polar and poorly absorbed compounds. In fact, microflora are located in the final parts of the intestinal tract and therefore may have only a minor influence on lipophylic compounds absorbed by passive diffusion in the upper gastrointestinal tract. With relation to the anaerobic conditions of the lower intestine, gut microflora afford mainly reductive and hydrolytic reactions, in contrast to the mostly oxidative and synthetic (conjugative) metabolism taking place in the tissues of the host organism. A peculiar phenomenon deriving from these metabolic differences is the enterohepatic circulation of chemicals. Indeed a lipophylic chemical that is excreted in the bile after glucuronidation in the liver, can be reabsorbed by passive diffusion in the lower intestine after hydrolysis of its glucuronide by gut microflora.

In this way the retention of a compound in the body can dramatically be prolonged.

An example of a particularly relevant influence of gut microflora on the absorption of an exogenous compound is represented by orally ingested stevioside, an intense natural sweetener used in Japan. This compound is a glycoside made up of one disaccharide and one monosaccharide residue linked to a diterpene structure, steviol. Metabolic studies revealed that stevioside administered orally (the route of human exposure) could not be absorbed as such in the intestine. However, it was extensively hydrolyzed by the gut microflora and the free steviol could be absorbed.[4] Subsequently the gut microflora could also act on the steviol conjugates excreted with the bile, giving rise also to enterohepatic circulation and causing a prolonged retention of the compound.

REFERENCES

1. **Ryan, J. A., Pahren, H. R., and Lucas, J. B.,** Controlling cadmium in the human food chain: A review and rationale based on health effects, *Environ. Res.,* 28, 251, 1982.
2. **Andersen, M. E., Clewell III, H. J., Gargas, M. L., Smith, F. A., and Reitz, R. H.,** Physiologically based pharmacokinetics and the risk assessment process for methylene chloride, *Toxicol. Appl. Pharmacol.,* 87, 185, 1987.
3. **Tephly, T. R., Watkins, W. D., and Goodman, J. I.,** The biochemical toxicology of methanol, in *Essays in Toxicology,* Vol. 5, Wayland, J., and Hayes, Jr., Eds., Academic Press, London, 1974, 149.
4. **Nakayama, K., Kasahara, D., and Yamamoto, F.,** Absorption, distribution, metabolism and excretion of stevioside in rats, *Shokuhin Eiseigaku Zasshi,* 27, 1, 1986.

CHAPTER 7

Comparative Aspects of the Metabolism and Toxicity of Xenobiotics in Fish

Luciano Vittozzi, Johannes Keizer, Maria A. Iannelli, and Stefania Soldano

The development of new and more selective pesticides has been the major reason for comparative studies of metabolism in nonmammalian species. Indeed the aim of the present pesticide research is the design of pesticides that are selectively toxic for a predetermined pest or group of pests. The most selective pesticides exploit species-specific receptors or metabolic pathways. However, the design of a pesticide usually takes advantage of quantitative interspecies differences of metabolism that modulate the toxicokinetics and the bioavailability of the substance at the target site, so that the desired species selectivity is attained. Mostly, similar factors are kept into consideration in order to endow pesticides with a low toxicity to fish.

A further reason of interest in comparative metabolic studies, raised from the growing concern for an environment endangered by chemicals, can also be found in the present International regulations for chemicals control, such as the EEC Directive 79/831. In this Directive a series of tests are envisaged that aim to assess the environmental impact of chemicals. Several toxicological tests on fish are required for a chemical already at the lower volumes of production (Table 1). The high consideration devoted to the protection of fish is mainly due to their use as food and because of their presence in the freshwater compartment, which is prone to chemical pollution. For the sake of standardization and of result comparability, a limited number of species are recommended for use in the ecotoxicological tests (Table 2); however, because of a realistic approach, tests on only one of these species are required. As a consequence, it is vital for the environment and essential for a correct evaluation of the impact of chemicals, that recommended test species be toxicologically representative of the species actually present in specific environmental compartments, and that recommended species be equivalent to each other as far as the ecotoxicity test endpoints are considered.

The general opinion attributes differences in fish toxicity test results to the influence of interlaboratory differences in experimental conditions and observation criteria, rather than to true interspecies differences. However, it is to be noted that another source of variability among the EEC test fish may be represented by their different developmental stages: for example, the guidelines recommend the use of juvenile trout and carp and, at variance, of adult guppy and zebra fish. Only recently a comprehensive review of literature data became available to soundly assess

Table 1. Data (with Particular Reference to Ecotoxicological Studies) to Be Presented before the Marketing of New Chemicals According to the VI Amendment to the Law 831/79 EEC

Basic dossier (obligatory)	
Chemical identity	Synonyms, composition, etc.
Chemical data	Utilization, precautions, etc.
Physicochemical properties	Melting point, etc.
Toxicological studies	Acute and prolonged toxicity, mutagenicity
Ecotoxicological studies	Acute toxicity (fish, Daphnia), biotic and abiotic degradation
Possibility to inactivate the chemical	
Level 1 (production > 100 t/year)	
Toxicological studies	Fertility, teratogenesis, prolonged toxicity
Ecotoxicological studies	Acute toxicity test on algae, plants, earth-worm
	Prolonged toxicity of Daphnia and fish
	Bioaccumulation in fish
	Biodegradation
Level 2 (production > 1000 t/year)	
Toxicological studies	Chronic toxicity, carcinogenesis, teratogenesis
Ecotoxicological studies	Bioaccumulation in different species
	Mobility
	Chronic toxicity including reproduction test on fish
	Toxicity to birds

Table 2. Fish Species Recommended by EEC for 96-h Acute Toxicity, 14-Day Prolonged Toxicity and Bioaccumulation Tests

Common name	Scientific name
Guppy	*P. reticulata*
Trout	*S. gairdneri*
Carp	*C. carpio*
Zebra fish	*B. rerio*
Bluegill	*L. macrochirus*
Golden orfe	*L. idus*
Red killifish	*O. latipes*
Fathead minnow	*P. promelas*

how different could be the susceptibilities of various fish species to toxicants.[1] This survey selected data on different EEC-recommended fish species appearing on the same publication, and therefore very likely obtained with minimal differences in test conditions and observation procedures. Comparative data of the acute toxicity (LC_{50}) of about 200 chemical substances were collected. They were produced mainly with the use of bluegill, trout, and fathead minnow and could be analyzed with statistical methods. It appeared that the most frequent value of the ratios between the LC_{50}s of the three species was lower than 2. However, the values of the LC_{50} ratios spread also toward higher values and a still relevant number of substances showed dramatic selectivity in their toxicity (with ratios of LC_{50}s between 10 and 300). Chemicals endowed with selective toxicity were very frequent among the organophosphorus compounds (Table 3),[1] although some cases of selectivity appeared in other chemical classes too. Usually fathead minnow resulted the most tolerant species with respect to trout and bluegill (Table 4).[1] This observation suggests that the action of such selective toxicants is not casual, but is modulated by similar species-dependent biological features. Based on the mechanism of toxicity of organophosphorus compounds, the relevant biological factors are likely to be found in their species-specific inhibition of acetylcholinesterases or to a species-specific balance of their toxifying (monooxygenation) and detoxifying (glutathione conjugation and hydrolysis) metabolic pathways.

Comparative information on the xenobiotic-metabolizing enzyme systems of the EEC test fish is becoming more and more extensive.[2-5] The results obtained with model substrates indicated

Table 3. Selective Toxicants in Different Chemical Classes

Chemical class	Total number of chemicals	Selective toxicants	Frequency
All chemicals	201	24	0.12
Chlorinated	73	6	0.08
Nonchlorinated	125	18	0.14
Carbamates	18	3	0.17
Phosphoesters	42	15	0.33
Others	147	6	0.04
Chlorinated	62	3	0.05
Nonchlorinated	85	3	0.03

Table 4. Relative Susceptibility of Bluegill, Trout, and Fathead Minnow to Selective Toxicants

Fish species	Susceptibility	
	Higher	Lower
Bluegill	20/26	6/26
Trout	13/20	7/20
Fathead minnow	3/26	23/26

Values indicate the number of comparisons in which the fish shows higher or lower susceptibility divided by the total number of comparisons between that fish and the other two species.

that guppy is endowed with monooxygenase activity levels about 10 times higher than all other tested fish independently of the substrate used (Figure 1).[2] Guppy is also endowed with higher GSH-transferase levels, although it seems that the most relevant interspecies differences are related to the isoenzyme qualitative composition.[3] Finally, it appeared that guppy is the most active test fish in hydrolyzing phenyl acetate, a typical A-esterase substrate, and that other types of esterase substrates are hydrolyzed at much lower rates in all test fish.[4] Additionally, oxonase (uncharged phosphoric ester hydrolase) activity does not correlate with any activity toward the model substrates used by Soldano et al.[4,6] More recent studies demonstrated that the differences in monooxygenase activities play a key role in the selective toxicity of diazinon, a phosphothionate ester insecticide, among fish species.[7,8] Furthermore, oxonase levels and the affinity between acetylcholinesterase and diazoxon are of major importance only when monooxygenase activity is low but detectable.[6]

At present the enzymological information obtained with the use of model substrates cannot be used easily to explain or predict the species-specific toxicity among fish. However, it appears that differences in xenobiotic-metabolizing enzymes may be strong enough to result in different susceptibility of fish species to toxicants.

Some of the observations described here deserve special consideration with regard to the evaluation of the impact of chemicals on fish. First, it is very likely that the environmental impact of selective toxicants on fish species is (or will be) underestimated if testing has been (or will be) performed with the use of fathead minnow, due to the comparatively low sensitivity of this species with respect to other commonly used species (Table 4). Second, the present regulatory requirement of testing on a single fish species only is clearly inadequate for those chemical classes that frequently show selective toxicity, i.e., phosphoric and thiophosphoric esters (OP) and, to a lesser degree, carbamates (CM). Indeed it is very important for environmental protection to know whether a compound is endowed with selective toxicity.

Finally, the occurrence of selective toxicants is very limited (3–5%) in chemical classes other than OP and CM chemicals. Therefore, for these compounds testing on a single species is the most realistic approach. However, since this proportion of selective compounds corresponds to a nonnegligible absolute number of chemicals, and for the other reasons described above, effective environmental protection should discourage the use of fathead minnow.

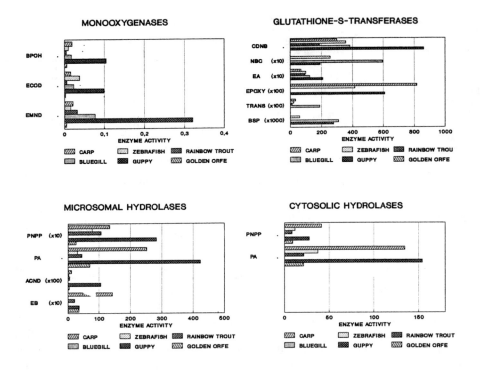

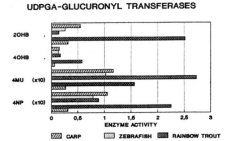

Figure 1. Activities of different hepatic enzymes devoted to the metabolism of xenobiotics in some fish species. Monooxygenases were assayed at their optimum temperatures, i.e., 25°C for trout microsomes and 30°C for microsomes from all the other species. Glutathione-S-transferases, hydrolases, and glucuronyl-transferases were assayed at the fish acclimation temperatures, i.e., 15°C for trout microsomes and 22°C for microsomes from all other fish. Activity is expressed as nmol/min/g body weight. Low activity data have been multiplied by the factor indicated in parentheses beside the substrate abbreviation. Substrate abbreviations and concentrations: BPOH, benzo[a]pyrene (hydroxylase activity), 0.2 mM; ECOD, ethoxy-coumarin (O-deethylase activity), 0.6 mM; EMND, ethylmorphine (N-demethylase activity), 5 mM; CDNB, 1-chloro-2,4-dinitrobenzene, 1 mM; NBC, 4-nitrobenzyl-chloride, 1 mM; EA, ethacrinic acid, 0.2 mM; EPOXY, 1,2-epoxy-3-(p-nitrophenoxy) propane, 0.5 mM; TRANS, $trans$-4-phenyl-3-buten-2-one, 0.05 mM; BSP, bromosulfophthalein, 0.03 mM; PNPP, 4-nitrophenyl-phosphate, 1 mM; PA, phenyl acetate, 0.5 mM; ACND, acetanilide, 10 mM; EB, ethyl butirrate, 0.5 mM; 2OHB, 2-hydroxybiphenyl, 0.25 mM; 4OHB, 4-hydroxybiphenyl, 0.3 mM; 4MU, 4-methylum-belliferone, 0.15 mM; 4NP, 4-nitrophenol, 1mM.

ACKNOWLEDGMENTS

This work has been partly supported by the Italian Research Council (CNR), in the frame of the Targeted Project "Prevention and Control of Disease Factors," Grant 92.00209.PF41.

REFERENCES

1. **Vittozzi, L., and De Angelis, G.,** A critical review of comparative acute toxicity data on freshwater fish, *Aquat. Toxicol.,* 19, 167, 1991.
2. **Funari, E., Zoppini, A., Verdina, A., De Angelis, G., and Vittozzi, L.,** Xenobiotic-metabolizing enzyme systems in test fish. I. Comparative studies of liver microsomal monooxygenases, *Ecotoxicol. Environ. Safety,* 13, 24, 1987.
3. **Donnarumma, L., De Angelis, G., Gramenzi, F., and Vittozzi, L.,** Xenobiotic-metabolizing enzyme systems in test fish. III. Comparative studies of liver cytosolic gluthathione-*S*-transferases, *Ecotoxicol. Environ. Safety,* 16, 180, 1988.
4. **Soldano, S., Gramenzi, F., Cirianni, M., and Vittozzi, L.,** Xenobiotic-metabolizing enzyme systems in test fish. IV. Comparative studies of liver microsomal and cytosolic hydrolases, *Comp. Biochem. Physiol.,* 101C, 117, 1992.
5. **Iannelli, M. A., Marcucci, I., and Vittozzi, L.,** Xenobiotic-metabolizing enzyme systems in test fish. V. Comparative studies of liver microsomal glucuronyl-transferases, *Ecotoxicol. Environ. Safety,* accepted.
6. **Keizer, J., D'Agostino, G., Volpe, M.T., Nagel, R., and Vittozzi, L.,** manuscript in preparation, 1993.
7. **Keizer, J., D'Agostino, G., and Vittozzi, L.,** The importance of biotransformation in the toxicity of xenobiotics to fish. I. Toxicity and bioaccumulation of diazinon in guppy (Poecilia reticulata) and zebra fish (Brachydanio rerio), *Aquat. Toxicol.,* 21, 239, 1991.
8. **Keizer, J., D'Agostino, G., Nagel, R., Gramenzi, F., and Vittozzi, L.,** Comparative diazinon toxicity in guppy and zebra fish: Different role of oxidative metabolism, *Environ. Toxicol. Chem.,* 12, 1243, 1993.

CHAPTER 8

The Role of Cytochrome P-450 in Drug Metabolism and Toxicity

Francesco De Matteis

INTRODUCTION

Man is exposed to a great variety of organic chemicals from the environment that cannot be used for energy production or as building blocks for his own body constituents. These foreign compounds, often referred to as *xenobiotics,* include drugs taken for therapeutic purposes, industrial chemicals that contaminate our food and environment, as well as natural products of vegetable and animal origin. The majority of these chemicals are lipid soluble and can easily cross biological membranes. Once they have gained access to the body they tend to persist as they distribute themselves in the fat and other tissues and are difficult to excrete. Although usually devoid of great chemical reactivity, they can nevertheless interact with biologically important molecules and give rise in this way to a number of pharmacological or toxicological responses.

Several organs and tissues, particularly the liver, contain enzymatic systems that are capable of metabolizing foreign chemicals, making them more water soluble and easier to excrete. This is usually accomplished by a two-phase metabolic process,[1] in which the lipid-like foreign compound is first modified so that a functional group (usually a hydroxyl group) is introduced into the molecule; this functional group is then coupled by conjugating enzymes to a polar molecule [such as glutathione, glucuronic acid, or sulfate (Figure 1)] and in this way the biliary or urinary excretion of the foreign chemical is greatly facilitated.

CYTOCHROME P-450 AND DRUG METABOLISM

Several enzymes contribute to the first phase of biotransformation, that concerned with the introduction of a functional group in the lipid-like foreign compound. Among these several hydrolytic enzymes capable of hydrolyzing various esters and amides, for example, the carboxyl esterases important in the detoxification of certain organophosphorus insecticides,[2] or the microsomal FAD-dependent monooxygenase active on several sulfur-or nitrogen-containing compounds.[3]

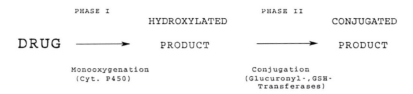

Figure 1. The two-phase metabolism of lipid soluble drugs and foreign chemicals (according to Williams[1]).

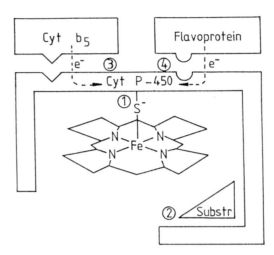

Figure 2. A simplified scheme of the interactions between the apoprotein of cytochrome P-450 and its heme prosthetic group (1), the substrate to be metabolized (2), the satellite components of the electron transport system [cytochrome b_5 (3) and the flavoprotein NADPH-cytochrome P-450 reductase (4)]. (Reproduced from De Matteis, F., *Selectivity and Molecular Mechanisms of Toxicity,* De Matteis, F., and Lock, E. A., Eds., Academic Press, London, 1980, 183–210.)

However, by far the most important enzyme system of "functionalization" is that represented by the group of cytochromes P-450, a family of hemoproteins localized in the membranes of the endoplasmic reticulum of the hepatocyte and most other animal cells. These hemoproteins, taken together, are probably one of the most versatile enzymic systems known, as they are capable of metabolizing a vast number of foreign compounds, compounds of very different size and configuration, some compounds in more than one way.

Before considering the main reasons for this versatility, it is necessary to clarify a few points of nomenclature and to discuss briefly the mechanism of action of this enzyme. Cytochrome P-450 owes its name to the wavelength (450 nm) of maximum absorbance of the spectrum of its carbon monoxide complex. As a hemoprotein, it consists of a protein moiety (apoprotein, or apocytochrome P-450) and of the heme prosthetic group.

The function of the apoprotein is to provide the region (or active site) where the xenobiotic becomes bound in order to be metabolized. To this end, the substrate-binding region of the various cytochromes accommodates compounds with different size and steric configuration—thus providing the basis for the substrate specificity of these enzymes—with the chemical grouping to be oxygenated presumably held in close proximity to the sixth coordination site of the heme prosthetic group (where oxygen is bound and activated). The apoprotein also establishes the structural and functional contacts with the other components of the membranes of the endoplasmic reticulum (microsomal membranes), among these, a flavoprotein and cytochrome b_5 (Figure 2), which participate in the mechanism of the enzymic reaction. These satellite components are involved in the donation of electrons to the cytochrome, two for catalytic cycle, ultimately resulting in one atom of the molecular oxygen being inserted into the drug substrate, while the other is reduced to water. For this reason the type of reaction most commonly associated with cytochrome P-450 is defined as *monooxygenation.* The enzyme can also function as an oxidase or a reductase in certain specialized cases, when no oxygen insertion takes place, but only electron transfer from the drug to the enzyme, or in the opposite direction, from the enzyme to the drug substrate, resulting in drug oxidation or reduction, respectively.

The heme prosthetic group of cytochrome P-450 is represented by protoheme IX (the iron complex of a porphyrin, protoporphyrin IX). The four equatorial positions of the heme iron are occupied by the pyrrole nitrogens of protoporphyrin IX, while the *fifth* coordination site is taken up by a sulfur ligand (thiolate anion) from a cisteinyl residue of the apoprotein. The iron of the heme group can oscillate from the oxidized (ferric) state to the reduced (ferrous) state. Once the heme iron has been reduced, oxygen becomes bound, in order to undergo reduction and activation. Oxygen binding takes place at the sixth coordination site of the heme iron (opposite the thiolate (S-)group, Figure 2), a position that is free or occupied by water in the native ferric hemoprotein.

The microsomal components that participate in the monooxygenation of xenobiotics have been solubilized from the microsomal membranes, purified, and characterized. An active system capable of metabolizing exogenous compounds in the presence of NADPH and oxygen can be *reconstituted in vitro,* by combining the following purified fractions:[4]

1. a flavoprotein (NADPH-cytochrome P-450 reductase);
2. a cytochrome P-450 (as discussed below, several such cytochromes exist, each exhibiting its own substrate specificity);
3. a phospholipid fraction, containing phosphatidylcholine.

Two additional microsomal components (another flavoprotein, NADH-cytochrome b_5 reductase, and cytochrome b_5) have also been shown to contribute to the overall rate of cytochrome P-450-dependent reactions, but the extent of their contribution depends on the particular cytochrome P-450 (and drug undergoing metabolism), being considerable in some cases and negligible or absent in others. These observations have a bearing on the source of the reducing equivalents for the monooxygenase reaction, where two one-electron reduction steps are required for catalytic cycle (see below). With most substrates the first electron to cytochrome P-450 is provided through NADPH-cytochrome P-450 reductase and the second preferentially through NADH-cytochrome b_5 reductase and cytochrome b_5, the favored electron donors being NADPH and NADH, respectively. With other substrates, both electrons appear to reach cytochrome P-450 exclusively through the first flavoprotein, NADPH-cytochrome P-450 reductase.

Activation of Oxygen at the Heme Iron Site of Cytochrome P-450

The cytochrome P-450-dependent drug monooxygenase reaction can be visualized as a cycle initiated by the binding of the substrate to the ferric cytochrome P-450 and concluding with the regeneration of the "uncombined" ferric hemoprotein. The following sequential steps can be considered (Figure 3).

(Step 1). Binding of the substrate to the ferric cytochrome. This is thought to be accompanied by a conversion of the spin state of the heme iron from a low to a high spin state, due to a change in the coordination number of the heme iron from six to five, with loss of an endogenous ligand (water) from the sixth coordination site and significant displacement of the iron from the plane of the four nitrogen atoms. An important associated change is an increase in the redox potential[5] from the strongly negative value of the low spin, native cytochrome (approximately -300 mV) to the value of approx. -200 mV for the high spin cytochrome–substrate complex of rat liver, an increase that will facilitate the reduction of the cytochrome (step 2), once the substrate has been bound.

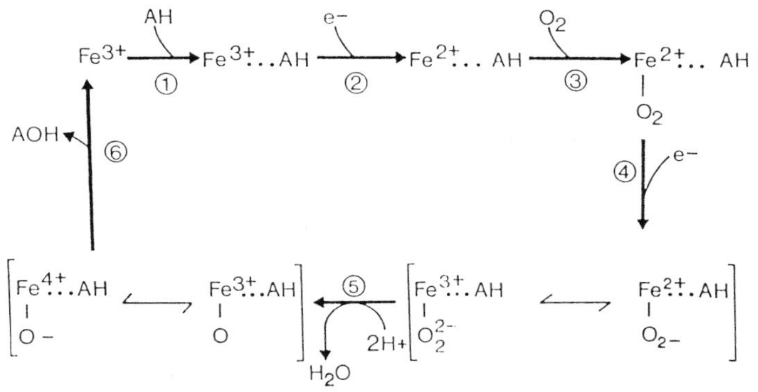

Figure 3. The cytochrome P-450-dependent drug monooxygenase cycle. The individual steps (1 to 6) are discussed in the text. (Modified from De Matteis, F., *Iron in Biochemistry and Medicine, Vol. 2*, Jacobs A., and Worwood, M., Eds., Academic Press, London, 1980, 293–294. With permission.)

(Step 2). The heme iron of the binary substrate–cytochrome P-450 complex undergoes a one-electron reduction from the ferric to the ferrous state.

(Step 3). The reduced hemoprotein then binds oxygen, so that the two reactants are brought together in a ternary complex consisting of substrate, molecular oxygen, and ferrous cytochrome.

(Step 4). The ternary complex undergoes a second one-electron reduction to a postulated peroxy-intermediate containing the substrate, ferric iron, and peroxide anion.

(Step 5). This intermediate decomposes, loses water, and generates an iron-bound monooxygen species (FeO) having a formal oxidation state of +5 {ferryl ion species (Fe^{4+} O^-)} and representing the "activated oxygen" form of the substrate-cytochrome P-450 complex.

(Step 6). The oxygen atom is then inserted into the substrate by a reaction, which is thought to operate through two discrete steps:[6,7] first, a carbon-centered radical intermediate is formed by hydrogen abstraction from the drug substrate to the activated ferryl oxygen of cytochrome P-450; in a subsequent step the carbon radical undergoes recombination with the ferryl oxygen, leading to insertion of a hydroxyl function in the substrate.

Multiplicity of Cytochromes P-450

The first indication that cytochrome P-450 could exist in more than one form was provided by the study of the effect of chemical inducers on the cytochrome P-450 system of the liver microsomes. Important qualitative differences were discovered between the "inducing" effects of 3-methylcholanthrene and phenobarbitone,[8] and this gradually led to the view, now conclusively established, that the two inducers stimulate the production of different hemoproteins. A very large number of cytochrome P-450 enzymes have been isolated from liver and other tissues of eukaryotes and prokaryotes, and their genes have been cloned and sequenced. A classification based on comparison of sequences of different cytochrome P-450 enzymes and their homology has been proposed in 1987[9] and the recommended nomenclature, updated in 1991,[10] recognizes three different levels of relationship: (1) families, comprising enzymes that are at least 40% homologous and designated by an Arabic number (e.g., 1, 2, etc.); and, within each family, (2) subfamilies, designated by a letter (e.g., A, B, C, etc.), and (3) individual members, denoted by a second Arabic numeral. Each individual cytochrome P-450 is therefore designated by the root symbol CYP (denoting cytochrome P-450), followed by an Arabic numeral, a letter, and a second Arabic numeral, in this order, specifying family, subfamily, and individual enzyme, respectively (e.g., CYP1A1, 1A2, this notation referring to both the enzyme *protein* and its mRNA). The same nomenclature is used for the corresponding *gene* (and cDNA) except that in this case italic characters are used (e.g., *CYP1A1, 1A2*, etc.).

Table 1. Main Cytochrome P-450 Enzymes Involved in Metabolism of Xenobiotics, Inducers Known to Stimulate Them, and Prototype Substrates Selectively Metabolized by Them

Enzyme	Species	Inducer	Prototype substrate (and reaction catalyzed)
CYP 1A1	Rat, mouse, rabbit, human	TCDD and 3-methyl-cholanthrene	7-Ethoxy resorufin (7-deethylase)
CYP 1A2	Rat, mouse, rabbit, human	TCDD and 3-methyl-cholanthrene	7-Ethoxy resorufin (7-deethylase)
CYP 2B1	Rat	Phenobarbital	7-Pentoxy resorufin (7-depentylase)
CPY 2B2	Rat	Phenobarbital	7-Pentoxy resorufin (7-depentylase)
CYP 2D6	Human, rat	—	Debrisoquine (4-hydroxylase)
CYP 2E1	Rat, rabbit, human	Ethanol, acetone	Aniline (p-hydroxylase)
CYP 3A1	Rat, human	Dexamethasone	Ethyl morphine (N-demethylase)
CYP 4A	Rat, rabbit	Clofibrate and other peroxisome proliferators	Lauric acid (11,12-hydroxylation)

A list of the most important cytochromes P-450 involved in the metabolism of xenobiotics is given in Table 1. The various enzymes differ in electrophoretic mobility on polyacrylamide gel (a function of their molecular size), in substrate specificity, including stereoselective metabolism for different regions of a complex molecule, such as a steroid, in their response to inducers and inhibitors (see later), in immunoreactivity to appropriately raised specific antibodies, and, most importantly, in their amino acid sequence and corresponding nucleotide sequence of their mRNA and cDNA. Since in most cases several of these enzymes are expressed together in the same tissue and their substrate specificity is not absolute, but overlapping, use is made of one or more of the properties described above to determine the "spectrum" of enzymes present in a tissue and their relative contribution to the overall metabolism of a drug substrate being studied. The use of purified individual enzymes in reconstituted systems is also very useful to ascertain their relative contribution to the metabolism of a given drug substrate.

Main Consequences of the Multiplicity of Cytochrome P-450 Enzymes

It is now appropriate to consider briefly some consequences of the multiplicity of the cytochromes in a given tissue, for example, in the liver.

The first important consequence is the extreme versatility of the monooxygenase system. Not only can each cytochrome accept, as substrates, a wide range of chemicals, but since different cytochromes coexist in the same tissue, each with a different substrate specificity, the system as a whole is one of the most versatile known. As already mentioned, it metabolizes organic substances of very different molecular size, structure, and shape: the chemical groups on which the oxidative attack is directed can also vary considerably, as can the induced modification of structure observed in the metabolites. The main monooxygenation reactions catalyzed by cytochromes P-450 are aliphatic and aromatic hydroxylation, alkene (and arene) epoxidation at carbon–carbon double bonds, and oxidative desulfuration. In same cases the products of hydroxylation are unstable and undergo spontaneous decomposition to deaminated and N- or O-dealkylated derivatives (oxidative deamination and N- or O-dealkylation). Less frequently the cytochrome may function as a reductase, as in azo and nitro reduction or in reductive metabolism of halogenated chemicals, such as CCl_4.

Another consequence of the multiplicity of the liver cytochromes P-450 is that each enzyme is under the control of a different gene, and may therefore vary in activity (and amount) independently of other members of this large family. This is encountered in the selective induction of certain cytochromes seen after administration of inducers (see below), and also in the so-called *genetic polymorphism* of drug oxidation, as will now be discussed.

An example well documented is that represented by debrisoquine, an antihypertensive drug metabolized principally by ring hydroxylation, the major urinary metabolite being the 4-hydroxy derivative. Most people convert between 50 and 70% of an oral dose of this drug into the hydroxylated metabolite, whereas in other individuals only 1 to 3% of the dose is metabolized.[11] Family studies suggest that people with the metabolic defect (poor metabolizers), between 1 and 15% of the population according to the ethnic background, are homozygous for an autosomal recessive character, which shows itself also as defective metabolism of spartein, phenformin, bufuralol, guanoxan, and several other drugs.[12] *In vitro* experiments with microsomes isolated from human livers have confirmed the suspicion that a single cytochrome with relatively high catalytic activity toward debrisoquine and the other drugs mentioned above is the enzyme involved in debrisoquine genetic polymorphism. In individuals who are homozygous for this genetic trait, this cytochrome (CYP 2D6) is absent, because of gene-inactivating mutations, three of which have been identified;[13,14] nevertheless these patients metabolize other unrelated drugs (antipyrine, acetanilid, tolbutamide, and amylobarbitone) perfectly normally both *in vivo* and *in vitro,* since the metabolism of these compounds is catalyzed by other cytochromes present in normal concentration (and activity). These individual variations and similar ones that concern, independently, the metabolism of other drugs, have great importance in medical practice, as the dose required to achieve the optimal plasma concentration of a drug will also vary considerably between individuals. Unless the dosage is adjusted to take into account the metabolic capacity of the patients, poor metabolizers will be at risk of excessive and potentially toxic concentrations of drug in their plasma. It is important to emphasize that similar variations in the metabolism of drugs—again based on genetic polymorphism of cytochrome P-450—have been described in rats, and are likely to be encountered in other species, particularly in the wild.

DRUG INTERACTIONS: DRUG-INDUCED INHIBITION AND STIMULATION OF THE CYTOCHROME P-450 SYSTEM

Foreign chemicals, beside serving as substrates of the cytochrome P-450 system, can also modulate the activity of this important system, decreasing it in some cases and stimulating it in others. These effects are at the basis of the phenomenon of *drug interaction,* where previous or concurrent administration of one drug influences the rate of metabolism of another, with the result that its pharmacological and toxicological properties are modified. These metabolic interactions are due to the fact that the specificity of substrate of most of the cytochromes that metabolize foreign chemicals is not very high. Because of this, any drug-induced inhibition or stimulation of a cytochrome may affect not only the subsequent metabolism of the very drug responsible, but also that of other foreign compounds, which happen to share the same metabolizing enzyme. This metabolic interaction between drugs explains why exposure to a foreign chemical may result in increased sensitivity or apparent tolerance to the pharmacological and toxicological properties of xenobiotics.

Inhibition of Cytochrome P-450

The most common case to be considered is the *competition* of two different foreign chemicals for the same enzyme. Provided that one chemical agent is present in a tissue, for example, in the liver, in a sufficiently high concentration, the metabolism of another substance sharing the same enzyme is significantly reduced, the degree of inhibition depending on the relative concentrations and affinities of the two compounds. This explains, for example, the competition for metabolism between debrisoquine and spartein, two drugs that utilize the same cytochrome. Other examples of inhibitory mechanisms are those provided by cimetidine, a drug with relatively high affinity for hepatic cytochrome P-450, responsible in this way for decreased clearance and prolongation

Table 2. Different Mechanisms by Which Foreign Chemicals May Cause Inhibition of Cytochrome P-450 Enzymes

Mechanism of inhibition	Example of foreign chemical responsible	Parameter affected
Competitive inhibition	Spartein	Metabolism of debrisoquine by CYP2B6
"Ligand" for the sixth coordination site of the heme iron	Cimetidine	Decreased liver metabolism of several drugs in human therapy
Suicidal inhibition	Ethylene	Loss of spectrally determined cytochrome P-450 and cytochrome-dependent activities
Heme depletion due to induction of heme oxygenase	COC_{12}	Loss of spectrally determined cytochrome P-450 and cytochrome-dependent activities

of the half-life of several drugs in human therapy; selective inhibitors of rat hepatic cytochrome P-450 enzymes, such as compound SKF 525-A, metyrapone, α-naphthoflavone, and diphenyloxazole. These compounds also compete with drugs substrates for the active site of cytochrome P-450, but some of them also act as *ligands* for the sixth coordination site of the heme iron (the position intended for oxygen binding) by possessing an atom with a lone electron pair (usually a nitrogen, or other hetero atoms), or by generating a ligand-type metabolite on metabolism[15] by cytochrome P-450 (Table 2). The enzyme inhibition resulting from these cytochrome–ligand complexes is not irreversible, however, since the ligand can be displaced[16] by lipophilic, substrate-type compounds and the cytochrome reactivated.

Another type of inhibition is that caused by foreign chemicals metabolized into very reactive derivatives, which inactivate irreversibly the metabolizing enzyme in a *suicidal* type of reaction. Inhibitors of this type can either damage preferentially the protein moiety of the cytochrome (carbon disulfide, parathion, and other thione-sulfur compounds,[2,17,18] or the heme prosthetic group, as with chemical agents possessing an unsaturated (olefinic or acetylenic) side chain.[19–22] Here too the inhibition is selective, as it will affect only those pathways of metabolism that are catalyzed by the inactivated cytochrome; however, unlike the strictly competitive type discussed above (type 1 in Table 2), with the suicidal inhibitors, the inhibition persists for some time after the agent responsible has been eliminated, as the lesion of the cytochrome will need to be repaired or some new cytochrome synthesized.

A final type of inhibition of cytochrome P-450 has also been described, which is of a more indirect type. It is mediated by a state of heme deficiency in a tissue, for example, in the liver, caused by a stimulation of the enzyme heme oxygenase. As a consequence, there is increased conversion of liver heme into bile pigments, a reduction of the heme available for the synthesis of the various hemoproteins, so that less cytochrome P-450 is formed, hence a reduction of its catalytic activity. This is the molecular basis for the depression of the hepatic drug metabolizing activity caused by cobalt and other metals,[23–25] a response also found in tumor-bearing animals[26] and after administration of stimulators of the immune response.[27]

Drug-Induced Stimulation of Cytochrome P-450

Chemical agents can stimulate the cytochrome P-450-dependent monooxygenase activities, by acting directly, possibly as allosteric effectors[28,29] on the hepatic cytochromes of man and animals. However, the most significant type of stimulation, both for its practical importance and its theoretical interest, is that mediated by *induction,* where the increased activity is accompanied by an increase in the *amount* of the cytochrome(s), usually because of an accelerated rate of biosynthesis. The concentration of one or more liver cytochromes can be induced by exposing an organism (or cells in culture) to foreign chemicals of very diverse chemical structure, both medicinal drugs and environmental pollutants, especially when the foreign compound is lipid soluble and difficult to eliminate. This phenomenon has now been documented for hundreds of chemicals, belonging to several distinct groups, according to the particular cytochromes that they induce. The first two classes of inducers to be identified were (1) the polycyclic aromatic hydrocarbons (typical example, 3-methylcholanthrene) described as inducers by Miller and co-workers[30–32] and (2) the barbiturates (typical example, phenobarbitone), studied by Remmer and co-workers[33,34] in their work on the mechanism of tolerance to the pharmacological action of barbiturates. Additional classes, identified more recently, are represented by (3) synthetic steroids such as pregnenolone 16α-carbonitrile and dexamethasone,[35] (4) clofibrate and other peroxisome proliferators,[36] and (5) ethanol and acetone.[37] The main enzymes induced by these and similarly active inducers are given in Table 1, although some inducers can also activate the synthesis of additional classes of enzymes (for example, phenobarbitone, in addition to its main effect on the cytochromes of the 2B class, can also stimulate the formation of enzymes of the 2C and 3A families). In most cases the induction appears to involve *de novo* synthesis of the enzyme protein through an increased rate of transcription of the corresponding gene and increased concentration of the mRNA encoding the specific cytochrome. Increased stabilization of the enzyme protein (and possibly of its mRNA) by the inducer itself or one of its metabolites may also contribute to the accumulation of the cytochrome in some cases (for example, the antibiotics erythromycin and troleandromycin reduce the rate of degradation of cytochrome P-450 3A protein by forming metabolic complexes with the cytochrome,[38] and in this way promote an apparent induction of the enzyme).

Recent work has addressed the mechanism by which the inducers activate the transcription of specific genes. Evidence for the involvement of specific receptors has been obtained in the induction of cytochromes P-450 1A by TCDD[39,40] and in the induction of cytochrome P-450 4A by clofibrate.[41] In both these cases the inducer appears to activate the receptor, so that it can bind specific recognition sequences in the corresponding gene, acting as a transcription activator, a mechanism shared by similarly acting receptors for steroid and thyroid hormones. In contrast, no receptor has so far been identified for phenobarbitone and, because of this, receptor-independent mechanisms have been postulated for P-450 induction by phenobarbitone and similarly acting inducers; these are discussed in a recent review.[42]

INFLUENCE OF METABOLISM ON THE TOXICITY OF FOREIGN CHEMICALS

The consequences of the metabolism of drugs and other foreign chemicals in terms of their toxicity vary according to whether (1) the parent compound, or (2) one of its metabolites is responsible for the toxic response. In the first case metabolism represents a mechanism of inactivation, as it affords protection, and a stimulation of the metabolic process, brought about, for example, by inducers, produces tolerance to doses that would otherwise be toxic and even lethal. Conversely, any inhibition of the metabolism of these compounds, whether genetically determined or acquired through previous or concurrent exposure to a chemical agent, will increase the risk of toxic symptoms. This situation is illustrated, for example, by poor metabolizers of debrisoquine and other drugs metabolized by cytochrome P-450 2D6, who are faced in general with longer and more intense pharmacological action of the drug they cannot metabolize, with a risk of toxic symptoms due to an exaggerated pharmacological response. Other drugs that are inactivated through hepatic metabolism are zoxazolamine and hexobarbitone, both widely used in experimental pharmacology in the past to monitor the liver cytochrome P-450 activity (by measuring the "paralysis" and "sleeping" time, respectively). Table 3 shows that when the metabolism of hexobarbitone is modified by previous treatment of the experimental animal with either an inducer or an inhibitor of the cytochrome P-450 system, its pharmacological properties are also markedly modified, obtaining either complete tolerance or an excessive pharmacological response, up to irreversible sedation and death.

In the case of a large number of potentially toxic chemicals, however, metabolism by the cytochrome P-450 system is an essential prerequisite for their toxicity. Compounds of this type

Table 3. Metabolic Interactions between Foreign Chemicals

Preliminary treatment	Duration of hypnotic action (min): average and range of at least 3 observations
None	35 (30–39)
Phenobarbital	5 (O–8)
Carbon disulfide	180 (160–OO)

Note: Tolerance or increased sensitivity to the hypnotic action of a barbiturate caused by previous treatment with either phenobarbital, a stimulator of liver drug metabolism, or carbon disulfide, an inhibitor of liver drug metabolism. Male rats (150 g) were given a short-acting barbiturate (hexobarbitone sodium, 150 mg/kg intraperitoneally) and the duration of their "sleeping time" was then measured. Some rats had been pretreated with phenobarbital (80 mg/kg intraperitoneally) or with carbon disulfide (1 ml/kg, orally) 24 h before the experiment. Note that "O" indicates lack of hypnosis, while "OO" denotes irreversible sedation leading to death.

Table 4. Liver Toxicity of Carbon Disulfide: Role of Carbon Disulphide Metabolism

Preliminary treatment	Increase in liver water content due to CS_2 treatment	$^{14}CO_2$ exhaled (% of dose)
None	30	4.7
Phenobarbital	650	13.1
Phenobarbital + CS_2	180	8.1

Note: Male rats (250 g), some pretreated with phenobarbital (80 mg/kg, ip) 48 and 24 h before the experiment [or with a small dose of CS_2 (5 µl, ip per rat) 5 h before the experiment] were all given a standard dose of CS_2 (20 µl, ip per rat) and their liver water content (an index of liver damage) was measured 24 h later. The radioactivity exhaled as $^{14}CO_2$ was collected for 6 h after administering $14CS_2$ (see De Matteis and Seawright,[48] 1973, for more details).

are themselves inactive, but are metabolized to reactive derivatives that interact covalently with biologically important molecules and initiate in this way a number of toxic reactions. The most insidious and threatening types of toxicity are in fact produced through this general mechanism, involving—as the initial step—the production of a reactive metabolite by cytochrome P-450. In this case, pretreatment with an inducer of the appropriate cytochrome P-450 will usually increase toxicity, whereas pretreatment with an inhibitor of the cytochrome will confer protection (the example of the liver toxicity of CS_2 is given in Table 4). The *reactivity* of a metabolite is an important factor in determining which target molecule will be preferentially attacked:[43] some are sufficiently stable and survive long enough to travel some distance from the site of production; others are so reactive that once produced, interact with suitable targets in their immediate environment. Accordingly the following cases can be distinguished.

1. Extremely reactive metabolites. These tend to interact with the very enzyme responsible for their production, as in the suicidal inactivation of cytochrome P-450, to which reference has already been made.
2. Very reactive metabolites. This group probably includes the great majority of cytotoxic drugs, with metabolites sufficiently stable to travel some distance from the activating enzyme, within the same cell where they are produced. It is against these metabolites that several lines of defense exist in the cell, particularly scavenging molecules, like reduced glutathione and conjugating enzymes. Once these various defenses have become depleted or saturated, then the reactive metabolites will attack biologically important molecules and structures, and initiate various types of toxicity. These include degenerative changes of increased severity in the hepatocyte (from hydropic degeneration to toxic necrosis), and also genotoxicity arising from covalent interaction of carcinogens with DNA.
3. Metabolites with intermediate reactivity. In some cases the metabolite is sufficiently stable to reach in a reactive form another organ close to the site of production, there to induce toxic manifestations. A relevant example is provided by the lung toxicity of monocrotaline, an alkaloid obtained from the seeds of *Crotalaria spectabilis* and metabolized to reactive pyrrole derivatives in the liver.[44] The hypothesis that these metabolites are produced in the liver and can then reach the lung through the blood circulation has been supported by

perfusion experiments with isolated liver and lung *in vitro*:[45] monocrotaline itself has no toxic effect on the isolated lung, but is metabolized by the liver into derivatives that damage the lung when the two organs are connected by a perfusion system.

4. Metabolites with limited reactivity. These metabolites are almost inert, but can nevertheless produce damage at a considerable distance from their site of production either because the target organ possesses particular structural or functional characteristics, with which the metabolite selectively interferes (for example, the axonal flow in the periferal nerve), or because the metabolite reaches high concentrations in the distal organ, or, finally, because a primary metabolite is subjected to a second activation step in the periphery. The first mechanism is illustrated by the toxicity of hexane, the agent responsible for peripheral neuropathy in workers of the leather and shoes industries and in glue-sniffers.[46] Hexane is metabolized in the liver to the immediate precursor (2,5-hexanediol) of a neurotoxic agent (2,5-hexanedione). The latter reacts covalently with lysine residues in the neurofilaments of the peripheral nerves and in this way appears to be responsible for protein crosslinking, which may in turn be responsible for a block in the axonal flow and for the development of the neuropathy.[47]

In conclusion, the relationships between metabolism and toxicity of foreign chemicals are very complex and the examples given in this chapter illustrate the considerable spectrum of mechanisms that is possible.

REFERENCES

1. **Williams, R. T.**, Pathways of drug metabolism, in *Handbook of Experimental Pharmacology*, Brodie B. B., and Gillette, J. R., Eds. Springer-Verlag, Berlin, 1971, 28, part 2: 226–242.
2. **De Matteis, F.**, Phosphorothionates, in *Sulphur-Containing Drugs and Related Organic Compounds*, Damani, L. A., Ed., Ellis Horwood, Chichester, 1989, 1, part B: 8–33.
3. **Ziegler, D. M.**, Flavin-containing monooxygenase: Catalytic mechanism and substrate specificities, *Drug Metab. Rev.*, 19, 1–32, 1988.
4. **Coon, M. J., Chiang, Y. L., and French, J. S.**, Chemical characterization of the enzymes involved in drug metabolism, in *The Induction of Drug Metabolism*, Vol. 14, Lindenlaub, R. W., and Estabrook, E., Eds. Schattauer Verlag, Stuttgard, 1979, Symposia Medica Hoechst 14, 201–211.
5. **Sliger, S. G., Cinti, D. L., Gibson, G. G., and Schenkman, J. B.**, Spin state control of the hepatic cytochrome P-450 redox potential, *Biochem. Biophys. Res. Commun.*, 90, 925–932, 1979.
6. **Groves, J. T.**, Biological strategies for the manipulation of dioxygen. The chemistry of cytochrome P-450, *Ann. N.Y. Acad. Sci.*, 471, 99–107, 1986.
7. **Ortiz de Montellano, P. R.**, Cytochrome P-450 catalysis: Radical intermediates and dehydrogenation reactions, *Trends Pharmacol. Sci.*, 10, 354–359, 1989.
8. **Conney, A. H.**, Pharmacological implications of microsomal enzyme induction, *Pharmacol. Rev.* 19, 317–366, 1966.
9. **Nebert, D. W., Adesnik, M., Coon, M. J., Estabrook, R. W., Gonzales, F. J., Guengerich, F. P., Gunsalus, I. C., Johnson, E. F., Kemper, B., Levin, W., Phillips, I. R., Sato, R., and Waterman, M. R.**, The P-450 gene superfamily: Recommended nomenclature, *DNA*, 6, 1–11, 1987.
10. **Nebert, D. W., Nelson, D. R., Coon, M. J., Estabrook, R. W., Feyereisen, R., Fujii-Kuriyama, Y., Gonzales, F. J., Guengerich, F. P., Gunsalus, I. C., Johnson, E. F., Loper, J. C., Sato, R., Waterman, M. R., and Waxman, J.**, The P-450 superfamily: Update on new sequences, gene mapping, and recommended nomenclature, *DNA Cell Biol.*, 10, 1–14, 1991.
11. **Idle, J. R., and Smith, R. L.**, Polymorphisms of oxidation of carbon centres of drugs and other compounds of clinical significance, *Drug Met. Rev.*, 9, 301–317, 1979.
12. **Boobis, A. B., Murray, S., Hampden, C. E., and Davies, D. S.**, Genetic polymorphism in drug oxidation: In vitro studies of human debrisoquine 4-hydroxylase and bufuralol 1'-hydroxylase activities, *Biochem. Pharmacol.*, 34, 65–71, 1985.

13. **Gough, A. C., Miles, J. S., Spurr, N. K., Moss, J. E., Gaedigk, A., Eichelbaum, M., and Wolf, C. R.,** Identification of the primary gene defect at the cytochrome P-450 CYP2D locus, *Nature (London),* 347, 773–776, 1990.
14. **Kagimoto, M., Heim, M., Kagimoto, K., Zeugin, T., and Meyer, U. A.,** Multiple mutations of the human cytochrome P-450IID6 (CYP2D6) in poor metabolizers of debrisoquine, *J. Biol. Chem.,* 265, 17209–17214, 1990.
15. **Franklin, M. R.,** Inhibition of mixed-function oxidations by substrates forming reduced cytochrome P-450 metabolic-intermediate complexes, in *Hepatic Cytochrome P-450 Monooxygenase System,* Schenkman, J. B., and Kupfer, D., Eds., Pergamon Press, Oxford, 1982, 763–783.
16. **Elcombe, C. R., Bridges, J. W., Gray, T. T. B., Nimmo-Smith, R. H., and Netter, K. J.,** Studies in the interaction of safrole with rat hepatic microsomes, *Biochem. Pharmacol.,* 24, 1427–1433, 1975.
17. **De Matteis, F.,** Covalent binding of sulfur to microsomes and loss of cytochrome P-450 during the oxidative desulfuration of several chemicals, *Molec. Pharmacol.,* 10, 849–854, 1974.
18. **Neal, R. A.,** Microsomal metabolism of thiono-sulfur compounds: Mechanisms and toxicological significance, *Rev. Biochem. Toxicol.,* 2, 131–171, 1980.
19. **De Matteis, F., and Cantoni, L.,** Alterations of the porphyrin nucleus of cytochrome P-450 caused in the liver by treatment with allyl-containing drugs. Is the modified porphyrin N-substituted?, *Biochem. J.,* 183, 99–103, 1979.
20. **Ortiz de Montellano, P. R., and Correia, M. A.,** Suicidal destruction of cytochrome P-450 during oxidative drug metabolism, *Annu. Rev. Pharmacol. Toxicol.,* 23, 481–503, 1983.
21. **White, I. N. H.,** Suicidal destruction of cytochrome P-450 by ethynyl substituted compounds, *Pharmaceut. Res.,* 141–188, 1984.
22. **Marks, G. S., McCluskey, S. A., Mackie, J. E., Riddick, D. S., and James, C. A.,** Disruption of hepatic heme biosynthesis after interaction of xenobiotics with cytochrome P-450, *FASEB J.,* 2, 2774–2783, 1988.
23. **Maines, M. D., and Kappas, A.,** Cobalt stimulation of heme degradation in the liver. Dissociation of microsomal oxidation of heme from cytochrome P-450, *J. Biol. Chem.,* 250, 4171–4177, 1975.
24. **De Matteis, F., and Unseld, A.,** Increased heme degradation caused by foreign chemicals: Comparison of the effects of 2-allyl-2-isopropilacetamide and cobaltous chloride, *Biochem. Soc. Trans,* 4, 205–209, 1976.
25. **De Matteis, F.,** Loss of microsomal components in drug-induced liver damage in cholestasis and after administration of chemicals which stimulate heme catabolism, in *Hepatic Cytochrome P-450 Monooxygenase System,* Schenkman, J.B., and Kupfer, D., Eds., Pergamon Press, Oxford, 307–340, 1982.
26. **Beck, W. T., Dedmon, M. L., and Ouellette, M. A.,** Biochemical basis for impaired drug metabolism in tumour-bearing rats. Evidence for altered regulation of hepatic microsomal hemoprotein synthesis, *Biochem. Pharmacol.,* 31, 1535–1543, 1982.
27. **El Azhary, R., Renton, K. W., and Mannering, G. J.,** Effect of interferon inducing agents (polyriboinisinic acid-polyribocytidylic acid and Tilorone) on the heme turnover of hepatic cytochrome P-450, *Mol. Pharmacol.,* 17, 395–399, 1980.
28. **Johnson, E. F., and Schwab, G. E.,** Constitutive forms of rabbit-liver microsomal cytochrome P-450: Enzymatic diversity, polymorphism and allosteric regulation, *Xenobiotica,* 14, 3–18, 1984.
29. **Boobis, A. R., and Davies, D. S.,** Human cytochromes P-450, *Xenobiotica,* 14, 151–185, 1984.
30. **Brown, R. R., Miller, J. A., and Miller, E. C.,** The metabolism of methylated aminoazo dyes. IV. Dietary factors enhancing demethylation, *J. Biol. Chem.,* 209, 211–222, 1954.
31. **Conney, A. H., Miller, E. C., and Miller, J. A.,** The metabolism of methylated aminoazo dyes. V. Evidence for induction of enzyme synthesis in the rat by 3-methylcholanthrene, *Cancer Res.,* 16, 450–459, 1956.
32. **Conney, A. H., Gillette, J. R., Inscoe, J. K., Trams, E. G., and Posner, H. S.,** 3,4-Benzpyrene-induced synthesis of liver microsomal enzymes which metabolize foreign compounds, *Science,* 130, 1478–1479, 1959.
33. **Remmer, H., and Asleben, B.,** Die Aktivierung der Entgiftung in den Lebermikrosomen Wahrend der Gewohnung, *Klin. Wochenschr.,* 36, 332–333, 1958.
34. **Remmer, H.,** Der beschleunigte Abbau von Pharmaka in den Lebermicrosomen unter dem Einflus von Luminal, *Arch. Exp. Pathol. Pharmakol.,* 235, 279–290, 1959.

35. **Schuetz, E. G., Wrighton, S. A., Safe, S. H., and Guzelian, P. S.,** Regulation of cytochrome P-450p by phenobarbital and phenobarbital-like inducers in adult rat hepatocytes in primary monolayer culture and in vivo, *Biochemistry,* 25, 1124–1133, 1986.
36. **Milton, M. N., Elcombe, C. R., and Gibson, G. G.,** On the mechanism of induction of microsomal cytochrome P-450 IVA1 and peroxisome proliferation in rat liver by clofibrate, *Biochem. Pharmacol.,* 40, 2727–2732, 1990.
37. **Thomas, P. E., Bandiera, S., Maines, S. L., Ryan, D. E., and Levin, W.,** Regulation of cytochrome P-450j, a high-affinity *N*-nitrosodimethylamine demethylase, in rat hepatic microsomes, *Biochemistry,* 26, 2280–2289, 1987.
38. **Watkins, P. B., Wrighton, S. A., Schuetz, E. G., Maurel, P., and Guzelian, P. S.,** Macrolide antibiotics inhibit the degradation of the glucocorticoid-responsive cytochrome P-450p in rat hepatocytes *in vivo* and in primary monolayer culture, *J. Biol. Chem.,* 261, 6264–6271, 1986.
39. **Poland, A., and Glover, E.,** Chlorinated dibenzo-p-dioxins: Potent inducers of delta-aminolevulinic acid synthetase and aryl hydrocarbon hydroxylase. II. A study of the structure-activity relationship, *Mol. Pharmacol.,* 9, 736–747, 1973.
40. **Hapgood, J., Cuthill, S., Denis, M., Poellinger, L., and Gustafsson, J. A.,** Specific protein-DNA interactions at a xenobiotic-responsive element: Copurification of dioxin receptor and DNA-binding activity, *Proc. Natl. Acad. Sci. U.S.A.,* 86, 60–64, 1989.
41. **Issemann, I., and Green, S.,** Activation of a member of the steroid hormone receptor superfamily by peroxisome proliferators, *Nature (London),* 347, 645–650, 1990.
42. **Waxman, D. J., and Azaroff, L.,** Phenobarbital induction of cytochrome P-450 gene expression, *Biochem. J.,* 281, 577–592, 1992.
43. **Gillette, J. R., Serrine, S. L., Monks, T. J., Satoh, H., and Pohl, L. R., Eds.,** *Chemically Reactive Metabolites: Introductory Remarks,* Taylor & Francis, London, 1985, 231–237.
44. **Mattocks, A. R.,** Mechanisms of pyrrolizidine alkaloid toxicity, in *Pharmacology and the Future of Man, Proceedings 5th Int. Cong. Pharmacol.,* Loomis, T.A., Ed., Karger, Basel, 1973, 2, Toxicological Problems: 114–123.
45. **Lafranconi, W. M., and Huxtable, R. J.,** Hepatic metabolism and pulmonary toxicity of monocrotaline using isolated perfused liver and lung, *Biochem. Pharmacol.,* 33, 2479–2484, 1984.
46. **Divincenzo, G. D., Krasavage, W. J., and O'Donoghue, J. L.,** Role of metabolism in hexacarbon neuropathy, in *The Scientific Basis of Toxicity Assessment,* Witschi, H., Ed., Elsevier, Amsterdam, 1980, 183–200.
47. **Graham, D. G., Anthony, D. C., Boekelheide, K., Maschmann, N. A., Richards, R. G., Wolfram, J. W., and Ramsay Shaw, B.,** Studies of the molecular pathogenesis of hexane neuropathy. II. Evidence that pyrrole derivatization of lysyl residues leads to protein crosslinking, *Toxicol. Appl. Pharmacol.,* 64, 415–422, 1982.
48. **De Matteis, F., and Seawright, A. A.,** Oxidative metabolism of carbon disulphide by the rat. Effect of treatments which modify the liver toxicity of carbon disulphide, *Chem.-Biol. Interact.,* 7, 375–388, 1973.
49. **De Matteis, F.,** Drugs as suicide substrates of cytochrome P-450, selectivity and molecular mechanisms of toxicity, De Matteis, F., and Lock, E. A., Eds., Macmillan, London, 1987, 183–210.
50. **De Matteis, F.,** Cytochrome P-450 and the metabolism of environmental chemicals, in *Iron in Biochemistry and Medicine, Vol. 2,* Jacobs. A., and Worwood, M., Eds., Academic Press, London, 1980, 293–324.

CHAPTER 9

Estimation of Induction of Enzymes That Metabolize Xenobiotics *In Vitro* and *In Vivo*

John R. Bend

INTRODUCTION

The Cytochrome P-450 Monooxygenase System

The most important enzymatic pathway for the metabolism of lipophilic compounds in animals is oxidation by the microsomal cytochrome P-450 (P-450) monooxygenase system which is composed of multiple forms or isozymes of the hemoprotein P-450 belonging, in mammals, to 10 distinct gene families and eighteen subfamilies,[1] and the flavoprotein, NADPH-P-450 reductase.[2] In 1991 Porter and Coon[3] characterized the P-450 monooxygenases as "the most versatile biological catalyst known" because they catalyze the oxidation of a multitude of lipophilic compounds including both endogenous compounds (bile acids, cholesterol, eicosanoids, fatty acids, lipid hydroperoxides, retinoids, steroid hormones) and exogenous or xenobiotic compounds (antioxidants, carcinogens, drugs, environmental pollutants, food additives, hydrocarbons, pesticides). With several classes of xenobiotic substrates, including chemical procarcinogens,[4] selected metabolites are more toxic than the parent chemical, a process termed metabolic activation or toxication. Endogenous compounds are also bioactivated by P-450; for example, arachidonic acid is metabolized to four isomeric epoxyeicosatrienoic acids (EETs) and to several other monohydroxy eicosatetraenoic acids (16-,17-,18-,19-, and 20-HETEs) formed only by this pathway.[5] Some EETs and HETEs have potent physiological and/or pathobiological effects in multiple tissues and cell types.[6,7] Isozymes of P-450 vary in their ability to metabolize chemicals: marked differences in regioselectivity and stereoselectivity of oxidation and apparent kinetic constants (K_m and V_{max} values) with the same substrate are common[7,9] and can be markedly altered by substitution of a single amino acid residue[10,11] at a critical site in the apoprotein. There are also large differences in the P-450 isozyme profile of different tissues and cell types. Liver contains the highest concentration of microsomal P-450, the greatest diversity of individual P-450 isozymes, and consequently this organ normally predominates in quantitative drug metabolism. P-450 is also found in many extrahepatic sites, however, including lung, gut, kidney, nasal and eye epithelia, and placenta. In lung, for example, the highest concentrations of P-450 are normally

Table 1. Cytochrome P-450 Subfamilies Important in Human Drug Metabolism[23,27-29]

P-450 Subfamily	Substrates oxidized
1A	Phenacetin, caffeine
2C	Mephenytoin, nirvanol, diazepam, tolbutamide
2D	Antiarrythmic drugs Encainide, flecanide, propranolol
	Tricyclic antidepressants Imipramine, nortriptyline
	Neuroleptics Haloperidol, perphenazine
2E	Ethanol, acetaminophen
3A	Cyclosporin A, nifedipine, lidocaine, quinidine, erythromycin, testosterone, cortisol

found in (epithelial) Clara and alveolar type II cells but lower amounts occur in ciliated, goblet, and vascular endothelial cells as well as alveolar macrophages.[12-14] The selective modulation of P-450 isozymes in a single tissue or cell type can have pronounced effects on the metabolism of both endogenous and exogenous substances, and on chemical-mediated target organ and/or cell toxicity by altering the balance between toxication and detoxication reactions.[15,16] Thus, the P-450 monooxygenase system plays a pivotal role in toxicology.

P-450 is also important in pharmacology because it serves as the rate-limiting step in the metabolism of many drugs. Those P-450 subfamilies that contribute to drug metabolism in humans are shown in Table 1, along with a few drugs whose metabolism is primarily catalyzed by these isozymes. (To be classified in the same subfamily P-450 isozymes must have at least 67% homology in amino acid sequence.[1]) P-450 2C, 2D, and 3A subfamilies are of particular interest in human drug metabolism for two reasons. First, they oxidize a wide array of commonly used drugs. Second, genetic polymorphism of isozymes of the P-450 2C[17], 2D,[18] and 3A[19] subfamilies are known to occur in humans and individuals may have a deficiency for the metabolism of some drugs. The molecular basis for the genetic deficiency can be that a particular isozyme is not expressed by some individuals or that a mutated protein with no or very little activity is expressed.[20] There are also differences in the ability of individual humans to metabolize some important drugs including cyclosporin A, 1,4-dihydropyridines such as nifedipine and felodipine, and erythromycin due to marked variations in the expression of P-450 3A4 in liver.[21]

The microsomal P-450 monooxygenase system is also important in physiology because it biosynthesizes (anabolism) and metabolizes (catabolism) endogenous compounds that function as intercellular and/or intracellular messengers.[22]

Several excellent reviews of the P-450 monooxygenases are available,[3,23-26] including those isozymes found in humans.[27-29] Since the P-450 system is essential for the metabolism of many lipophilic xenobiotics, it is well studied at the structural and regulatory levels, and much is known about its induction,[30-32] it is the only enzyme system discussed here. However, many of the comments made are also applicable to other inducible enzymes that participate in the metabolism (and ultimate excretion) of lipophilic chemicals. These include the microsomal UDP-glucuronosyltransferases,[33] the cytosolic glutathione S-transferases,[34] and the microsomal epoxide hydrolases.[35]

Enzyme Induction

Induction refers to an increase in the steady-state concentration of an enzyme, in this case a specific P-450 isozyme, subsequent to the administration of or exposure to a chemical compound. Induction most frequently results from an increase in gene transcription (increased protein synthesis) but it can also occur as a result of decreased degradation of a protein or its mRNA. Induction of P-450 isozymes quantitatively important in the metabolism of a drug or other chemical normally increases its rate of biotransformation and can result in increased, decreased, or unaltered pharmacological activity of a drug or toxicity of a chemical, depending upon whether the induced metabolic pathways are toxication or detoxication in nature.

Table 2. Inducers of Various Subfamilies of Cytochrome P-450 Involved in Drug Metabolism in Humans[23,24,27–32]

Cytochrome P-450 subfamily	Prototype inducer
1A	Polycyclic hydrocarbons[a]
	Charbroiled meat[a]
	Dioxins[a]
	Bioflavanoids[a]
2B	Phenobarbital
2C	Phenobarbital
2E	Ethanol[a]
	Isoniazid[a]
	Fasting[a]
3A	Phenobarbital[a]
	Dexamethasone[a]
	Erythromycin[a]
	Rifampicin[a]

[a]Predominant subfamily induced.

Many different classes of compounds including drugs, pesticides, environmental pollutants, and certain food constituents can induce P-450 isozymes but induction is P-450 subfamily (and isozyme) selective for a given chemical class (Table 2). This is not surprising, since genes closely related in their coding regions are also closely related in their regulatory regions.[24]

The P-450 1A subfamily is strongly induced by polycyclic aromatic hydrocarbons (PAH) such as benzo[a]pyrene and 3-methylcholanthrene, polychlorinated dioxins (e.g., TCDD) and dibenzofurans, as well as planar polychlorinated (PCB) and polybrominated biphenyl (PBB) congeners. Enough PAH are formed during the charcoal broiling of beef to induce P-450 1A in humans that eat four such meals a week; a similar response occurs in smokers. Bioflavanoids, such as β-naphthoflavone, found in foods will also induce P-450 1A isozymes.[30,31] In mammals, there are two P-450 1A isozymes, P-450 1A1 (which is induced in both liver and extrahepatic tissues) and P-450 1A2, which is expressed and induced only in liver.[24,36]

The P-450 1A subfamily is of special interest as a biomarker of environmental pollution since induction occurs in fish and invertebrates as well as mammals and because many of the chemicals that are inducers of this system are extremely toxic and occur as environmental contaminants. Some of the advantages of using this system for environmental surveillance are that the mechanism of P-450 1A induction is well understood,[24,37] it occurs via the Ah receptor with a dose–response relationship parallel to that of the toxicity of these chemicals, which is also dependent upon interaction with this receptor, the response is extremely sensitive and returns to normal when the chemical pollutants are removed. My colleagues and I[38] and Stegeman[39–40] have used P-450 1A induction in liver of marine fish to demonstrate their exposure to varying degrees of pollutants that induce 1A1 in their natural environment.

The induction of pulmonary P-450 1A1 is also of major interest in humans because of the relationship between the up-regulation of this isozyme by PAH and the conversion of these compounds to ultimate carcinogens by 1A1.[41]

Phenobarbital is a prototypical inducer for several other compounds including chlorcyclizine, chlordane, DDT, phenytoin, phenylbutazone, and tolbutamide. In mammals, phenobarbital is a potent inducer of the P-450 2B subfamily and a weaker inducer of P-450 2C and 3A isozymes. In humans, the predominent P-450 subfamily induced by phenobarbital is 3A. Several important drugs including dexamethasone, erythromycin, and rifampicin also induce P-450 3A isozymes in humans.

Repeated, large doses of ethanol or isoniazid will induce P-450 2E1 in experimental animals and humans. This isozyme is also up-regulated by starvation.

For the reasons discussed above, including the possibility of enhanced toxicity from exposure to drugs and environmental pollutants, it is often important to be able to determine whether or not an enzyme system is induced (or deficient) and the biochemical nature of the induction (or

Table 3. Cytochrome P-450 Subfamily and/or Isozyme Selective Substrates[23,24,27–29]

P-450 Subfamily or isozyme	Monooxygenase activity
1A1	7-Ethoxyresorufin O-deethylation
1A2	Phenacetin O-deethylation
2B1	7-Pentoxyresorufin O-depentylation
2C	Testosterone hydroxylation
2E	Aniline 4-hydroxylation
3A	Erythromycin N-demethylation

Table 4. Cytochrome P-450 Subfamily Selective Mechanism-Based Inhibitors

P-450 Subfamily	Inhibitor
1A	2-Ethynylnaphthalene[42]
	N-(2-p-Nitrophenyl)dichloracetamide[43]
2B	Secobarbital[44]
	N-α-Methylbenzyl-1-aminobenzotriazole[45–47]
2C	Cimetidine[48]
	21,21-Dichloropregnenolone[49]
2E	Diallyl sulfone[50]
3A	Gestodene[51]
	Thiosteroids[52]
4A	10-Undecynoic acid[53]

deficiency), which includes the isozyme selectivity of the response and the type of compound responsible.

METHODS TO ASSESS INDUCTION OF P-450 *IN VITRO*

Dependent upon Cytochrome P-450 Monooxygenase Activity

Some of the methods commonly used to demonstrate enzyme induction *in vitro* rely upon enzyme activity with substrates that are highly selective for individual isozymes or isozymes of individual subfamilies of P-450 (Table 3). Depending upon the degree of induction, increases in turnover of these substrates by microsomes from hepatic or extrahepatic tissues can be very marked (up to several hundred fold). Monooxygenase activities that are commonly used for this purpose include benzo[*a*]pyrene hydroxylation and 7-ethoxyresorufin O-deethylation for P-450 1A1; phenacetin O-deethylation or aromatic amine (e.g., 4-aminobiphenyl) N-hydroxylation for P-450 1A2; 7-pentoxyresorufin O-dealkylation for P-450 2B1/2 (rat) and 2B4 (rabbit) and orthologues of these isozymes in other species; the regioselective and stereoselective oxidation of testosterone for P-450 2C and 3A isozymes; aniline 4-hydroxylation for P-450 2E; and erythromycin N-demethylation for P-450 3A isozymes.

The discovery of P-450 subfamily selective/specific mechanism-based inhibitors (Table 4) provides another way to monitor monooxygenase activities for induction *in vitro*. Comparison of monooxygenase activities with the isozyme/subfamily selective substrates mentioned above with microsomes preincubated with or without NADPH and the appropriate concentration of the P-450 selective mechanism-based inhibitor will give further assurance that an elevated monooxygenase activity is catalyzed by one specific P-450 isozyme or subfamily. Such verification is especially worthwhile if working in a species whose P-450 monooxygenases have not been characterized.

One advantage of including a functional assay when evaluating whether or not induction has occurred is that immunochemical and cDNA hybridization assays (for P-450 and mRNA concentrations, respectively) do not differentiate between apoenzyme (inactive, does not contain protoporphyrin IX) and holoenzyme (active monooxygenase that contains heme).

Dependent upon Content of Cytochrome P-450 Isozyme or mRNA

The amazing progress made over the last decade in investigating the structures of different P-450 isozymes at the amino acid and nucleotide sequence levels[1,24,29] has provided the necessary reagents (monospecific polyclonal and monoclonal antibodies; cDNA and synthetic oligonucleotide probes) to specifically assay for individual P-450 isozymes and their mRNA. There is sufficient homology of individual isozymes that these antibodies and cDNA probes can be successfully used across species, at least to decipher whether induction has occurred and the relative increase at the protein and/or mRNA levels.

For example, antibodies to P-450 are used to quantitate (semiquantitate across species) the amount of specific P-450 isozymes present in microsomal preparations by immunostaining of proteins after separation by SDS-polyacrylamide gel electrophoresis (i.e., western blotting) and mRNAs encoding specific P-450 isozymes are quantitated by hybridization to a radiolabeled cDNA or oligonucleotide probe (under conditions of lower stringency across species).

These techniques have been successfully used in many recent investigations[42,54–58] to follow changes in isozyme expression during development as well as induction subsequent to administration or exposure to a chemical or complex mixtures of chemicals.

However, as mentioned above, these methods are best coupled with functional measurements of monooxygenase activity to get an accurate assessment of the status of the microsomal P-450 monooxygenase system in liver and extrahepatic tissues of experimental or wild animals that can be sacrificed for these purposes. Unfortunately, due to its invasive nature, this methodology cannot be applied to humans although tissues removed during surgery and various populations of easily isolated and enriched cells (lymphocytes, alveolar macrophages) are routinely studied in this manner.

METHODS TO ASSESS INDUCTION OF P-450 *IN VIVO*

Pharmacokinetic Studies

Initial studies of *in vivo* enzyme induction in humans relied on analysis of the concentration of an indicator drug in plasma (area under the plasma concentration curve or AUC measurements) over time. Enzyme induction is indicated by marked decreases in AUC values or by a markedly decreased plasma half-time subsequent to drug administration or exposure to various environmental contaminants. Drugs that have been used as *in vivo* probes in studies of this type include aminopyrine, antipyrine, phenytoin, theophylline, and caffeine. Similar experiments have also been performed by monitoring drug concentrations in saliva over time so that repeated blood sampling is unnecessary.

The Breath Test

An excellent noninvasive procedure that can be used for compounds that are metabolized to CO_2, or other volatile metabolites, is the so-called breath test. This is best used for compounds that have an *O*-methyl or *N*-methyl substituent that is demethylated to formaldehyde by P-450 monooxygenase activity. The formaldehyde is rapidly converted to formate and bicarbonate *in vivo*, and then 1 mol of CO_2 is exhaled for each 2 mol of drug demethylated.[59] Drugs containing a radioactive ($^{14}CH_3$) or stable ($^{13}CH_3$) isotope in an appropriate position of the molecule are used and the rate at which $^{14}CO_2$ or $^{13}CO_2$ is exhaled over time is determined by liquid scintillation spectroscopy (for ^{14}C) or mass spectrometry (for ^{13}C). The most detailed research conducted with the breath test in humans has used *N*-[^{14}C]methylerythromycin as substrate, and has verified that this procedure selectively measures monooxygenase activity of P-450 3A isozymes.[60–62] Marked changes in the rate of CO_2 expiration have also been noted with substrates including aminopyrine, caffeine, phenacetin, and ethanol in humans following exposure to various drugs and environmental contaminants. However, the relative contributions that various P-450 subfamilies are making to the increased turnover of these substrates is yet to be definitively demonstrated. In any case this is an extremely important area and additional isozyme selective/specific substrates will certainly be validated and/or developed in the near future.

HPLC Analysis of Urinary Metabolites

It is possible to analyze the concentrations of parent drug and its metabolites (both conjugated and unconjugated) excreted in urine by high-performance liquid chromatography (HPLC). This methodology has excellent potential for assessment of *in vivo* induction of the P-450 monooxygenase system in humans, particularly now that many recombinant human P-450 isozymes have been expressed in cells and their ability to metabolize common drug and carcinogen substrates assessed.[21,63,64] Consequently, metabolic pathways of drugs catalyzed by a single isozyme (or a few closely related isozymes) will continue to be identified and provide additional probes for assessment of *in vivo* P-450 metabolism at the isozyme level. Phenotyping of individuals for genetic deficiencies by HPLC analysis of metabolites excreted in urine over a fixed time period after administration of a specific dose of drug (e.g., debrisoquine or bufuralol for P-450 2D6 deficiency) is also common and economical (as opposed to genotyping by molecular biology approaches).

CONCLUSION

There are now a wide array of sophisticated *in vitro, invasive* techniques available to assess whether or not the P-450 monooxygenase system of an individual human being or a laboratory or wild animal is induced and the specific P-450 isozyme that has been up-regulated. Recently, progress has also been made in the development of noninvasive procedures that provide similar information. Given the importance of such assessments further developments will occur rapidly in this area.

ACKNOWLEDGMENT

This work was supported by Medical Research Council of Canada Grant MT 9722 to JRB.

REFERENCES

1. **Nebert, D. W., Nelson, D. R., Coon, M. J., Estabrook, R. W., Feyereisen, R., Fujii-Kuriyama, Y., Gonzalez, F. J., Guengerich, F. P., Gunsalus, I. C., Johnson, E. F., Loper, J. C., Sato, R., Waterman, M. R. and Waxman, D. J.,** The P450 superfamily: Update of new sequences, gene mapping, and recommended nomenclature. *DNA Cell Biol.,* 10, 1, 1991.
2. **Black, S. D. and Coon, M. J.,** Comparative structures of P-450 cytochromes, in *Cytochrome P-450: Structure, Mechanism, and Biochemistry,* Ortiz de Montellano, P. R., Ed., Plenum Press, New York, 1986, 161.
3. **Porter, T. D., and Coon, M. J.,** Cytochrome P-450: Multiplicity of isoforms, substrates, and catalytic and regulatory mechanisms. *J. Biol. Chem.,* 266, 13469, 1991.
4. **Guengerich, F. P.,** Roles of cytochrome P450 enzymes in chemical carcinogenesis and cancer chemotherapy, *Cancer Res.,* 48, 2946, 1988.
5. **Capdevila, J. H., Falck, J. R., and Estabrook, R. W.,** Cytochrome P450 and the arachidonate cascade, *FASEB J.,* 6, 731, 1992.

6. **McGiff, J. C.,** Cytochrome P-450 metabolism of arachidonic acid, *Annu. Rev. Pharmacol. Toxicol.,* 31, 339, 1991.
7. **Fitzpatrick, F. A., and Murphy, R. C.,** Cytochrome P-450 metabolism of arachidonic acid: Formation and biological actions of "epoxygenase"-derived eicosanoids. *Pharmacol. Rev.,* 40, 229, 1989.
8. **Harris, C., Philpot, R. M., Hernandez, O., and Bend, J. R.,** Rabbit pulmonary cytochrome P-450 monooxygenase system: Isozyme differences in the rate and stereoselectivity of styrene oxidation, *J. Pharmacol. Exp. Ther.,* 236, 144, 1986.
9. **Harada, N., and Negishi, M.,** Substrate specificities of cytochrome P-450, C-P-450$_{16\alpha}$ and P-450$_{15\alpha}$, and contribution to steroid hydroxylase activities in mouse liver microsomes, *Biochem. Pharmacol.,* 37, 4778, 1988.
10. **Juvonen, R. O., Iwasaki, M., and Negishi, M.,** Structural function of residue-209 in coumarin 7-hydroxylase (P450coh): Enzyme-kinetic studies and site-directed mutagenesis, *J. Biol. Chem.,* 266, 16432, 1991.
11. **Matsunaga, E., Zeugin, T., Zanger, U. M., Aoyama, T., Meyer, U. A., and Gonzalez, F. J.,** Sequence requirements for cytochrome P-450IID1 catalytic activity: A single amino acid change (ILE^{380} PHE) specifically decreases V_{max} of the enzyme for bufuralol but not debrisoquine hydroxylation, *J. Biol. Chem.,* 265, 17196, 1990.
12. **Domin, B. A., Devereux, T. R. and Philpot, R. M.,** The cytochrome P-450 monooxygenase system of rabbit lung: Enzyme components, activities, and induction in the nonciliated bronchiolar epithelial (Clara) cell, alveolar type II cell, and alveolar macrophage, *Mol. Pharmacol.,* 30, 296, 1986.
13. **Serbjit-Singh, C. J., Nishio, S. J., Philpot, R. M., and Plopper, C. G.,** The distribution of cytochrome P450 monooxygenase in cells of the rabbit lung: An ultrastructural immunocytochemical characterization. *Mol. Pharmacol.,* 33, 279, 1988.
14. **Overby, L. H., Nishio, S., Weir, A., Carver, G. T., Plopper, C. G., and Philpot, R. M.,** Distribution of cytochrome P450 1A1 and NADPH-cytochrome P-450 reductase in lungs of rabbits treated with 2,3,7,8-tetrachlorodibenzo-*p*-dioxin: Ultrastructural immunolocalization and *in situ* hybridization. *Mol. Pharmacol.,* 41, 1039, 1992.
15. **Bend, J. R., Serabjit-Singh, C. J., and Philpot, R. M.,** The pulmonary uptake, accumulation and metabolism of xenobiotics, *Annu. Rev. Pharmacol. Toxicol.,* 25, 97, 1985.
16. **Bend, J. R. and Serabjit-Singh, C. J.,** Xenobiotic metabolism by extrahepatic tissues: Relationship to target organ and cell toxicity, in *Drug Metabolism and Drug Toxicity,* Mitchell, J.R., and Horning, M. G., Eds., Raven Press, New York, 1984, 99.
17. **Nakamura, K., Goto, F., Ray, W. A., McCallister, C. B., Jacqz, E., Wilkinson, G. R., and Branch, R. A.,** Interethnic differences in genetic polymorphism of debrisoquine and mephenytoin between Japanese and Caucasian populations, *Clin. Pharmacol. Ther.,* 38, 402, 1985.
18. **Meyer, U. A., Skoda, R. C., and Zanger, U. M.,** The genetic polymorphism of debrisoquine/sparteine metabolism-molecular mechanisms, *Pharmacol. Ther.,* 46, 297, 1990.
19. **Wrighton, S. A., Ring, B. J., Watkins, P. B., and Vandenbranden, M.,** Identification of a polymorphically expressed member of the human cytochrome P-450III family, *Mol. Pharmacol.,* 36, 95, 1989.
20. **Kagimoto, M., Heim, M., Kagimoto, K., Zeugin, T., and Meyer, U. A.,** Multiple mutations of the human cytochrome P-450IID6 gene (CYP2D6) in poor metabolizers of debrisoquine: Study of the functional significance of individual mutations by expression of chimeric genes, *J. Biol. Chem.,* 265, 17209, 1990.
21. **Aoyama, T., Yamano, S., Waxman, D. J., Lapenson, D. P., Meyer, U. A., Fischer, V., Tyndale, R., Inaba, T., Kalow, W., Gelboin, H. V., and Gonzalez, F. J.,** Cytochrome P-450 hPCN3, a novel cytochrome P450IIIA gene product that is differentially expressed in adult human liver: cDNA and deduced amino acid sequence and distinct specificities of cDNA-expressed hPCN1 and hPCN3 for the metabolism of steroid hormones and cyclosporine, *J. Biol. Chem.,* 264, 10388, 1989.
22. **Nebert, D. W.,** Proposed role of drug-metabolizing enzymes: Regulation of steady state levels of the ligands that effect growth, homeostasis, differentiation, and neuroendocrine functions, *Mol. Endocrinol.,* 5, 1203, 1991.
23. **Guengerich, F. P., Ed.,** *Mammalian Cytochromes P-450,* Vols. I and II, CRC Press, Boca Raton, FL, 1987.
24. **Gonzalez, F. J.,** Molecular genetics of the P-450 superfamily, *Pharmacol. Ther.,* 45, 1, 1990.

25. **Ortiz de Montellano, P. R., Ed.,** *Cytochrome P-450: Structure, Mechanism and Biochemistry,* Plenum Press, New York, 1986, 1–556.
26. **Waterman, M. R. and Johnson, E. F., Eds.,** Cytochrome P450, *Methods Enzymol.,* 206, 1–655.
27. **Guengerich, F. P.,** Characterization of human microsomal cytochrome P-450 enzymes, *Annu. Rev. Pharmacol. Toxicol.,* 29, 241, 1989.
28. **Guengerich, F. P.,** Human cytochrome P-450 enzymes, *Life Sci.,* 50, 1471, 1992.
29. **Gonzalez, F. J., and Gelboin, H. V.,** Human cytochromes P450: Evolution, catalytic activities and interindividual variations in expression, *Prog. Clin. Biol. Res.,* 372, 11, 1991.
30. **Okey, A. B., Roberts, E. A., Harper, P. A., and Denison, M. A.,** Induction of drug-metabolizing enzymes: Mechanisms and consequences. *Clin. Biochem.,* 19, 132, 1986.
31. **Conney, A. H.,** Induction of microsomal cytochrome P-450 enzymes: The first Bernard B. Brodie lecture at Pennsylvania State University, *Life Sci.,* 39, 2493, 1986.
32. **Sotaniemi, E. A., and Pelkonen, R. O., Eds.,** *Enzyme Induction in Man,* Taylor & Francis, London, 1987, 1–245.
33. **Burchell, B., and Coughtrie, M. W. H.,** UDP-glucuronosyltransferases, *Pharmacol. Ther.,* 43, 261, 1989.
34. **Mannervik, B., and Danielson, U. H.,** Glutathione transferases-structure and catalytic activity, *CRC Crit. Rev. Biochem.,* 23, 283, 1988.
35. **Guengerich, F. P.,** Epoxide hydrolase: properties and metabolic roles, *Rev. Biochem. Toxicol.,* 4, 5, 1984.
36. **Goldstein, J. A., and Linko, P.,** Differential induction of two 2,3,7,8-tetrachlorodibenzo-p-dioxin inducible forms of cytochrome P450 in extrahepatic versus hepatic tissue, *Mol. Pharmacol.* 25, 185, 1984.
37. **Whitlock, J. P., Jr,** The regulation of cytochrome P-450 gene expression, *ISI Atlas Sci.: Pharmacol.,* 351, 1988.
38. **Foureman, G. L., White, N. B. Jr. and Bend, J. R.,** Biochemical evidence that winter flounder (*Pseudopleuronectes americanus*) have induced hepatic cytochrome P450 dependent monooxygenase activities, *Can. J. Fish. Aquat. Sci.,* 40, 854, 1983.
39. **Elskus, A. A., and Stegeman, J. J.,** Induced cytochrome P450 in *Fundulus heteroclitus* associated with environmental contamination by polychlorinated biphenyls and polynuclear aromatic hydrocarbons, *Mar. Environ. Res.,* 27, 31, 1989.
40. **Monosson, E., and Stegeman, J. J.,** Cytochrome P450 E (P450 1A) induction and inhibition in winter flounder by 3, 3′, 4, 4′ tetrachlorobiphenyl: Comparison of response in fish from Georges Bank and Narragansett Bay, *Environ. Toxicol. Chem.,* 10, 763, 1991.
41. **McLemore, T. L., Adelberg, S., Liu, M. C., McMahon, N. A., Yu, S. J., Hubbard, W. C., Czerwinski, M., Wood, T. G., Storeng, R., Lubet, R. A., Eggleston, J. C., Boyd, M. R., and Hines, R. N.,** Expression of CYP1A1 gene in patients with cancer: Evidence for cigarette smoke-induced gene expression in normal lung tissue and for altered gene regulation in primary pulmonary carcinomas, *J. Natl. Cancer Inst.,* 82, 1333, 1990.
42. **Hammons, G. F., Alworth, W. L., Hopkins, N. E., Guengerich, F. P., and Kadlubar, F. F.,** 2-Ethynylnaphthalene as a mechanism-based inactivator of the cytochrome P450 catalyzed N-oxidation of 2-naphthylamine, *Chem. Res. Toxicol.,* 2, 367, 1989.
43. **Miller, N. E., and Halpert, J. R.,** Mechanism-based inactivation of the major β-naphthoflavone-inducible isozyme of rat liver cytochrome P-450 by the chloramphenicol analog N-(2-p-nitrophenethyl) dichloroacetamide, *Drug Metab. Disp.,* 15, 846, 1987.
44. **Lunetta, J. M., Sugiyama, K., and Correia, M. A.,** Secobarbital-mediated inactivation of rat liver cytochrome P-450$_b$: A mechanistic reappraisal, *Mol. Pharmacol.,* 35, 10, 1989.
45. **Mathews, J. M., and Bend, J. R.,** N-Alkylaminobenzotriazoles as isozyme-selective suicide inhibitors of rabbit pulmonary microsomal cytochrome P-450, *Mol. Pharmacol.,* 30, 25, 1986.
46. **Woodcroft, K. J., and Bend, J. R.,** N-Aralkylated derivatives of 1-aminobenzotriazole as isozyme selective, mechanism-based inhibitors of guinea pig hepatic cytochrome P-450-dependent monooxygenase activity, *Can. J. Physiol. Pharmacol.,* 68, 1278, 1990.
47. **Woodcroft, K. J., Szczepan, E. W., Knickle, L. C., and Bend, J. R.,** Three N-aralkylated derivatives of 1-aminobenzotriazole as potent and isozyme-selective, mechanism-based inhibitors of guinea pig pulmonary cytochrome P450 in vitro, *Drug Metab. Disp.,* 18, 1031, 1990.

48. Chang, T., Levine, M., and Bellward, G. D., Selective inhibition of rat hepatic microsomal cytochrome P450. II. Effect of the *in vitro* administration of cimetidine, *J. Pharmacol. Exp. Ther.*, 260, 1450, 1992.
49. Halpert, J., Jaw, J. Y., Balfour, C., Mash, E. A., and Johnson, E. F., Selective inactivation by 21-chlorinated steroids of rabbit liver and adrenal microsomal cytochromes P450 involved in progesterone hydroxylation, *Arch. Biochem. Biophys.*, 264, 462, 1988.
50. Brady, J. F., Ishizaki, H., Fukoto, J. M., Lin, M. C., Fadel, A., Gapac, J. M., and Yang, C. S., Inhibition of cytochrome P-450 2E1 by diallyl sulfide and its metabolites, *Chem. Res. Toxicol.*, 6, 642, 1991.
51. Guengerich, F. P., Mechanism-based inactivation of human liver microsomal cytochrome P450 IIIA4 by gestodene, *Chem. Res. Toxicol.*, 3, 363, 1990.
52. Underwood, M. C., Cashman, J. R., and Correia, M. A., Specifically designed thiosteroids as active-site-directed probes for functional dissection of rat liver cytochrome P450 3A isozymes, *Chem. Res. Toxicol.*, 5, 42, 1992.
53. Muerhoff, A. S., Williams, D. E., Reich, N. O., CaJacob, C. A., Ortiz de Montellano, P. R. and Masters, B. S., Prostaglandin and fatty acid ω- and (ω-1)-oxidation in rabbit lung: Acetylenic fatty acid mechanism-based inactivators as specific inhibitors. *J. Biol. Chem.*, 264, 749, 1989.
54. Friedberg, T., Siegert, P., Grassow, M. A., Bartlomowicz, B., and Oesch, F., Studies of the expression of cytochrome P450IA, P450IIB, and P450IIC gene family in extrahepatic and hepatic tissues, *Environ. Health. Perspect.*, 88, 67, 1990.
55. Omiecinski, C. J., Redlich, C. A., and Costa, P., Induction and developmental expression of cytochrome P450IA1 messenger RNA in rat and human tissues: Detection by the polymerase chain reaction, *Cancer Res.*, 50, 4315, 1990.
56. Raval, P. R., Iversen, P. L., and Bresnick, E., Induction of cytochromes P450IA1 and P450IA2 as determined by solution hybridization, *Biochem. Pharmacol.*, 41, 1719, 1991.
57. Pineau, T., Daujat, M., Pichard, L., Girard, F., Angevain, J., Bonfils, C., and Maurel, P., Developmental expression of rabbit cytochrome P450 *CYP1A1, CYP1A2* and *CYP3A6* genes: Effect of weaning and rifampicin, *Eur. J. Biochem.*, 197, 145, 1991.
58. Marcus, C. B., Wilson, N. M., Jefcoate, C. R., Wilkinson, C. F., and Omiecinski, C. J., Selective induction of cytochrome P450 isozymes in rat liver by 4-n-alkyl-methylenedioxybenzenes, *Arch. Biochem. Biophys.*, 277, 8, 1990.
59. Watkins, P. B., Breath tests as noninvasive assays of P450s, *Methods Enzymol.*, 206, 517, 1991.
60. Watkins, P. B., Murray, S. A., Winkelman, L. G., Heuman, D. M., Wrighton, S. A., and Guzelian, P. S., Erythromycin breath test as an assay of glucocorticoid-inducible liver cytochromes P-450, *J. Clin. Invest.*, 83, 688, 1989.
61. Watkins, P. B., Hamilton, T. A., Annesley, T. M., Ellis, C. N., Kolars, J. C., and Voorhees, J. J., The erythromycin breath test as a predictor of cyclosporin A blood levels, *Clin. Pharmacol. Ther.*, 48, 120, 1990.
62. Lown, K., Kolars, J., Turgeon, K., Merion, R., Wrighton, S. A., and Watkins, P. B., The erythromycin breath test selectively measures P450IIIA in patients with severe liver disease, *Clin. Pharmacol. Ther.*, 51, 229, 1992.
63. Nhamburo, P. T., Gonzalez, F. J., McBride, O. W., Gelboin, H. V., and Kimura, S., Identification of a new P450 expressed in human lung: Complete cDNA sequence, cDNA-directed expression, and chromosome mapping, *Biochemistry*, 28, 8060, 1989.
64. Czerwinski, M., McLemore, T. L., Philpot, R. M., Nhamburo, P. T., Korzekwa, K., Gelboin, H. V., and Gonzalez, F. J., Metabolic activation of 4-ipomeanol by complementary DNA-expressed human cytochromes P-450: Evidence for species specfic metabolism, *Cancer Res.*, 51, 4636, 1991.

SECTION III

Toxic Effects of Contaminants

CHAPTER **10**

Environmental Impact of Pollutants: Biochemical Responses as Indicators of Exposure and Toxic Action (Biochemical Biomarkers)

Colin H. Walker

INTRODUCTION

A number of biochemical responses to sublethal doses of organic pollutants (henceforward simply ''pollutants'') may be useful for measuring exposure of vertebrates to them in the field.[13,17,22] In this account the emphasis will be upon responses that are reasonably stable and are detectable by simple, sensitive, and specific procedures following nondestructive sampling procedures. Methods involving the use of blood samples will be given particular consideration, since this type of nondestructive sampling is readily accomplished in the field.[8]

For the purpose of this discussion, a biochemical biomarker will be defined as ''a biochemical response to an environmental chemical that provides a measure of exposure, and sometimes, also, of toxic effect.'' By this definition, any biochemical response that is related to the ''dose'' of chemical received is a biochemical biomarker of exposure.[22] However, particular interest centers upon biochemical responses that also give measures of toxic effect. This is most clearly the case where the biomarker is itself part of the molecular mechanism that underlies toxicity (e.g., the inhibition of brain acetylcholinesterase activity by organophosphates). It may also be the case where there is no direct involvement in the mechanism of toxic action—where a biomarker response at a site removed from the site of action is closely correlated with the degree of interaction of the same chemical at the site of action.

Biochemical responses may be subdivided into five different categories. ''Inhibition of enzymes'' and ''adduct formation'' are direct effects of either original pollutants or of their active forms. ''Effects on metabolic pathways'' and ''release of enzymes into serum'' are secondary effects. ''Induction'' is less easy to classify since little is known about the way in which pollutants act as inducers. All of these approaches will be discussed here with the exception of adduct formation, which is dealt with elsewhere (Shugart, Chapter 3).[17]

One of the most difficult problems in field studies is getting reliable estimates of degree of exposure. In what follows, reference will be made to the relationship between dose and response under laboratory conditions. There is not, however, a simple relationship between the laboratory and field situations. The formulation of the chemical and the nature and duration of exposure in the field are factors that will influence the pattern of uptake by and distribution in the exposed organisms and thereby the toxic effect that will be produced. Extrapolation from the laboratory to the field is not a simple matter.

INHIBITION OF ENZYMES

Where a pollutant causes the stable inhibition of an enzyme (i.e., irreversible or nearly so) this may provide the basis for a useful biochemical biomarker to be used in the field. The degree to which enzyme activity is lost provides a measure of the level of exposure. The most widely exploited interaction of this type is the inhibition of cholinesterases and other "B"-type esterases by organophosphorus insecticides (ops) (and in certain cases by carbamate insecticides). The inhibition of "B"-type esterases by organophosphorus insecticides has been studied in depth and detail, and well illustrates the strengths and limitations of the approach. The remainder of this section will be restricted to this topic.

As defined by Aldridge[1] "B" esterases are inhibited by organophosphates. The organophosphates function as "suicide" substrates, causing phosphorylation of the enzyme, which is slowly reversible. If the phosphorylated enzyme undergoes aging, no reactivation occurs. Cholinesterases constitute a subclass of "B"-esterases, which function as both sites of action and sites of detoxification for ops. The acetylcholinesterase of the nervous system is held to be the principal site of action of organophosphorus insecticides in vertebrates and in insects. In mammalian blood, two main forms of cholinesterase are recognized—(1) the butyrylcholinesterase (EC 3.1.1.8) of the serum and (2) the acetylcholinesterase of erythrocytes (EC 3.1.1.7).[5] In avian blood there are relatively large quantities of butyrylcholinesterase in the serum but relatively small—and variable—amounts of acetylcholinesterase. In addition to cholinesterases, the blood of mammals and birds contain a group of "B" esterases, usually referred to as unspecific carboxylesterases (EC 3.1.1.1) since they can hydrolyze a wide range of lipophilic carboxylesterases, both naturally occurring and man-made. Both cholinesterases and unspecific carboxylesterases have proved to be of value in monitoring exposure of vertebrates to pesticides in the field.[4,8,10,12,14,20]

A fundamental problem in using inhibition of esterases as an indication of exposure to organophosphates is finding a suitable control. With high levels of exposure, it is relatively easy to demonstrate substantial reductions in blood esterase activities compared to the normal range of values for the species.[12] However, lower levels of exposure are more difficult to monitor. In birds, several studies have shown that the cholinesterases or carboxylesterases of blood can increase following exposure to low levels of pesticides.[20] This may be due, at least in part, to a release of esterase into the blood. In birds, the picture is further complicated by the possibility of diurnal variation in carboxylesterases activity.[21] The underlying problem is that activities are usually measured on the basis of rate of metabolism of substrate per unit volume of blood. Measured in this way, activities are determined by two factors: (1) the quantity of enzyme present and (2) the specific activity of the enzyme. Ops can cause changes in both factors and at low levels of exposure one effect tends to cancel out the other. An increased activity due to more enzyme being present in blood is offset by a decrease in specific activity due to inhibition. Attempts to restore the activity of inhibited blood esterases using, e.g., pyridine-2-aldoxime methiodide (P_2AM), have been of limited value due to aging of the enzyme.

These problems can be overcome by developing assay procedures that will determine the level of a particular esterase protein in blood. This can be achieved by producing antibodies to specific esterases, which can then be used to develop immunochemical assays (e.g., ELISA assays) that will enable the measurement of both the *quantity* of esterase protein and the *specific activity* of that protein in blood. Such an approach is currently under investigation in the Starling (*Sturnus vulgaris*), which has a serum carboxylesterase responsible for some 65% of the total activity of the blood toward α-naphthyl-acetate. This esterase is being purified so that antibodies can be raised to it; the antibodies can then be used in an immunochemical assay.[20] By this approach it should be possible to develop simple, sensitive, and specific assays for blood esterases. The specific activities of particular esterases, which should be fairly constant, can act as controls. This should resolve not only the problems referred to above, but also the problem of interindividual variation, and variations related to sex and age. In the case of the Starling, it has been estimated that immunochemical methods should detect exposures down to the equivalent of 1% of the LD_{50} for certain ops.[20] In birds, serum butyrylcholinesterases are more sensitive to inhibition than are the ''unspecific'' serum carboxylesterases. This suggests that, by taking several different esterases of contrasting sensitivity to organophosphates, a wide range of different levels of exposure can be monitored.

With an improved assay procedure, the inhibition of blood esterases can give a useful indication of level of exposure to ops. However, these enzymes do not represent the site of action, and there does not appear to be a very good correlation between the extent to which they are inhibited and the toxic action of an organophosphate.[10] A better correlation is found between degree of inhibition of brain cholinesterase and toxic action, this enzyme being a site of action for the compounds.[6,7] It remains to be seen whether improved assays for blood esterases can give a better correlation between the level of inhibition and degree of toxic effects than has been achieved with the present methods.

INDUCTION

Several types of enzymes are readily induced following exposure to pesticides, notably microsomal monooxygenases (MOs) and certain other enzymes concerned with the metabolism for foreign compounds. The following discussion will focus on MOs because of the sensitivity of some of them to induction and because they have been well characterized.

The catalytic sites of MOs are provided by various forms of the hemprotein cytochrome P-450. MOs are found in the endoplasmic reticulum of most cell types of vertebrates, but are especially abundant in the liver (hepatic microsomal monoxygenases, HMOs). Certain P-450 forms are readily inducible by environmental agents. One group, called ''phenobarbitone-inducible'' cyto P-450s, are induced by barbiturates and by some lipophilic insecticides (e.g., DDT, DDE).[25] This group shows a wide range of substrate specificities. Another type of cytochrome P-450, P-450 IAI/IAII (sometimes called cytochrome P-448) is induced by polycyclic aromatic compounds, e.g., 3-methylcholanthrene, naphthoflavone, and by some planar organochlorine compounds, e.g., dioxin (TCDD) and planar PCB congeners.[15] Other forms are induced by clofibrate and ethanol. In some birds prochloraz, imazalil, and related fungicidal ergosterol biosynthesis inhibitors are potent inducers of hepatic cytochrome P-450. The induction is of mixed type, involving several different forms of cytochrome P-450.[11]

The induction of cytochrome P-450, may provide a valuable indication of exposure of vertebrates to pesticides. In a recent study with marine fish, Stegeman and Kloepper-Sams[19] produced strong evidence for induction of a hepatic cytochrome P-450 IAI (orthologue of rat P-450) as a consequence of exposure to aromatic hydrocarbons. Evidence has also been produced for the environmental induction of a similar P-450 in sea birds collected in British and Irish coastal waters during 1978–1984.[18]

The identification of specific induced forms of cytochrome P-450 can be accomplished using very small amounts of tissue, if an appropriate antibody is available (e.g., by Western blotting). The main problem in the present context is that the liver is the most appropriate tissue to sample. It may be possible to obtain samples of liver by needle biopsy, although this is not easy with many wild vertebrate species on account of their small size. Two other strategies are worthy of consideration. Some tissues than can be obtained by nondestructive sampling (e.g., white blood cells) may contain inducible forms of cytochrome P-450. A further possibility is that inductions caused in the liver may result in changes in levels of some of their natural substrates (e.g., steroids) in the blood. Monitoring could, in this case, focus on changes that occur in blood components following induction.

RELEASE OF ENZYMES INTO BLOOD

When chemicals cause damage to tissues, certain enzymes are released into blood. The consequent increase in the activity of the enzymes in blood can provide a useful indication of tissue damage. The measurement of activities of such enzymes has been widely used in clinical medicine for the diagnosis of liver damage.

A few studies have attempted to adapt this approach to provide an indication of sublethal effects of pesticides upon birds. In a study with Japanese quail, *pp'*-DDMU, a metabolite of *pp'*-DDT, caused an increase in the level of glutamate oxaloacetate aminotransferase (GOT) in plasma.[24] There were also some changes in plasma enzyme activities following dosing with carbophenothion, but species differed in their response.[23]

On the limited evidence so far available, changes in serum or plasma enzymes can occur after exposure to relatively high levels of pesticides, but appear to be of only limited usefulness in the present context.

EFFECTS OF METABOLIC PATHWAYS

When pesticides and other biologically active substances have toxic effects upon living organisms there are likely to be wide ranging secondary biochemical effects. Where effects are the clear consequence of a particular biochemical lesion, they may be useful as indicators of toxic effect.

One case in point is the effect of anticoagulant rodenticides upon clotting factors. Rodenticides such as warfarin, diphenacoum, and brodifacoum act as vitamin K antagonists. Their action inhibits the operation of vitamin K-dependent enzymes which function as carboxylases for clotting factors. Exposure to these compounds can lead to a change in the balance of clotting factors in the blood.[16] In humans, administration of anticoagulants leads to an increased level of preprothrombin (PIVKA-protein induced by vitamin K absence). The measurement of PIVKA levels in human blood can be achieved by an immunochemical method, and can provide a useful way of monitoring exposure to anticoagulants and of indicating the degree if their toxic or pharmacological action.[3]

CONCLUSIONS

A number of different types of responses to sublethal doses of pollutants have been described, all of which (with the possible exception of induction) are detectable in blood, and apparently lend themselves to nondestructive sampling procedures. Methods are particularly valuable where they relate directly to toxic effect. If appropriate methods are developed, enzyme inhibition, binding to macromolecules, and some secondary effects on metabolic pathways can give indications of levels of exposure well below the LD_{50} values. On the other hand, the release of liver enzymes such as GOT and induction appear to be relatively insensitive.

Only limited information can be gained by the use of one approach. In designing monitoring procedures it is clearly desirable to adopt an appropriate combination of methods, in addition to whatever residue analysis may be possible. A major challenge is to link biochemical biomarker responses, to responses at higher levels of organization—at the physiological and at the population and community level. This is a critical area that has, as yet, received little attention.

Methods should be carefully chosen, having regard for the compound and the species under investigation, to ensure (1) that a suitable wide range of exposures can be effectively monitored, and (2) that, as far as possible, early toxic effects are measured. This approach will be strengthened aided by related laboratory studies where sublethal doses are given to the species that is to be studied in the field. There is a clear need for the development of more procedures that will provide a measure of sublethal toxic effects.

REFERENCES

1. **Aldridge, W. N.**, Serum esterase I, *Biochem. J.*, 53, 110–117, 1953.
2. **Bailey, E., Farmer, P., and Shuker, D. E. G.**, Estimation of exposure to alkylating carcinogens by the GC-MS determination of adducts to haemoglobin and nucleic acid bases in urine, *Arch. Toxicol.*, 60, 187–191, 1987.
3. **Blanchard, R. A., Furie, B. C., Kruger, S. F., and Furie, B.**, Plasma prothrombin and abnormal prothrombin antigen: The correlation with bleeding and thrombosis complications in patients treated with warfarin, *Blood*, 58 (Suppl.), 235, 1981.
4. **Bunyan, P. J., and Taylor, J.**, Esterase inhibition in birds poisoned by Thimet, *J. Agric. Food Chem.*, 14, 132–137, 1966.
5. **Dixon, M., Webb, E. C., Thorne, C. J. R., and Tipton, K. F.**, Enzymes, 3rd ed., Longman, New York, 1979.
6. **Fleming, W. J.**, Recovery of brain and plasma cholinesterase activities in ducklings exposed to organophosphorus pesticides, *Arch. Environ. Contam. Toxicol.*, 10, 215–229, 1981.
7. **Fleming, W. J., and Bradbury, S. P.**, Recovery of cholinesterase activity in mallard ducklings administered organophosphorus pesticides, *J. Toxicol. Environ. Health*, 8, 885–897, 1981.
8. **Fossi, M. C., and Leonzio, C., Eds.**, *Non-Destructive Biomarkers in Vertebrates*, Lewis Publishers, Boca Raton, FL, 1993.
9. **Gupta, R. C.**, Enhanced sensitivity of ^{32}P postlabelling analysis of aromatic carcinogen-DNA adducts, *Cancer Res.* 455, 5656–5662, 1985.
10. **Hill, E. F., and Fleming, W. J.**, Anticholinesterase poisoning of birds: Field monitoring and diagnosis of acute poisoning, *Environ. Toxicol. Chem.*, 1, 27–38, 1982.
11. **Walker, C. H., and Johnston, G. O.**, Potentiation of pesticide toxicity in birds; the role of cytochrome P-450, *Biochem. Soc. Trans.*, 21, 1066–1068, 1993.
12. **Ludke, J. L., Hill, E. F., and Dieter, M. P.**, ChE responses and related mortality among birds fed ChE inhibitors, *Arch. Environ. Contam. Toxicol.*, 3, 1–21, 1975.
13. **McCarthy, J. F., and Shugart, L. R.**, Eds., *Biomarkers of Environmental Contamination*, Lewis Publishers, Boca Raton, FL, 1990.
14. **Mineau, P., Ed.**, *Cholinesterase-Inhibiting Insecticides: Their Impact on Wildlife and the Environment*, Elsevier, Amsterdam, 1991.
15. **Nebert, D. W., and Gonzalez, F. J.**, P-450 genes: Structure, evaluation and regulation, *Annu. Rev. Biochem.*, 56, 945–993, 1987.
16. **Olson, R. E.**, The function and metabolism of Vitamin K, *Annu. Rev. Nutr.*, 4, 281–337, 1984.
17. **Peakall, D.**, *Animal Biomarkers as Pollution Indicators*, Chapman and Hall, 1992.
18. **Ronis, M. J. J., Hansson, T., Borlakoglu, J. T., and Walker, C. H.**, Cytochrome P-450 forms of sea birds: Cross reactivity studies with purified rat cytochromes, *Xenobiotica*, 19, 1167–1174, 1989.
19. **Stegeman, J. J., and Kloepper-Sams, P. J.**, Cytochrome P-450 isosomes and monooxygenase activity in aquatic animals, *Environ. Health Perspect.*, 71, 87–95, 1987.
20. **Thompson, H, M., Walker, C. H., and Hardy, A. R.**, Esterases as indicators of avian exposure to insecticides. Proceedings of the Symposium on Field Methods for the Study of Environmental Effects of Pesticides, *British Crops Protection Council*, 40, 39–45, 1988.
21. **Thompson, H. M., Walker, C. H., and Hardy, A. R.**, Avian esterases as indicators of exposure to insecticides—the factor of diurnal variation, *Bull. Environ. Contam. Toxicol.*, 41, 4–11, 1988.

22. **Walker, C. H.,** Biochemical responses as indicators of toxic effects of chemicals in ecosystems, *Toxicol. Lett.,* 65/66, 527–533, 1992.
23. **Westlake, G. E., Bunyan, P. J., and Stanley, P. I.,** Variation in the response of plasma enzyme activities in avian species dosed with carbophenothion, *Ecotoxicol. Environ. Safety,* 2, 151–159, 1978.
24. **Westlake, G. E., Bunyan, P. J., Stanley, P. I., and Walker, C. H.,** The effect of 1,1-di (p-chlorophenyl)-2-chloroethylene on plasma enzymes and blood constituents on the Japanese quail, *Chem. Biol. Interact.,* 25, 197–210, 1979.
25. **Yoshioka, H., Miyata, T., and Omura, T.,** Induction of phenobarbital-inducible form of cytochrome P-450 in rat liver microsomes by pp'DDE, *J. Biochem. (Jpn.),* 95, 937–947, 1984.

CHAPTER 11

Reactions of Molluscan Lysosomes as Biomarkers of Pollutant-Induced Cell Injury

Michael N. Moore

INTRODUCTION

Marine molluscs are widely used in monitoring the marine environment for the impact of pollutant chemicals.[1-3] These animals frequently accumulate both metals and organic xenobiotics in their tissues, making them particularly suitable for analysis of chemical residues.[1] Another approach has involved the development of indicators of adverse biological effect at the biochemical, cellular, and physiological levels of biological organization.[2,3] Such indicators or "biomarkers" have potential for use as early warning distress signals of impending environmental damage.[2-4] Cellular pathological reactions in molluscs have proven to be a fruitful area of study in the context of biomarkers of pollutant-induced cell injury, and the lysosomal system has been identified as a particular target for the toxic effects of many contaminants.[4,5] Many of the tissues in molluscs are extremely rich in lysosomes, which makes them highly suitable for purposes of detecting the adverse reactions in these organelles.[4] The aim of this chapter is to briefly describe some of the reactions of lysosomes to toxic pollutants, how these are related to the process of cell injury and its higher level consequences, and, also, how such reactions can be used as monitoring tools. The focus of the chapter is on pollutant-induced cell injury in the hepatopancreas or digestive gland, which is the molluscan liver analogue and a major interface with the environment. Many toxic chemicals enter this organ in association with ingested foods where they are then taken up by the endocytotically active tubular epithelial cells and induce cellular damage.[4,5]

CELL INJURY

Alterations in specific aspects of cellular organization offer a means of identifying and characterizing adaptive responses or reactions to cell injury by metals and lipophilic xenobiotics. It should be possible to observe structural–functional alterations in individual cell types or groups of cells at an early stage of a response, before an integrated cellular alteration would manifest itself at the level of organ or whole-animal physiological processes.[6] Some of these cellular reactions may be generalized, whereas others are likely to be specific for toxic chemicals.[6,7]

When cells are injured they undergo a series of biochemical and structural alterations. These can be classified into two phases: a reversible phase preceding cell death at which time an irreversible phase commences.[8] Trump and Arstila have defined the "point of no return" or point of cell death as the point beyond which changes are irreversible even if the injurious stimulus is removed and the cell is returned to a normal environment.[8] This chapter is largely concerned with the reversible phase associated with sublethal cell injury. Cells are able to continue their existence following many types of injury by means of adaptive physiological responses. Examples of such adaptations in mammalian cells include hypertrophy, atrophy, fatty changes, proliferations of smooth endoplasmic reticulum, increased lysosomal autophagy, aging, neoplastic transformations, and accumulation of materials such as lipofuscin.[8] A number of these changes involve functional and structural alterations in intracellular membranes, particularly those of the lysosomal system in, for example, autophagy and the endoplasmic reticulum in induction of cytochrome(s) P-450.[3,7–11]

Xenobiotic-induced cellular pathology reflects disturbances of structure–functions at the molecular level of biological organization. In many instances, the earliest detectable changes or "primary events" are associated with a particular class of subcellular organelle such as the lysosome or endoplasmic reticulum.[12] Investigations in mammals have revealed that much of the damaging action of xenobiotics is produced by highly reactive derivatives. It is these activated chemical forms that are responsible for the initiation of what may be termed the primary intracellular disturbances. These may spread rapidly into a complex network of associated secondary and higher order disturbances that become progressively more difficult for the cell to reverse or modify.

There are numerous ways in which the structure and function of organelles and cells can be disturbed by toxic chemical contaminants and these have been grouped by Slater into four main categories:[12]

1. Depletion or stimulation of metabolites or co-enzymes
2. Inhibition or stimulation of enzymes and other specific proteins
3. Activation of a xenobiotic to a more toxic molecular species
4. Membrane disturbances

It is primarily with categories 2 and 4 that we will be concerned in considering injury induced by xenobiotics, such as PAHs to the lysosomal system, where membrane damage results in major changes in the structure and function of this cellular compartment.

PATHOLOGICAL REACTIONS OF LYSOSOMES

Mammalian lysosomes are noted for their responsiveness to many types of cell injury and molluscan lysosomes have also been shown to respond to a wide range of injurious environmental agents including metals and polycyclic aromatic hydrocarbons.[13–17] The cells of many molluscan tissues are especially rich in lysosomes.[7,18,19] Such tissues include the digestive gland, pericardial gland, kidney, adipogranular cells in the mantle, and ovarian eggs (oocytes).[18–26] Tissues in many other invertebrates also have numerous lysosome-rich cells such as in the gastrodermis of coelenterates.[27,28]

There are a number of ways that lysosomes can react to cellular injury; these can be divided into basically three categories.[16] Lysosomal responses can be considered as decreases or increases in (1) lysosomal contents such as hydrolytic degradative enzymes, (2) rate of membrane fusion events with either the cell membrane or other components of the vacuolar system, and (3) lysosomal membrane permeability.[16] For a variety of reasons, however, lysosomal reactions to cell

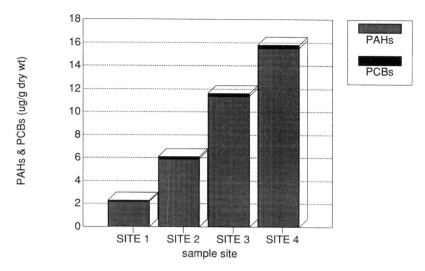

Figure 1. Contaminant gradient showing concentrations of total representative polycyclic aromatic hydrocarbons (PAHs) and total polychlorinated biphenyls (PCBs) in mussels sampled from four sites in Langesundfjord, in Norway. From Klungsøyr et al.[42] and Moore.[5]

injury are not well understood due to the many forms of cell injury and wide variety of organisms studied. Furthermore, one type of lysosomal change may assume the appearance of another type.[16,17]

Lysosomal injury may be either primary or secondary and it has been difficult generally to decide which situation applies for many injurious stimuli.[16] Even in mammals, which are the most studied systems, the extent and ultimate cause of lysosomal reactions to injury are frequently unknown. Many exogenous and endogenous chemical agents are known that influence lysosomal fusion and motility, as well as modifying membrane permeability, and these have aided considerably in achieving a better understanding of lysosomal response patterns.[16,29]

Lysosomal reactions to cellular injury have been categorized above into three groups. Reactions to particular stimuli, such as PAHs, may be considered as appropriate adaptive responses leading to the re-establishment of cellular homeostasis, a conservative response in which the cell is protected from further injury (e.g., detoxication and/or excretion) or an inappropriate response leading to further cellular deterioration and ultimately cell death.[8,16,17] The remainder of this section will attempt to describe observed reactions of molluscan lysosomes, induced by metals and PAHs in terms of alterations (increases or decreases) in lysosomal contents, rate of fusion events, or membrane permeability.

Many environmental xenobiotics (e.g., PAHs) and metals (e.g., copper) are known to be sequestered within lysosomes under certain conditions.[6,13–15,30–35] These may in some circumstances be accompanied by lysosomal damage.[30–35] Exposure of marine molluscs to PAHs has been demonstrated to result in increases in the activities of certain lysosomal enzymes, notably β-glucuronidase and acid phosphatase. A consequence of this pattern would be to prepare the cell for the degradation of particular macromolecules hence making the products available for maintenance of the cell.[16,38]

Other lysosomal contents such as lipofuscin have also been observed to accumulate following exposure to copper, PAHs, or oil-derived hydrocarbons in several species of molluscs (Figures 1 and 2).[3,4,39–42] Lipofuscin or aging pigment is a product of free-radical peroxidative reactions derived from the autophagy of lipoprotein membranes.[43] In molluscan digestive cells this pigment accumulates in residual bodies or tertiary lysosomes and can be excreted by exocytosis.[19] Exposure of marine snails (*Littorina littorea*) to phenanthrene has resulted in an elevated accumulation of lipofuscin in these cells, although this material is rapidly lost following removal of the injurious

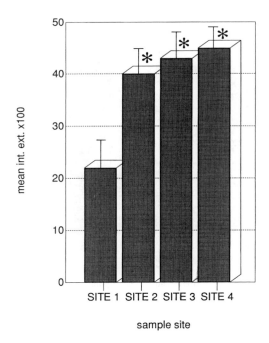

Figure 2. Midrodensitometric determination of lipofuscin (mean ± 95% Cl) in sectioned mussel digestive cells from the sites indicated in Figure 1. Difference from the reference site (1) indicated by * $p < 0.001$, $n = 10$, U-test. From Moore.[3]

stimulus; this build-up of lipofuscin is probably indicative of enhanced autophagy.[39,40] The consequences of this accumulation are not fully understood, although the indications of elevated production of free radicals within the lysosomal environment can hardly be considered beneficial.

Lysosomal accumulation of unsaturated neutral lipid has been described in digestive cells following exposure to organic xenobiotics (Figure 3).[42] This reaction is related to fatty change and the excess lipid is apparently transferred to enlarged lysosomes (Figure 4) where it is not degraded, possibly due to functional failure of the intralysosomal environment and inhibition of acid lipases.[3]

The regulation of the lysosomal system is dependent on controlled fusion with other components of the vacuolar system such as phagosomes, primary lysosomes, and the plasma membrane.[16,17,29] The digestive cells of molluscs are largely concerned with heterophagy and the digestion of food material.[19] Disturbances of the fusion processes involved could have marked consequences for the nutritional status of the organism by perturbing "normal" intracellular digestion and the balance of autophagy to heterophagy.[6]

There are a number of indications that experimental exposure to PAHs such as anthracene and phenanthrene, as well as oil-derived PAHs in both the field and laboratory, induces profound alterations in the rate of fusion events in the lysosomal–vacuolar system of the digestive cells.[7,36,44,45] Ultrastructural studies show that the large secondary lysosomes (2–5 μm diameter approximately) in the digestive cells show marked increases in the presence of internalized membrane-bound vesicular components and that these secondary lysosomes become abnormally enlarged (up to 15 μm in diameter) in mussels, oysters, and periwinkles (Figure 4).[3,40,44,46] Quantitative image analysis of these enlarged lysosomes demonstrates that both lysosomal volume and surface area within the cells are significantly increased, while numerical density is decreased.[44] This is perhaps indicative of fusion of vacuolar components to produce these abnormally enlarged lysosomes. This type of response has also been observed to occur when mussels are exposed to

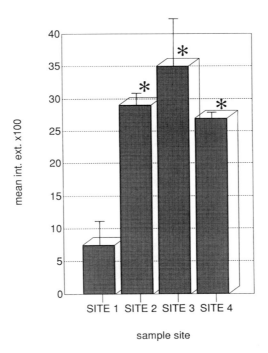

Figure 3. Microdensitometric determination of neutral lipid (mean ± 95% CI) in sectioned mussel digestive cells from the sites indicated in Figure 1. Difference from the reference site (1) indicated by * $p < 0.001$, $n = 10$, U-test. From Moore.[3]

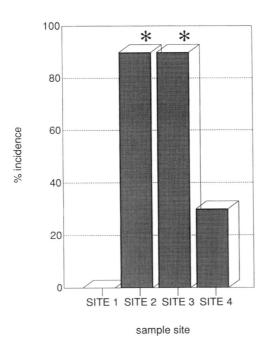

Figure 4. Incidence of pathological enlargement of lysosomes in sectioned mussel digestive cells from the sites indicated in Figure 1. Difference from the reference site (1) indicated by * $p < 0.001$, $n = 10$, U-test. From Moore.[3]

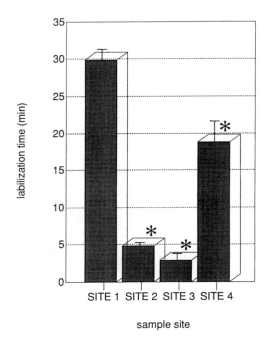

Figure 5. Lysosomal membrane stability (mean ± 95% CI) based on membrane permeabilization (37°C at pH 4.5) in sectioned mussel digestive cells from the sites indicated in Figure 1. Difference from the reference site (1) indicated by * $p < 0.001$, $n = 10$, U-test. From Moore.[3]

an abrupt increase in salinity and has been linked to increased fusion of lysosomal vacuoles and autophagy.[47] These alterations are also associated with elevated protein catabolism and formation of amino acids as measured within the lysosomal cellular compartment.[48] The formation of these "giant" lysosomes has also been linked with atrophy of the digestive tubule epithelium, which is largely comprised of digestive cells.[44-47] This latter relationship will be discussed in more detail in a subsequent section.

The third category of lysosomal disturbance involves membrane permeability. This property can be investigated both biochemically and cytochemically. It is, however, more realistic to employ cytochemical procedures in molluscan digestive cells as the large secondary lysosomes do not readily lend themselves to the trauma of homogenization and subsequent fractionation.[49] Cytochemical procedures for the determination of lysosomal permeability are well established for a number of enzyme substrates in molluscs and these are all conceptually based on the Bitensky Fragility Test for lysosomal stability.[6,7,21,50,51] Lysosomal destabilization is measured as increased permeability to certain enzyme substrates whose products can be used to give a measurable final cytochemical reaction product.[21,49]

Destabilization of lysosomes has been demonstrated in molluscan digestive cells as a result of injury by copper, 2-methyl naphthalene, 2,3-dimethyl naphthalene, anthracene, phenanthrene, diesel oil emulsion, water accommodated fraction of crude oil (North Sea - Auk), and PAH and PCB contaminated field samples (Figure 5).[4,6,7,10,11,30,31,36,37,44,52] Assessment of this type of injury has been confirmed as an extremely sensitive index of cellular condition and the destabilization of the lysosomal membrane appears to bear a quantitative relationship to the magnitude of the stress response and this presumably contributes to the intensity of catabolic or degradative effects, as well as to the level of pathological change that results.[7]

The consequences of destabilization of the secondary lysosomal compartment have been investigated in several experimental studies. Subcellular fractions rich in destabilized secondary lysosomes have been shown to contain significantly increased concentrations of amino acids as compared to stable lysosomes.[48] This is indicative of enhanced intralysosomal protein catabolism. Ultrastructural investigations of cells with destabilized lysosomes indicated increased secondary lysosomal volume with evidence of increased autophagy and possible heterophagy of apoptotic cell fragments, thus providing further indications of elevated catabolic activity.[40,47]

A quantitative cytochemical approach such as the measurement of lysosomal permeability or stability, based on substrate penetrability (hydrolase latency), may also provide insight into the mechanisms of pollutant-induced cell injury in molluscs. Caution, is required, however, in the interpretation of lysosomal damage as a primary event, when it may in fact be a secondary or higher order alteration. Recent investigations of lysosomal responses to specific PAHs have demonstrated that the lysosomal disturbances are complex and differ markedly for PAHs that are structurally dissimilar, such as the isomeric three-ring forms anthracene and phenanthrene.[53] The biochemical evidence of relatively low activity for cytochrome(s) P-450 monooxygenase in molluscan digestive gland cells, when considered together with the ability of these cells to accumulate and retain very high concentrations of PAH, indicates that their loss by metabolic transformation is limited.[9] The fact that the secondary lysosomes are often lipid-rich, particularly following exposure to PAHs and PCBs (Figure 5), would tend to argue for a direct effect on the lysosomes by these xenobiotics, rather than a secondary effect.[3] Aromatic hydrocarbons and PAHs have been shown to penetrate synthetic phospholipid membranes and alter their physicochemical properties including membrane fluidity and permeability.[54,55]

The possibility also exists that reactive species could be produced by oxidation of PAHs within the lysosomes themselves; many molluscan lysosomes contain lipofuscin, which is probably being produced *in situ* by peroxidation processes.[56,57] These lysosomes are noted for their accumulation of metal ions including those of iron and copper that are known to be involved in the generation of oxygen radicals.[23,30,43,58] If reactive species are formed from PAHs by such a process then these could also contribute to lysosomal membrane injury; and in fact the lysosomes of digestive cells have been shown to be a significant site of oxyradical generation.[51,58] There is evidence of enhanced formation of lipofuscin-rich lysosomes and residual bodies in digestive cells following exposure of the snail *L. littorea* to phenanthrene, although the major factor here is probably increased autophagy of cytoplasmic lipoprotein membranes.[3,39,40]

Further evidence of lysosomal destabilization comes from an ultrastructural study of the effects of phenanthrene on secondary lysosomes in digestive cells.[40,59] This has demonstrated the presence of corrugation of the bounding membrane with possible associated microautophagic activity.[59] Increased frequency of membrane breaks has also been described, and while these breaks may be artifacts of fixation and tissue processing, their relative infrequency in control lysosomes is indicative of the greater fragility of lysosomes from cells exposed to phenanthrene.[40] Apparent leakage of lysosomal β-glucuronidase has been demonstrated in the case of lysosomes from the digestive cells of phenanthrene-exposed *L. littorea*; extracellular release of lysosomal β-glucuronidase was also observed in these cells.[40] Limited release of lysosomal enzymes into the cytosol and nucleoplasm has been described by Szego following treatment of rat preputial gland cells with 17β-estradiol, which destabilizes the lysosomal membrane.[60]

The consequences of such a release of lysosomal enzymes is uncertain but is believed to lead to enhanced cell damage and possibly cell death.[16,17] Evidence from rat preputial gland cells indicates increased protein catabolism following estrogen treatment, although enzyme release in these cells is noninjurious and precedes initiation of cell division.[60] This may represent a fundamental difference in the lysosomal response to physiological agonists as opposed to xenobiotics.

In summary, the evidence of both ultrastructural, quantitative cytochemical and morphometric approaches indicates that chemical contaminants induce profound alterations in both structure and function. These involve all three categories of lysosomal response and there are some grounds for suggesting that membrane destabilization may represent the primary injury that could lead to the other events described above. Cytochemical demonstration of lysosomal membrane destabilization has proven to be a useful investigative tool for PAHs, metals, and other injurious agents.[4,7,20] That this procedure does in fact measure membrane destabilization is further supported by evidence of reversibility and restabilization by treatment with hydrocortisone, an established membrane stabilizer.[21,48]

CELLULAR CONSEQUENCES OF LYSOSOMAL DAMAGE

Exposure of marine molluscs to single PAHs and petroleum hydrocarbons results in atrophy of the epithelium of the digestive tubules.[44-46] This atrophy or epithelial "thinning" involves structural changes in the digestive cells, the major component of the epithelium. These changes have been quantified in the mussel *Mytilus edulis* using image analysis of histological sections and in the clam *Mercenaria mercenaria* using morphometry.[61] There is evidence in both mussels and periwinkles that this may be a generalized response to toxic xenobiotics and other stressors such as starvation.[36,44,47,62]

Investigation of digestive cell atrophy has revealed that there is a significant increase in lysosomal volume as described in the previous section.[44,45] This increase in volume of the lysosomal compartment involved the formation of enlarged or "giant" lysosomes and this alteration is associated with membrane destabilization and increased permeability.[44] There is also evidence for increased lysosomal fusion events leading to the formation of the enlarged lysosomes.[40,47] As discussed in the preceding section, the consequences of these lysosomal disturbances would be increased autolytic and autophagic activity presumably leading to atrophy of the digestive cells.

Physiological investigation of mussels has shown that scope for growth is significantly reduced (and in some cases becomes negative) following exposure to PAHs and petroleum hydrocarbons.[51,63] This situation is indicative of relatively enhanced tissue catabolism. Samples from these experiments demonstrate digestive cell atrophy and lysosomal disturbances as described above, arguing strongly for a mechanistic link from the lysosomal events through to the whole animal.[44,45,51] In fact, increased protein catabolism has been demonstrated in digestive glands with low lysosomal stability.[64]

Turning to considerations of effects on reproduction at the cellular level, mussels exposed to PAHs derived from diesel oil showed a reduction in the volume of storage cells in the mantle tissue, a reduction in volume of ripe gametes, and increased degeneration or atresia of oocytes.[65] These data indicate a direct impairment of the reproductive processes and the implication is that reproductive capability would be reduced, both by degeneration of oocytes and reduction in ripe gametes, as well as by a reduction in the energy reserves available for gametogenesis as supplied by the connective tissue storage cells.[24] Ultrastructural investigations designed to explore the mechanisms of oocyte degeneration have revealed that degradative lysosomal enzymes are associated with yolk granules and with pinocytotic phenomena that occur along the basal membrane of developing oocytes.[26] Lysosomal enzymes are also associated with the degradation (atresia) and resorption of oocytes, as well as the resorption of adipogranular storage cells.[24-26]

CONCLUSIONS

The advantage of using an integrated cellular biomarker approach to the problem of assessment of sublethal effects of pollutants in molluscs is clearly evident from the preceding sections, particularly in facilitating the development of conceptual linking of responses at several levels of biological organization. Study of the responses to metals and PAHs has extended our understanding of certain aspects of physiological functions and has also highlighted the capacity for pollutants to disturb biochemical and cellular processes, particularly those involving the lysosomes (Table 1).

Although lysosomal function in molluscs is reasonably well understood, there are still considerable gaps in our knowledge as is also the case in mammals. Regulation of lysosomal catabolism

Table 1. Pathological Reactions of Lysosomes Associated with Toxic Metals and Lipophilic Xenobiotics

Type of pollutant	Lysosomal reaction
Heavy metals	Loss of membrane integrity
	Lysosomes abnormally enlarged
	Lipofuscin accumulation
Lipophilic xenobiotics	Loss of membrane integrity
	Lysosomes abnormally enlarged
	Lipid accumulation
	Lipofuscin accumulation
	Induction of specific acid hydrolases

is obviously critical in maintaining cellular homeostasis and this is evidently perturbed by exposure to toxic chemicals. The nature of these control mechanisms, which often involve membrane–membrane interactions, are complex and unclear.[27] Lysosomally related processes involve endocytosis, phagosome–lysosome interaction, autophagy, and exocytosis.[66] In each situation, lysosomes must fuse with discrete compartments of the cytoplasm in order to carry out their function and this appears to be dependent on the composition of the membrane involved.[66] The regulation of lysosomal function and the effect of membrane disturbance by pollutants clearly represent important areas for future investigation.

There is also considerable scope for the investigation of the underlying processes involved in the uptake, loss, and bioaccumulation of xenobiotics and metals tissues. A better understanding of lysosomal involvement in these processes and the mechanisms of toxicity should provide a toxicological interpretation of the chemical data concerning complex chemical residues in tissues of molluscs.

Prediction of ecological consequences is probably one of the most difficult problems facing those involved in biological effects measurements. However, this does not invalidate the use of cellular biomarkers as an "early warning system" for detection of environmental deterioration in terms of damage to the health of individuals. The rationale for the development of such environmental safeguards is that it is first essential to understand the mechanisms of toxicity and how the animal responds to toxic insult. Once this is established by laboratory experimentation it then becomes possible to develop effects indices or biomarkers based on the pathological cellular reactions and physiological responses. In this context, certain marine molluscs such as mussels (*M. edulis*) and periwinkles (*L. littorea*) have been found to be highly responsive indicator species in investigations of pollutant toxicology.[7,9,67]

A major difficulty in the use of marine molluscs in environmental impact assessment involves our current limited understanding of the mechanisms of toxicity of pollutant chemicals in these animals; consequently this raises problems in the interpretation of the significance or specificity of any biological responses measured, whether at the molecular, cellular, or whole animal levels of organization. Attempts to draw parallels with mechanisms in mammalian toxicology can be misleading and may be of limited value. A further complicating factor arises from the fact that contamination from a single chemical pollutant seldom occurs in environmental isolation, as has already been mentioned above, and this, therefore, poses the question of interactive biological effects resulting from multiple xenobiotic and metal challenge.[68,69]

A partial solution to this problem may lie in the development of two types of biomarker: first, those that indicate exposure to metals or lipophilic xenobiotics; and second, those that indicate cell damage.[70]

Despite these problems, pathological reactions involving lysosomes do appear to have applicability to monitoring and this has been demonstrated for contaminant effects of both metals and xenobiotics.[3,71–74] Lysosomal reactions to pollutants have also been described in fish liver, demonstrating that their use as biomarkers of cell injury can be transferred across the taxonomic divide.[75–78]

A final cautionary statement is necessary: the use of biomarkers based on biochemical changes or cellular pathology must avoid the use of single tests in isolation. It is essential to use batteries of linked tests (cf. clinical testing in human and veterinary medicine) in assessment of chemical impact on health of animals in the environment, otherwise there is a serious risk of interpretative error. In essence, there is no simplistic solution or "litmus paper" test.

REFERENCES

1. **Goldberg, E. D., Bowen, V. T., Farrington, J. W., Harvey, G., Martin, J. H., Parker, P. L., Risebrough, R. W., Robertson, W., Schneider, E., and Gamble, E.,** The mussel watch, *Environ. Conserv.,* 5, 101, 1978.
2. **Bayne, B. L., Livingstone, D. R., Moore, M. N., and Widdows, J.,** A cytochemical and biochemical index of stress in *Mytilus edulis* (L.), *Mar. Pollut. Bull.,* 7, 221, 1976.
3. **Moore, M. N.,** Cytochemical responses of the lysosomal system and NADPH-ferrihemoprotein reductase in molluscan digestive cells to environmental and experimental exposure to xenobiotics, *Mar. Ecol. Prog. Ser.,* 46, 81, 1988.
4. **Moore, M. N.,** Lysosomal cytochemistry in marine environmental monitoring. *Histochem. J.,* 22, 187, 1990.
5. **Moore, M. N.,** The Robert Feulgen Lecture 1990. Environmental distress signals: Cellular reactions to marine pollution, *Prog. Histo-Cytochem.,* 23, 1, 1990.
6. **Moore, M. N.,** Cytochemical determination of cellular responses to environmental stressors in marine organisms, in *Biological Effects of Marine Pollution and the Problems of Monitoring,* McIntyre, A. D., and Pearce, J. B., Eds., *Rapp. P.-V. Reun. Cons. Int. Explor. Mer.,* 179, 7, 1980.
7. **Moore, M. N.,** Cellular responses to pollutants, *Mar. Pollut. Bull.,* 16, 134, 1985.
8. **Trump, B. F., and Arstila, A. V.,** Cell membranes and disease processes, in *Pathobiology of Cell Membranes,* Vol. 1, Trump, B. F., and Arstila, A. V., Eds., Academic Press, New York, 1975, 1.
9. **Livingstone, D. R.,** Responses of the detoxication/toxication enzyme systems of mollusks to organic pollutants and xenobiotics, *Mar. Pollut. Bull.,* 16, 158, 1985.
10. **Moore, M. N.,** Cellular responses to polycyclic aromatic hydrocarbons and phenobarbital in *Myitlus edulis, Mar. Enivron. Res.,* 2, 255, 1979.
11. **Moore, M. N., Lowe, D. M., and Fieth, P. E. M.,** Lysosomal responses to experimentally injected anthracene in the digestive cells of *Mytilus edulis, Mar. Biol.,* 48, 297, 1978.
12. **Slater, T. F.,** *Biochemical Mechanisms of Liver Injury,* Slater, T. F., Ed., Academic Press, New York, 1978, 44.
13. **Allison, A. C.,** Lysosomes and cancer, in *Lysosomes in Biology and Pathology,* Vol. 2, Dingle, J.T., and Fell, H.B., Eds., Elsevier, Amsterdam, 1969, 178.
14. **Tzartsidze, M. A., Lomadze, B. A., and Shengelia, M. G.,** Characteristics of benzo(a)pyrene binding with lysosomal membrane, *Vop. Med. Khim.,* 30, 17, 1984.
15. **Baccino, F. M.,** Selected patterns of lysosomal response in hepatocytic injury, in *Biochemical Mechanism of Liver Injury,* Slater, T. F., Ed., Academic Press, New York, 1978, 581.
16. **Hawkins, H. K.,** Reactions of lysosomes to cell injury, in *Pathobiology of Cell Membranes,* Vol. 2, Trump, B.F., and Arstila, A.V., Eds., Academic Press, New York, 1980, 252.
17. **Ericsson, J. L. E., and Brunk, U. T.,** Alterations in lysosomal membranes as related to disease processes, in *Pathobiology of Cell Membranes,* Vol. 1, Trump, B. F., and Arstila, A. V., Eds., Academic Press, New York, 1975, 217.
18. **Sumner, A. T.,** The distribution of some hydrolytic enzymes in the cells of the digestive gland of certain lamellibranchs and gastropods, *J. Zool. (London),* 158, 277, 1969.
19. **Owen, G.,** Lysosomes, peroxisomes and bivalves, *Sci. Prog. Oxf.,* 60, 299, 1972.
20. **Moore, M. N.,** Lysosomes and environmental stress, *Mar. Pollut. Bull.,* 13, 42, 1982.
21. **Moore, M. N.,** Cytochemical demonstration of latency of lysosomal hydrolases in digestive cells of the common mussel, *Mytilus edulis,* and changes induced by thermal stress, *Cell Tissue Res.,* 175, 279, 1976.

22. **Moore, M. N., Bubel, A., and Lowe, D. M.,** Cytology and cytochemistry of the percardial gland cells of *Mytilus edulis* and their lysosomal responses to injected horseradish peroxidase and anthracene, *J. Mar. Biol. Assoc. U.K.,* 60, 135, 1980.
23. **Lowe, D. M., and Moore, M. N.,** The cytochemical distributions of zinc (Zn II) and iron (Fe III) in the common mussel *Mytilus edulis,* and their relationship with lysosomes, *J. Mar. Biol. Assoc. U.K.,* 59, 851, 1979.
24. **Lowe, D. M., Moore, M. N., and Bayne, B. L.,** Aspects of gametogenesis in the marine mussel *Mytilus edulis, J. Mar. Biol. Assoc. U.K.,* 62, 133, 1982.
25. **Bayne, B. L., Bubel, A., Gabbott, P. A., Livingstone, D. R., Lowe, D. M., and Moore, M. N.,** Glycogen utilization and gametogenesis in *Mytilus edulis* L., *Mar. Biol. Lett.,* 3, 89, 1982.
26. **Pipe, R. K., and Moore, M. N.,** The ultrastructural localization of acid hydrolases in developing oocytes of *Mytilus edulis, Histochem. J.,* 17, 939, 1985.
27. **Moore, M. N., and Stebbing, A. R. D.,** The quantitative cytochemical effects of three metal ions on a lysosomal hydrolase of a hydroid, *J. Mar. Biol. Assoc. U.K.,* 56, 995, 1976.
28. **Tiffon, Y., Rasmont, R., de Vos, L., and Bouillon, J.,** Digestion in lower metazoa, in *Lysosomes in Biology and Pathology,* Vol 3, Dingle, J. T., Ed., Elsevier, Amsterdam, 1973, 49.
29. **Szego, C. M., and Pietras, R. J.,** Lysosomal functions in cellular activation: Propagation of the actions of hormones and other effectors, in *International Review of Cytology,* Vol. 88, Bourne, G. M., Danielli, J. F., and Jeon, K. W., Eds., Academic Press, New York, 1984, 1.
30. **Viarengo, A., Zanicchi, G., Moore, M. N., and Orunesu, M.,** Accumulation and detoxication of copper by the mussel *Mytilus galloprovincialis* Lam: a study of the subcellular distribution in the digestive gland cells, *Aquatic Toxicol.,* 1, 147, 1981.
31. **Viarengo, A., Pertica, M., Mancinelli, G., Orunesu, M., Zanicchi, G., Moore, M. N., and Pipe, R. K.,** Possible role of lysosomes in the detoxication of copper in the digestive gland cells of metal-exposed mussels, *Mar. Environ. Res.,* 14, 469, 1984.
32. **George, S. G.,** Heavy metal detoxication in the mussel *Mytilus edulis*—composition of Cd-containing kidney granules (tertiary lysosomes), *Comp. Biochem. Physiol.,* 76C, 53, 1983.
33. **George, S. G.,** Heavy metal detoxication in *Mytilus* kidney—an in vivo study of Cd- and Zn-binding to isolated tertiary lysosomes, *Comp. Biochem. Physiol.,* 76C, 59, 1983.
34. **Harrison, F. L., and Berger, R.,** Effects of copper on the latency of lysosomal hexosaminidase in the digestive cells of *Mytilus edulis, Mar. Biol.,* 68, 109, 1982.
35. **Sternlieb, I., and Goldfischer, S.,** Heavy metals and lysosomes, in *Lysosomes in Biology and Pathology,* Vol. 5, Dingle, J. T., and Dean, R. T., Eds., Elsevier, Amsterdam, 1976, 185.
36. **Moore, M. N., Pipe, R. K., Farrar, S. V., Thomson, S., and Donkin, P.,** Lysosomal and microsomal responses to oil-derived hydrocarbons in *Littorina littorea,* in *Oceanic Processes in Marine Pollution —Biological Processes and Waste in the Ocean,* Vol. 1, Capuzzo, J. M., and Kester, D. R., Eds., Krieger Publishing, Melbourne, FL, 1986, 89.
37. **Moore, M. N., Pipe, R. K., and Farrar, S. V.,** Lysosomal and microsomal responses to environmental factors in *Littorina littorea* from Sullom Voe, *Mar. Pollut. Bull.,* 13, 340, 1982.
38. **Moore, M. N., Koehn, R. K., and Bayne, B. L.,** Leucine aminopeptidase (aminopeptidase-1), N-acetyl-β-hexosaminidase and lysosomes in the mussel, *Mytilus edulis* L., in response to salinity changes, *J. Exp. Zool.,* 214, 239, 1980.
39. **Moore, M. N., Mayernik, J. A., and Giam, C. S.,** Lysosomal responses to a polynuclear aromatic hydrocarbon in a marine snail: Effects of exposure to phenanthrene and recovery, *Mar. Environ. Res.,* 17, 230, 1985.
40. **Pipe, R. K., and Moore, M. N.,** An ultrastructural study on the effects of phenanthrene on lysosomal membranes and distribution of the lysosomal enzyme β-glucuronidase in digestive cells of the periwinkle *Littorina littorea, Aquatic Toxicol.,* 8, 65, 1986.
41. **Wolfe, D. A., Clark, R. C., Foster, C. A., Hawkes, J. W., and Macleod, W. D.,** Hydrocarbon accumulation and histopathology in bivalve mollusks transplanted to the Baie de Morlaix and the Rade de Brest, in *Amoco Cadiz: Fates and Effects of the Oil Spill,* CNEXO, Paris, 1981, 599.
42. **Klungsϕyr, J., Wilhelmsen, S., Westrheim, K., Saetvedt, E., and Palmork, K. H.,** The GEEP Workshop: organic chemical analysis, *Mar. Ecol. Prog. Ser.,* 46, 19, 1988.
43. **Brunk, U. T., and Collins, V. P.,** Lysosomes and age pigment in cultured cells, in *Age Pigments,* Sohal, R. S., Ed., Elsevier/North-Holland Biomedical Press, Amsterdam, 1981, 243.

44. **Lowe, D. M., Moore, M. N., and Clarke, K. R.,** Effects of oil on digestive cells in mussels: Quantitative alterations in cellular and lysosomal structure, *Aquat. Toxicol.,* 1, 213, 1981.
45. **Moore, M. N., and Clarke, K. R.,** Use of microstereology and quantitative cytochemistry to determine the effects of crude oil-derived aromatic hydrocarbons on lysosomal structure and function in a marine bivalve mollusk *Mytilus edulis, Histochem. J.,* 14, 713, 1982.
46. **Couch, J. A.,** Atrophy of diverticular epithelium as an indicator of environmental irritants in the oyster *Crassostrea virginica, Mar. Environ. Res.,* 14, 525, 1984.
47. **Pipe, R. K., and Moore, M. N.,** Ultrastructural changes in the lysosomal-vacuolar system in digestive cells of *Mytilus edulis* as a response to increased salinity, *Mar. Biol.,* 87, 157, 1985.
48. **Bitensky, L., Butcher, R. S., and Chayen, J.,** Quantitative cytochemistry in the study of lysosomal function, in *Lysosomes in Biology and Pathology,* Vol. 3, Dingle, J. T., Ed., Elsevier, Amsterdam, 1973, 465.
50. **Moore, M. N., and Lowe, D. M.,** Cytological and cytochemical measurements, in *The Effects of Stress and Pollution on Marine Animals,* Bayne, B. L., et al., Eds., Praeger Scientific, New York, 1985, 46.
51. **Lowe, D. M., and Moore, M. N.,** Cytological and cytochemical procedures, in *The Effects of Stress and Pollution on Marine Animals,* Bayne, B. L., et al., Eds., Praeger Scientific, New York, 1985, 179.
52. **Widdows, J., Bakke, T., Bayne, B. L., Donkin, P., Livingstone, D. R., Lowe, D. M., Moore, M. N., Evans, S. V., and Moore, S. L.,** Responses of *Mytilus edulis* L. on exposure to the water accommodated fraction of North Sea oil, *Mar. Biol.,* 67, 15, 1982.
53. **Moore, M. N., and Farrar, S. V.,** Effects of polynuclear aromatic hydrocarbons on lysosomal membranes in mollusks, *Mar. Environ. Res.,* 17, 222, 1985.
54. **Roubal, W. T., and Collier, T. K.,** Spin-labelling techniques for studying mode of action of petroleum hydrocarbons on marine organisms, *Fish. Bull.,* 73, 299, 1975.
55. **Nelson, A.,** Membrane-mimetic electro-chemistry in the marine sciences, *Mar. Environ. Res.,* 17, 306, 1985.
56. **Grossman, J. C., and Khan, M. A. Q.,** Metabolism of naphthalene by pigeon liver microsomes, *Comp. Biochem. Physiol.,* 63C, 251, 1979.
57. **Allison, A. C., and Young, M. R.,** Vital staining and fluorescence microscopy of lysosomes, in *Lysosomes in Biology and Pathology,* Vol. 2, Dingle, J. T., and Fell, H. B., Eds., Elsevier, Amsterdam, 1969, 600.
58. **Winston, G. W., Moore, M. N., Straatsburg, I., and Kirchin, M.,** Decreased stability of digestive gland lysosomes from the common mussel *Mytilus edulis* L. by *in vitro* generation of oxygen-free radicals, *Arch. Environ. Contam. Toxicol.,* 21, 401, 1991.
59. **Nott, J. A., and Moore, M. N.,** Effects of polycyclic aromatic hydrocarbons on molluscan lysosomes and endoplasmic reticulum, *Histochem. J.,* 19, 357, 1987.
60. **Szego, C. M.,** Lysosomal function in nucleocytoplasmic communication, in *Lysosomes in Biology and Pathology,* Vol. 4, Dingle, J.T., and Dean, R., Eds., Elsevier, Amsterdam, 1975, 385.
61. **Tripp, M. R., Fries, C. R., Craven, M. A., and Grier, C. E.,** Histopathology of *Mercenaria mercenaria* as an indicator of pollutant stress, *Mar. Environ. Res.,* 14, 521, 1984.
62. **Bayne, B. L., Holland, D. L., Moore, M. N., Lowe, D. M., and Widdows, J.,** Further studies on the effects of stress in the adult on the eggs of *Mytilus edulis, J. Mar. Biol. Assoc. U.K.,* 58, 825, 1978.
63. **Bayne, B. L., Moore, M. N., Widdows, J., Livingstone, D. R., and Salkeld, P.,** Measurement of the responses of individuals to environmental stress and pollution, *Phil. Trans. R. Soc. London Ser. B.,* 286, 563, 1979.
64. **Moore, M. N., and Viarengo, A.,** Lysosomal membrane fragility and catabolism of cytosolic proteins: Evidence for a direct relationship, *Experientia,* 43, 320, 1987.
65. **Lowe, D. M., and Pipe, R. K.,** Cellular responses in the mussel Mytilus edulis following exposure to diesel oil emulsions: Reproductive and nutrient storage cells, *Mar. Environ. Res.,* 17, 234, 1985.
66. **Wells, W. W., and Collins, C. A.,** Phosphorylation of lysosomal membrane components as a possible regulatory mechanism, in *Lysosomes in Biology and Pathology,* Vol. 7, Dingle, J. T., Dean, R. T., and Sly, W., Eds., Elsevier, Amsterdam, 1984, 119.
67. **Widdows, J.,** Physiological responses to pollution, *Mar. Pollut. Bull.,* 16, 129, 1985.
68. **Malins, D. C., and Collier, T. K.,** Xenobiotic interactions in aquatic organisms: Effects on biological systems, *Aquat. Toxicol.,* 1, 257, 1981.

69. **Moore, M. N., Widdows, J., Cleary, J. J., Pipe, R. K., Salkeld, P. N., Donkin, R., Farrar, S. V., Evans, S. V., and Thomson, P. E.,** Responses of the mussel *Mytilus edulis* to copper and phenanthrene: Interactive effects, *Mar. Environ. Res.,* 14, 167, 1984.
70. **Moore, M. N.,** Molecular cell pathology of pollutant induced liver injury in flatfish: Use of fluorescent probes, *Mar. Ecol. Prog. Ser.,* 91, 127, 1992.
71. **Lowe, D. M.,** Alterations in cellular structure of *Mytilus edulis* resulting from exposure to environmental contaminants under field and experimental conditions, *Mar. Ecol. Prog. Ser.,* 46, 91, 1988.
72. **Axiak, V., George, J. J., and Moore, M. N.,** Petroleum hydrocarbons in the marine bivalve *Venus verrucosa*: Accumulation and cellular responses, *Mar. Biol.,* 97, 225, 1988.
73. **Regoli, F.,** Lysosomal responses as a sensitive index in biomonitoring heavy metal pollution, *Mar. Ecol. Prog. Ser.,* 84, 63, 1992.
74. **Krishnakumar, P. K., Asokan, P. K., and Pillai, V. K.,** Physiological and cellular responses to copper and mercury in the green mussle *Perna viridis* (Linnaeus), *Aquat. Toxicol.,* 18, 163, 1990.
75. **Köhler, A.,** Cellular effects of environmental contamination in fish from the River Elbe and the North Sea, *Mar. Environ. Res.,* 28, 417, 1989.
76. **Köhler, A.,** Lysosomal perturbations in fish liver as indicators for toxic effects of environmental pollution, *Comp. Biochem. Physiol.,* 100C, 123, 1991.
77. **Köhler, A., Deisemann, H., and Lauritzen, B.,** Ultrastructural and cytochemical indices of toxic injury in dab liver, *Mar. Ecol. Prog. Ser.,* 91, 141, 1992.
78. **Lowe, D. M., Moore, M. N., and Evans, B.,** Contaminant impact on interactions of molecular probes with lysosomes in living hepatocytes from Dab (*Limanda limanda*), *Mar. Ecol. Prog. Ser.,* 91, 135, 1992.

CHAPTER 12

Biochemical Approach to Toxicity Evaluation of the Rodenticides, Norbormide and α-Naphthyl Thiourea

R. Radhakrishanamurty

Both the rodenticides are effective against albino rats. Norbormide has potent hyperglycemic effects. This compound affects some of the liver mitochondrial energy linked properties including ATPases. The microsomal enzymes are unaffected in norbormide fed animals.

Administration of α-naphthyl thiourea to albino rats caused excessive urinary excretions of albumin, inorganic phosphorus, and urea in 22 h. In α-naphthyl thio[^{14}C]urea-injected rats, considerable radioactivity was observed in liver and kidney. About 80% of the activities present in serum and pleural effusion were found in the respective albumin fractions. Approximately 40% of the dose administered was excreted in urine and less than 1% in feces in 20 h. Decreases in cytochrome P-450 and activities of microsomal drug-metabolizing enzymes were observed in treated rats. Mitochondrial properties were not affected.

INTRODUCTION

Both the rodenticides norbormide [5-(α-hydroxy-α-2-pyridylbenzyl)-7-(α-2-pyridylbenzylidene)-5-norborene-2,3-dicarboxymide)] and α-naphthyl thiourea (l-naphthyl 2-thiourea) are selective specific rodenticides. Norbormide is toxic to the genus *Rattus* at 5–15 mg/kg body weight[1] and recommended by various international agencies as a safe single dose rodenticide. The pharmacological properties of this compound as a peripheral vasoconstrictor[1] and its cardiovascular effect[2] have been reported. However, there are no reports on the biochemical changes induced by this compound. α-Naphthyl thiourea, another specific rodenticide, is toxic to *Rattus norvegicus* at 5–10 mg/kg body weight.[3] It causes accumulation of pleural effusion and pulmonary edema.[4] The compound was reported to cause inhibition of tyrosinase and copper-catalyzed oxidation of ascorbic acid,[5] hyperglycemia in rats,[6] and hyperchlolesterolemia in dogs.[7] In view of the importance of these two compounds, the biochemical changes induced by them in rats and mice have been studied in some detail and a summary of the results obtained are described in this review.

LABORATORY STUDIES

Administration of norbormide (1 to 1.5 mg/100 g body weight) to rats by oral feeding, stomach tube, or intraperitoneal injection resulted in a significant increase of about 100 to 150% in blood glucose levels, when the rats were killed after reaching the "coma" state, which normally took 1 to 3 h. When the compound was fed at this dose to rats starved for 48 h, the blood glucose levels were not much different from controls. Feeding of the compound at subacute doses (0.5 mg/100 g body weight) for 8 days caused some increases (60%). When the rats were pretreated with 1 to 1.25 g of glucose prior to norbormide, both males and females showed similar blood glucose response. However, the toxic dose for the male rats was slightly higher (1.5 mg/100 g body weight). The results pointed to a hyperglycemic condition. In both male and female rats receiving norbormide, the glycogen levels of the liver were significantly low (50% of control). The glycogen levels of skeletal muscle and glutathione levels of the livers were also low.[8]

In view of the high glucose levels and the concomitant changes in glycogen and glutathione levels of liver of rats given norbormide insulin (2 to 5 IU) was able to counteract the hyperglycemic condition, although the lethal action persisted. Hyperglycemic effects induced by norbormide do not appear to be the major cause for the lethal action of the compound.

Since norbormide was reported to be a specific rodenticide to the rat species, the effects of the compound on blood glucose levels in two strains of mice were investigated. Administration of norbormide even 7.34 mg/100 g body weight to Wistar and tumorigenic strains did not produce any effect on blood glucose levels. The mice did not enter into a "coma"-type condition as was observed with rats.

In order to investigate further the cause of the increase in blood glucose, experimental conditions were standardized for performing bilateral adrenalectomy in both male and female rats. The rats lived for 15–20 days after the operation. In adrenalectomized animals, the increases were only of the order of 36% indicating a major role for adrenaline in blood glucose increase. Adrenaline levels in norbormide-ingested animals were elevated in blood and adrenals when compared with controls.

Norbormide favored the liver mitochondrial swelling, which was not prevented by 2,4-dinitrophenol or reversed by a combination of ATP, Mg^{2+}, and bovine serum albumin. In rats receiving norbormide, Mg^{2+}-activated ATPase showed an increase whereas dinitrophenol-activated ATPase did not show any differences. Also, the $^{32}P_i \rightleftharpoons$ ATP exchange reaction and ^{45}Ca uptake by liver mitochondria were adversely affected, and the liver mitochondrial functions were inhibited to some extent. *In vitro* studies with the intestinal mucosa of rats have shown Mg^{2+}- and Ca^{2+}-activated ATPase were significantly increased.[9]

Studies with the acetylcholinesterase enzyme system have shown that there was an increase of this enzyme in one fraction (cerebellum) of the brain, or in the intestine and plasma depending on the mode of administration of the compound. The alkaline and acid phosphatases have shown significant increases in the serum of norbormide-treated rats. Slight increases in alanine and aspartate aminotransferase were also observed. The lactate dehydrogenase activity of serum of rats given norbormide was found to increase from 428 to 1180 units in treated rats. Gel electrophoresis of the serum samples indicated considerable increases in the activity in most of the bands.

α-Naphtlyl thiourea, another safe single dose rodenticide, was selected for a comparative study. In the literature this compound was reported to increase blood glucose levels. In the present study, a progressive increase culminating in a total increase of 60 to 80% of blood glucose was observed during the 22-h period after oral feeding of the compound to both sexes of rats. Bilateral adrenalectomy did not influence the increase, as was observed with norbormide, indicating that the adrenals have no function on blood glucose levels. Insulin (1 I.U.) lowered glucose levels but the lethal action could not be reversed. α-Naphthyl thiourea feeding to albino mice did not have any effect on blood glucose levels, indicating its specificity to rats.[10]

In α-naphthyl thiourea-fed rats, phosphorus, creatinine, and urea contents of blood and urine were higher. The blood urea in rats increased with time. Increases of lesser magnitude were observed with albino mice within 4–5 h when the compound was fed at 6 mg/100 g body weight. The arginase activities of liver and serum were significantly elevated. General increases in phospholipids, triglycerides, and cholesterol contents in serum of treated rats were observed. They indicate the possibility of kidney damage. This was further demonstrated in histopathological studies, which showed degenerative changes indicating acute glomerular nephritis of the lipoidal type. The levels of albumin contents of α-naphthyl thiourea-treated rats were decreased by 50% over the controls. The electrophoretic patterns revealed increases in α- and β-globulins. It is known that pleural effusion accumulated when the compound was given to rats. In the present study, an analysis of pleural effusion has shown a considerable amount of glucose, urea and triglycerides.[11]

The L-aspartate:2-oxoglutarate aminotransferase activities of liver supernatant and plasma were significantly elevated in α-naphthyl thiourea-treated rats. The L-alanine:2-oxoglutarate aminotransferase activities of these tissues did not show any significant changes.[12] The total and oubain-sensitive Na^+-K^+-ATPases of membrane components were significantly decreased under the influence of α-naphthyl thiourea. Inhibition of Na^+-K^+-ATPase of kidney microsomes is substantially reversed by lecithin and cephalin and the enzyme activity is protected by ATP. The mechanism of inhibition revealed that the compound strongly inhibited Na^+-activated ATPase, culminating in overall inhibition of the phosphorylation step of catalysis of Na^+-K^+-ATPase.[13] The amounts of the microsomal enzymes, N'N-dimethylanilinedemethylase, aryl 4-hydroxylase, and reduced NAD^+ dehydrogenase, and cytochrome P-450 of α-naphthyl thiourea-treated rats showed a decrease of 20–50% compared to controls. The levels of these enzymes and cytochrome P-450 showed increases of 2 to 3 times in phenobarbitone-treated animals.

Radioactive α-naphthyl thio[^{14}C]urea was prepared in the laboratory using recrystallized α-naphthylamine and potassium thiocyanate [KS^{14}CN]. Temperature of incubation was found to be very critical for obtaining a biologically active product. At incubation temperatures ranging from 110 to 130°C, the products obtained had melting points of about 280°C and did not possess biological activity. Only the compound obtained on incubation at 100 to 105°C and a melting point of 180–185°C and exhibited all the desired properties of the authentic compound. Within 4–8 h of oral administration to rats, considerable radioactivity was detected in the kidney and liver samples. Even after 20 h, about 36% of the activity was present in the stomach. Radioactivity was detected in intestine and skeletal muscle also. In the serum, about 80% of the activity was associated with the albumin fraction. Approximately 40% of the dose administered was excreted in the urine and less of urinary excretory products showed that about 36% of the compound was in unchanged form. The remainder was resolved into 3–4 components with 5–25% activities. These components had lower R_f values when compared with α-naphthyl thiourea.[14]

In conclusion, the investigation has confirmed that both norbormide and α-naphthyl thiourea are effective against albino rats. The mechanism of action of these compounds appeared to be different as revealed by the results. Norbormide has potent hyperglycemic effects as a result of which interdependent reactions are seriously affected. This compound affects some of the liver mitochondrial energy-linked properties. α-Naphthyl thiourea, on the other hand, is a potent nephrotoxic compound in addition to having hyperglycemic activity. The effects of this compound, however, are confined to the functional properties of the microsomes, and have no effects on the mitochondrial properties tested.

ACKNOWLEDGMENT

The work presented in this review was done in collaboration with Dr. T. N. Patil and formed a major part of his Ph.D. dissertation to Nagpur University, India.

REFERENCES

1. **Raszkowski, A. P., Poos, G. I., & Mohrbacher, R. J.,** *Science,* 144, 412, 1964.
2. **Yelnosky, J., and Lawlor, R.,** *Eur. J. Pharmacol.,* 16, 117, 1971.
3. **Richter, C. P.,** *J. Am. Med. Assoc.,* 129, 927, 1945.
4. **Richter, C. P.,** *J. Thorac. Surg.,* 23, 66, 1952.
5. **DuBois, K. P., and Erway, W. F.,** *J. Biol. Chem..* 165, 711, 1946.
6. **DuBois, K. P., Horman, R. G., and Erway, W. F.,** *J. Pharmacol.,* 89, 186, 1947.
7. **Chanutin, A., Gjessing, E. C., and Ludweig, S.,** *Proc. Soc. Exp. Biol. Med.,* 64, 174, 1947.
8. **Patil, T.N., and Radhakrishnamurty, R.,** *Ind. J. Biochem. Biophys.,* 10, 206, 1973.
9. **Patil, T.N., and Radhakrishnamyrty, R.,** *Ind. J. Biochem. Biophys.,* 14, 68, 1977.
10. **Patil, T.N., and Radhakrishnamurty, R.,** *Ind. J. Biochem. Biophys.,* 15, 108, 1978.
11. **Patil, T.N., and Radhakrishnamurty.,** *Pest. Biochem. Physiol.,* 8, 217, 1978.
12. **Patil, T.N., and Radhakrishnamurty.,** *Pest. Biochem. and Physiol.,* 11, 74, 1979.
13. **Patil, T.N., and Radhakrishnamurty.,** *Ind. J. Biochem. Biophys.,* 15, 294, 1978.
14. **Patil, T.N., and Radhakrishnamurty.,** *Ind. J. Biochem., Biophys.,* 14, 275, 1977.

CHAPTER 13

Detection of Xenobiotic–Protein Adducts: Electrophoretic and Immunochemical Approaches

Barbara Magi, Barbara Marzocchi, Cristina Lazzeri, Luca Bini, Annalisa Santucci, and Vitaliano Pallini

Foreign molecules, xenobiotics, synthesized and spread by human activity, react sometimes unpredictably with preexisting biological molecules. Xenobiotics include drugs, food additives, as well as indoor and outdoor pollutants.

Reactions involving DNA have attracted much attention for their mutagenic and carcinogenic consequences. Interaction with proteins, although less studied, appear to be complicated by the wide variety of protein structures and functions.

Interference with protein function may result from reversible binding without the formation of covalent adduct. For example, chlorinated benzene compounds reduce the plasma level of the thyroid hormone thyroxine. Thyroxine contains iodinated benzene rings and is displaced from the binding site of carrier proteins on the basis of structural similarity. The binding of chlorinated benzene compounds is not covalent and is eventually reversed by the hormone.[1]

Covalent modifications and damage to proteins occur even without direct binding, as in the case of xenobiotics whose metabolism activates lipid peroxidation. Aldehydes and ketones are formed from fatty acids and react with proteins.[2,3] In this case, xenobiotics enhance a phenomenon related to spontaneous aging.[4]

This chapter deals with protein modification through the formation of covalent adducts with xenobiotics or with their metabolites. These adducts have long lives and are usually destroyed when the protein is broken down by intracellular proteases. Covalent modification sometimes abolishes protein function and cell degeneration and death may occur. In other cases, toxicity is thought to be mediated by immune responses to the modified proteins.[3] Bound xenobiotics play the role of haptenes and confer to the target proteins new antigenic properties. Such interactions with the vertebrate immune system are frequent, complex, and poorly understood. Studies on the mechanism of toxicity cannot but benefit from the development of detection methods for modified proteins.

Detection of protein adducts is useful also in the case of xenobiotics feared essentially for their mutagenic activity. The damage can be monitored at the level of DNA adducts, which tend, however, to be removed by DNA repair enzymes. On the other hand, adducts with proteins with a long life span, e.g., hemoglobin, are not repaired and provide an integrated measure of exposure over prolonged time periods. Examples of this approach are found in References 5 to 7.

The characterization of protein adducts is based on procedures of protein chemistry, such as specific cleavage to oligopeptides, chromatographic separation, localization of isotopically labeled xenobiotic molecules, and, in more recent years, mass spectrometry of modified proteins.[3]

Xenobiotics that generate multiple metabolic products or react at high concentration may involve more than one type of amino acid, however, quite frequently the reaction occurs with specific amino acid side chains.[3] For example, histidine side chains are preferentially alkylated by ethylene and propylene oxide and by styrene 7,8-oxide, a product of oxidation of styrene by microsomal enzymes. In human hemoglobin, histidines exposed at the protein surface predominantly react with styrene oxide.[6]

Cysteine sulfhydryl groups bind a product of the metabolism of acetaminophen, a common analgesic sometimes responsible for liver necrosis and kidney damage.[3] Quite a few products of the xenobiotic oxidative metabolism bind to cysteine residues. In addition, cysteine sulfhydryl groups can form mixed disulfides with disulfiram and other thiol compounds.

Lysine side chain amino groups are also mentioned. n-Hexane and n-butyl-ketone, common industrial solvents, are both metabolized to hexanedione, a neurotoxic compound. It forms a Schiff base with lysine amino groups, followed by other reactions that covalently cross-link polypeptide chains. Cross-linked proteins have been found in neuron and erythrocyte cytoskeleton.

Lysine amino groups also undergo an acylation reaction, e.g., those with the metabolites of halothane, an inhalation anesthetic, with aspirin, and with penicillin. The therapeutic purpose of aspirin is esterification of one serine in prostaglandin synthetase. However, at common doses aspirin also forms amide bonds with several lysines at a specific position in the sequence of hemoglobin chains and with one preferred residue (Lys-199) out of the 59 lysines in human serum albumin.[3] Treatment with penicillin (aimed at inactivating vital enzymes of bacterial pathogens) results in a blood concentration high enough to acylate six specific lysine residues in serum albumin. Penicilloylated serum albumin is an allergene.

The formation of adducts is not only amino acid specific: it is protein specific. Identification of target proteins can be achieved even in complex systems such as cell organelles, whole cells, and plasma after electrophoretic separation. SDS-polyacrylamide gel electrophoresis is frequently used to separate polypeptides with covalently bound xenobiotic moieties. The latter are evidenced by radioactive labeling or by staining with specific antibodies by immunoblotting. Antibodies to xenobiotics can be raised by artificial immunization of experimental animals.

Antibodies to the trifluoroacetyl haptene, a product of halothane metabolism, have been employed to identify modified proteins in liver microsomes and in kidney mitochondria. In both cases a high specificity for binding to proteins is observed, i.e., only a few bands in SDS-polyacrylamide electrophoretic gels are stained by the antibody. These proteins conceivably represent target molecules that may play a role in the mechanism of toxicity. Indeed, immunomicroscopy of kidney slices indicates an association between modified proteins and cell death.[9]

As we have mentioned above, acetaminophen may produce serious liver damage and its metabolites form cysteine adducts. With the aid of an antibody, one main target protein in mice liver has been identified, purified, and partially sequenced. The data indicate high homology with the sequence of a cDNA that corresponds to a 56-kDa selenium-binding protein.[10] The inactivation of this protein by covalent modification may be related to the aminophen-induced death of liver cells.

Immunoblotting can be combined with high resolution two-dimensional electrophoresis. In human plasma, some 800 polypeptides are separated, and only 4–5 among them, including albumin, can be shown to have undergone modification during therapy with penicillins.[10a]

The same approach can be employed in the mapping of xenobiotic-binding sites in protein sequences. Figure 1a shows the electrophoretic separation of fragments obtained by cleavage of

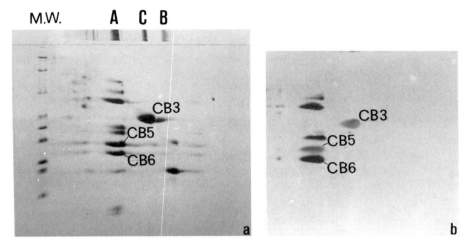

Figure 1. (a) Diagonal gel electrophoretic pattern of CNBr cleavage products from *in vitro* penicilloylated albumin. The first step (separation of fragments) is performed according to Laemmli,[11] the second step (separation of peptides) according to Schäger and Von Jago.[12] M. W., molecular weight standards. From the top of the gel: serum albumin (66,000), ovalbumin (43,000), carbonic anhydrase (31,000), soybean trypsin inhibitor (21,500), myoglobin (17,000), lysozyme (14,400), myoglobin fragment 56-153 (11,000), myoglobin fragment 56-131 (8000), myoglobin fragment 1-55 (6000). A, C, B indicate the three fragments originated from unreduced albumin by cleavage with CNBr. On the top of the diagonal gel they are run only in the first step. (b) Immunoblotting[13] with anti-penicillin antiserum of the peptides released from fragments A, C, and B, by reduction. Immunoreactive spots have been identified as peptides CB3, CB5, and CB6 on the basis of their molecular weights and NH$_2$-terminal sequences, determinated after electroblotting onto PVDF membranes. Other immunopositive spots derived from fragment A contain CB5 and CB6 sequences incompletely cleaved by CNBr.

human serum albumin with CNBr by "diagonal electrophoresis." Three fragments consisting of peptides held together by disulfide bridges are first separated in a nonreducing step (A, C, B in Figure 1). A second electrophoretic step is performed after reduction and dissociation of the fragments. The resulting peptides are separated and related to the amino acid sequence of albumin on the basis of their molecular weights and by their amino-terminal sequences, determined by automated Edman sequencing. Immunoblotting with anti-ampicilloyl antibodies stains selectively the cyanogen bromide peptides, which carry covalently bound antibiotic (Figure 1b). The immunopositive spots CB3, CB5, and CB6 (so named according to current nomenclature) correspond, respectively, to the sequence Cys-124–Glu-297, Phe-330–Arg-445, and Pro-447–Val-547 of albumin. Ampicillin-binding sites are to be looked for in these regions of the albumin sequence. The mapping of reactive sites in the protein sequence is a clue to the description of the structural interaction with the xenobiotic, if the 3D structure of the protein is known.

Adducts can be finely studied by isoelectric focusing if the binding alters the electric charge of the protein. This phenomenon occurs when charged groups of the protein, e.g., lysine amino groups, are involved and/or the xenobiotic is a charged molecule.

Figure 2A shows the focusing of normal human serum albumin, which evidentiates three main isoforms, the predominant being the most cathodic. pI values have been measured with a flat electrode. Covalent binding with β-lactame antibiotics (penicillin, ampicillin, ceftriaxone) generates a series of anodic isoforms, as expected, on the basis of the loss of lysine amino groups. Bands in the series correspond to albumin molecules with different levels of modification, i.e., with different numbers of antibiotic molecules per protein molecule. The higher the ratio, the greater the anodic shift. Sulbactam is evidently the least reactive among the antibiotics we have tested. The method can be used to assess the level of modification as a function of the xenobiotic concentration.

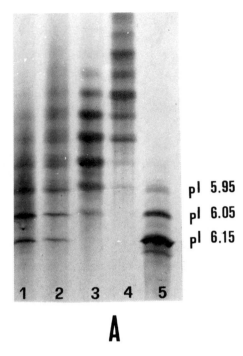

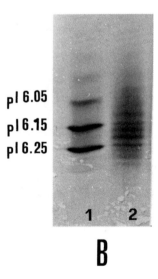

Figure 2. Albumin pattern obtained after isoelecric focusing in the presence of 8 M urea and 60 mM serine in the 5 to 8 pH range according to Rocha et al.[14] In A, lanes, 1, 2, 3, and 4 show human serum albumin incubated *in vitro* with sulbactam, ceftriaxone, ampicillin, and penicillin under the conditions reported.[8] Isoforms of native human serum albumin are shown in lane 5. In B, lane 1 shows rabbit serum albumin and lane 2 rabbit serum albumin after incubation with styrene oxide under the conditions described.[6]

Figure 2B reports data on the alkylation of rabbit serum albumin by styrene oxide. Sharp bands corresponding to protein adducts are observed, slightly shifted toward the cathode. This behavior is probably related with an alteration of the properties of the histidine imidazole, a weak acid with pK_a 6, induced by alkylation.

On the basis of the occurrence of specifically reacting amino acids in specific target proteins, one may conclude that, in general, damage to proteins by xenobiotics is more precisely site-directed than that to DNA. It is interesting that xenobiotic-binding sites have formed over a long time during evolution of proteins. The blind watchmaker has preceded the quick xenobiotic synthesizer, the latter, too, blind about the consequences of his syntheses. The understanding of the rules of xenobiotic–protein recognition may help open our eyes. The description of the 3D structure of the binding sites is already possible for some proteins, and it can be a tool for a "rational xenobiotic design."

REFERENCES

1. **Van Der Berg, K. J.** Interaction chlorinated phenols with thyroxine binding sites in human transthyretin, albumin and thyroid binding globulin, *Chem. Biol. Interact.,* 76, 63–75, 1990.
2. **Benedetti, A., Comporti, M., Fulceri, R., and Esterbauer, H.,** Cytotoxic aldehydes originating from the peroxidation of liver microsomal lipids. Identification of 4,5-dihydroxydecenal, *Biochim. Biophys. Acta,* 792, 172–181, 1984.
3. **Harding, J. J.,** Non enzymatic covalent post-translation modification of proteins *in vivo, Adv. Prot. Chem.,* 37, 247–334, 1985.
4. **Stadtman, E. R.,** Covalent modification reactions are marking steps in protein turnover, *Biochemistry,* 29, 6323–6331, 1990.
5. **Kautianen, A., Törnquist, M., Svensson, K., and Osterman-Golkar, S.,** Adducts of malonaldehyde and a few other aldehydes to hemoglobin, *Carcinogenesis,* 10, 2123–2130, 1989.
6. **Kaur, S., Hollander, D., Haas, R., and Burlingame, A.,** Characterization of structural xenobiotic modifications in proteins by high sensitivity tandem mass spectrometry. Human hemoglobin treated *in vitro* with styrene, 7,8-oxide, *J. Biol. Chem.,* 264, 16981–16984, 1989.
7. **Bailey, E., Brooks, A. G., Bird, I., Farmer, P. B., and Street, B.,** Monitoring exposure to 4,4′-methylenedianiline by the gas chromatography-mass spectrometry determination of adducts to hemoglobin, *Anal. Biochem.,* 190, 175–181, 1990.
8. **Yvon, M., Anglade, P., and Wal, J. M.,** Identification of the binding sites of benzyl penicilloyl, the allergenic metabolite of penicillin, on the serum albumin molecule, *FEBS Lett.,* 263, 237–240, 1990.
9. **Hayden, P. J., Ichimura, T., McCann, D. J., Pohl, L. R., and Stevens, J. L.,** Detection of cysteine conjugate metabolite adduct formation with specific mitochondrial proteins using antibodies raised against halothane metabolite adducts, *J. Biol. Chem.,* 266, 18415–18418, 1991.
10. **Pumford, N. R., Martin, B. M., and Hinson, J. A.,** A metabolite of acetaminophen covalently binds to the 56 KDa selenium binding protein, *Biochem. Biophys. Res. Commun.,* 182, 1348–1355, 1992.
10a. **Bini, L., et al.,** 1993. Submitted.
11. **Laemmli, U. K.,** Cleavage of structural proteins during the assembly of the head of bacteriophage T4, *Nature (London),* 227, 680–685, 1970.
12. **Schägger, H., and Von Jagow, G.,** Tricine-sodium dodecyl sulfate-polyacrylamide gel electrophoresis for the separation of proteins in the range from 1 to 100 kDa, *Anal. Biochem.,* 166, 368–379, 1987.
13. **Towbin, H., Staehelin, T., and Gordon, J.,** Electrophoretic transfer of proteins from polyacrylamide gels to nitrocellulose sheets: Procedure and some applications, *Proc. Natl. Acad. Sci. U.S.A.,* 76, 4350–4354, 1979.
14. **Rocha, J., Kämpt, J., Ferrand, N., Amorim, A., and Ritter, H.,** Separation of human alloalbumin variants by isoelectric focusing, *Electrophoresis,* 12, 313–314, 1991.

CHAPTER 14

Structural and Biochemical Alterations in the Gills of Copper-Exposed Mussels

A. Viarengo, N. Arena, L. Canesi, F. A. Alia, and M. Orunesu

In marine organisms heavy metal accumulation can alter the physiology of the gill tissue. Data are presented here on the structural and biochemical alterations occurring in the gill cells of Cu-exposed mussels. Immunohistochemical data obtained with antitubulin antibodies indicate that Cu^{2+} causes microtubule disassembly in the gill ciliated epithelium. Moreover, biochemical data demonstrate a decrease in the glutathione content and increased accumulation of lipid peroxidation products in the tissue. These results suggest that Cu^{2+} ions may affect the microtubule structure of the gills either by directing binding to tubulin SH groups or, indirectly, by inducing alterations of redox balance and oxidative stress conditions in the tissue. These results may explain the mechanisms by which an excess of Cu^{2+}, an essential metal, can alter the activity of gill cells and, consequently, the physiology of mussels.

INTRODUCTION

Heavy metals, such as Cd^{2+}, Hg^{2+}, Cu^{2+}, and Zn^{2+}, represent a major class of contaminants in the marine environment.[1,2] Among these copper, an "essential" metal, can be extremely toxic to organisms, when present in the cells in excess amounts.[3,4]

In both marine vertebrates and invertebrates the gills are the main site of accumulation of the soluble forms of the metal, this often resulting in morphological and physiological alterations.[5-14] This is of particular importance in the case of lamellibranch molluscs, such as *Mytilus* sp., whose gills are involved in both respiration and feeding processes. Mussel gills are covered with a ciliated epithelium whose continuous beat produces a current that draws water between the individual gill filaments. Particles from this current are sorted and trapped by specialized cilia and conveyed to the digestive system. Therefore, in these organisms, the physiological processes of respiration and feeding are intimately associated.

Copper exposure has been shown to affect mussel physiology, altering some essential parameters such as oxygen uptake, heart and filtration rate, and scope for growth.[7,8,13,16,17] However, little is known about the cellular and biochemical mechanisms involved in copper toxicity to mussel gill structure and functions.

Due to the extremely high affinity of copper ions for sulfhydryl groups, the toxic effects of the metal may be related to binding to SH-containing proteins and enzymes.[3,4,18] Moreover, in aerobic conditions, copper ions are potentially involved in redox reactions leading to the production of oxyradicals and to oxidative stress conditions in the tissues.[19]

In this work the possible effects of Cu^{2+} on ciliary microtubules and on the distribution of other cytoskeletal proteins, such as α-actinin and filamin, of mussel gill cells were evaluated. Tubulin, the main component of the cilia structure, is an SH-containing protein,[20,21] which could represent a potential target for both heavy metal cations and oxidizing species. Therefore, in an attempt to evaluate a possible relationship between Cu-induced gill damage and oxidative stress, the level of glutathione (GSH), the main soluble thiol involved in the maintenance of the cellular redox balance,[22] was estimated. Moreover, the concentration of malondialdehyde (MDA), which is the main lipid peroxidation product,[23] was evaluated.

METHODS

Animals and Treatments

Mussels (*Mytilus galloprovincialis* Lam.) 4–5 cm long were collected from La Spezia (Italy). Before treatment, they were kept in an aquarium in static tanks containing aerated, EDTA-free artificial sea-water[24] (1 L/mussel) at 15°C for 3 days. During treatment, mussels were exposed to copper (20 µg/L/mussel) for 1, 3, and 6 days. The metal was added daily in the form of standard solutions of $CuCl_2$. The seawater was changed daily.

Histochemistry

Whole gills were carefully dissected, placed in Petri dishes, and repeatedly washed with cytoskeleton-stabilization buffer (CSS).[25,26] In the experiments with antitubulin antibodies $CaCl_2$ in CSS was replaced by 2 mM EGTA. After washing, samples were fixed in a freshly prepared *p*-formaldehyde solution (0.75% *p*-formaldehyde, 2.5% glutaraldehyde in CSS buffer) for 1 h. Tissues were then rinsed in PBS buffer, dehydrated, and embedded in paraffin at 60°C. Sections (6 µm thick) were cut with a rotating Reichert microtome and dried at 37°C.

Immunofluorescence

For immunofluorescence microscopy, sections were deparaffinized, rehydrated, and then incubated with primary antibodies (mouse anti-β-tubulin, anti-α-actinin, antifilamin, and normal swine serum) in the following sequence:[28,29] (1) normal swine serum (1:10 dilution) for 30 min at room temperature; (2) anti-β-tubulin (1:200 dilution), antifilamin, and anti-α-actinin antibodies (1:250 dilution) at 4°C overnight, followed by washing in CSS (in a parallel set of control experiments, the primary antibody was omitted); (3) FITC antimouse for 30 min at room temperature followed by washing in phosphate-buffered saline (PBS).

Sections were mounted in 90% glycerol in PBS and examined using a Leitz fluorescence microscope Dialux 20 EB equipped with phase contrast (PC), epifluorescence (filter Ex = 450–490 nm; Em = 520 nm) and automatic microphotography.

Immunoenzymatic Techniques

Immunoenzymatic techniques were based on the peroxidase–antiperoxidase procedure (PAP), using horseradish peroxidase conjugates for detection. Rehydrated sections were separately incubated with two blocking reagents (3% hydrogen peroxide and normal 1:10 serum in PBS) to reduce the nonspecific background staining. Incubation with the primary antibody (diluted in PBS as in the immunofluorescence procedure) was performed for 1 h at 37°C. Sections were washed three times in PBS for 5 min and incubated with the secondary antibody (tubulin:sheep antimouse IgG, α-actinin, and filamin:goat antirabbit IgG, 1:50 in PBS) for 1 h at room temperature, washed again three times, and incubated with antiperoxidase antibodies (labeling antibody: tubulin antisheep PAP conjugated; α-actinin and filamin antigoat PAP conjugated, 1:80 in PBS) for 1 h at room temperature.

Sections were subsequently immersed for 5 min in a filtrated, freshly prepared 10% diaminobenzidine solution in 0.1 M Tris-HCl buffer (pH 7.1) with 125 µL 3% hydrogen peroxide. Samples were washed three times in distilled water and stained with Harris' hematoxilin for 5 min. After washing in running water, sections were mounted in aqueous mounting medium (Dako DK) and stored in the dark at 4°C. Microscopic observations were carried out within 24 h. Blanks were prepared by replacing the primary antibody with PBS. No nonspecific reactions were observed.

Glutathione Assay

Gills were homogenized in 5 vol of cold 1 N PCA/2 mM EDTA and centrifuged at 30,000g at 4°C for 20 min. Aliquots of the supernatant were neutralized with 2 M KOH/0.3 M MOPS and centrifuged at 1000g for 15 min at 4°C. The glutathione content (GSH + 1/2 GSSG) in 100–200 µL of neutralized supernatant was evaluated by the GSH reductase enzymatic method.[30]

Evaluation of the Malondialdehyde (MDA) Content

Gills were homogenized in 30 mM Tris-HCl buffer, pH 7.4, at 4°C. Aliquots of the homogenate were added with an equal volume of acetonitril and subsequently centrifuged at 5000g for 15 min at 4°C. The supernatants (20 µl) were utilized for the evaluation of the MDA content by HPLC on a Waters carbohydrate analysis column (3.5 mm × 30 cm; n. 84038).[31]

Analysis for Metal Content

Gill copper concentrations were determined by inductively coupled plasma-atomic emission spectroscopy (ICP-AES).[32]

CHEMICALS

All reagents were of analytical or HPLC grade. Immunofluorescence antibodies: primary antibodies: monoclonal anti-β-tubulin (cod. 6068), anti-α-actinin (cod. 6587), and antifilamin (cod. 1083) were from Bio-Makor (Rehovot, Israel). Indirect immunofluorescence secondary antibodies (conjugated with fluorescein isothiocyanate): antimouse FITC conjugate (Sera Lab, cod. SBA-6120–02) and antirabbit FITC conjugate (Bio-Makor, cod. 4271). Immunoperoxidase antibodies: Sera Lab monoclonal (cod. RPX-2000) and polyclonal (cod. RPX-3000) antibodies were utilized.

RESULTS

Data reported in Table 1 show the copper content in the gills of mussels exposed to the metal (20 µg/L/mussel) for 1, 3, and 6 days. Copper was progressively accumulated in the tissue up to an extremely high concentration in 6 day-exposed mussels (about 24 µg/g wet weight tissue) with a 50-fold increase with respect to controls (0.45 µg/g).

Microscopic examination revealed that copper accumulation results in alteration of the gill structure (Figures 1–3). In particular, the cilia appeared shorter and did not show the regular

Table 1. Copper Content in the Gills of Control and Copper-Exposed Mussels

Treatment	Copper (μg/g wet weight)
Control	0.45 ± 0.05
Cu 1 day	3.16 ± 0.25[a]
Cu 3 days	11.30 ± 0.91[a]
Cu 6 days	24.20 ± 1.45[a]

Note: Mussels were exposed to Cu^{2+} (20 μg/L/animal) for 1, 3 and 6 days. Data represent the mean ± SD of 5 experiments each involving 12 animals.
[a] = $p < 0.01$ (Mann–Whitney U-test).

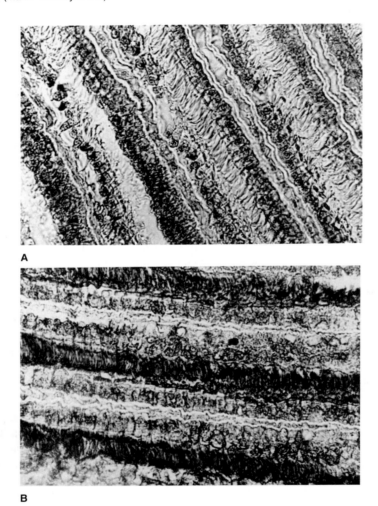

Figure 1. (A,B) Gill tissue of control mussels (400×); (A) blank of the immunoperoxidase reaction (hematoxylin) performed without the primary antibody; (B) peroxidase–antiperoxidase (PAP) antitubulin reaction plus hematoxylin; (C) gill tissue of mussels exposed to copper for 3 days. PAP antitubulin reaction plus hematoxylin; (D) gill tissue of mussels exposed to copper for 6 days. PAP antitubulin reaction plus hematoxylin.

arrangement observed in the gills of control animals (Figures 1B, C, and D). Moreover, in Cu-exposed animals hypertrophy of gill filaments was observed (Figure 3). This kind of morphological alteration has been previously reported in the gills of metal-exposed fish.[5,6]

The immunohistochemical analysis indicated that copper leads to alterations of the cytoskeletal structure of the gill cells. In fact, the antitubulin PAP reaction seemed to be proportionally decreased with the time of exposure and, therefore, with the accumulation of the metal in the

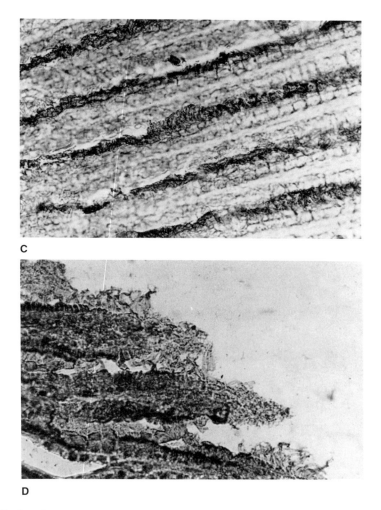

Figure 1. Continued.

tissue (Figure 1). Similarly, as evidenced by indirect immunofluorescence utilizing antitubulin antibodies, the distribution of tubulin was drastically altered in the cells of 6-day-exposed mussels (Figure 2).

The intensity of the α-actinin immunofluorescent reaction did not show great differences with respect to controls, even though the distribution of the protein seemed to be slightly more dispersed in the gills of 6-day-exposed mussels (Figure 3). No alterations in the distribution of filamin were observed (data not shown).

The results shown in Table 2 demonstrate that copper exposure resulted in a significant decrease of the gill GSH content (of about 50% in the gills of 3- and 6-day-exposed animals), indicating alteration of the redox balance in the tissue. Moreover, in the gills of 6-day-exposed mussels, the concentration of malondialdehyde (MDA), which represents the main product of membrane lipid peroxidation, was significantly increased. These data demonstrate that copper accumulation induces oxidative stress conditions in the tissue.

DISCUSSION AND CONCLUSIONS

Many physiological, biochemical, histological, and ultrastructural studies have shown that heavy metal cations interfere with respiration and osmoregulation processes in fish, crustaceans,

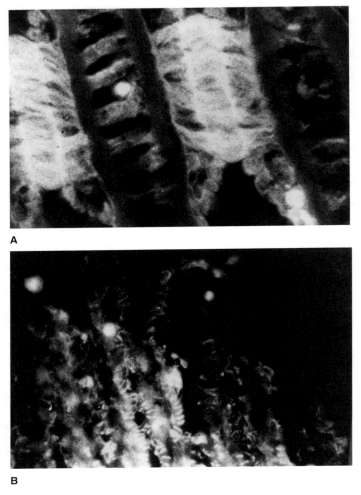

Figure 2. Gill tissue of mussels. Indirect immunofluorescence reaction with anti-β-tubulin primary antibodies (400×). (A) Control mussels; (B) mussels exposed to copper for 6 days.

and molluscs by disrupting the structure of the gill cells.[5–7,9,14,33] In mussels, it has been demonstrated that copper can affect gill function by inhibiting ciliary activity and decreasing oxygen consumption and filtration rate.[7,8,16,34] These sublethal effects are reflected by altered feeding and growth rates,[13,35] as indicated by a net reduction in scope for growth.[17]

There are different hypothesis to explain the mechanisms by which copper ions can alter the physiology of mussel gills. An immediate response to the presence of copper in the seawater is a copious secretion of mucus,[8] this possibly leading to clogging of the cilia rather than to damage to gill structure.[7] Moreover, it has been suggested that the effects of copper may be the result of both separation of gill filaments and depressed lateral cilia activity.[16,36] This latter effect has been attributed to a neuronal mechanism of inhibition rather than to a direct effect of copper on interfilamentar ciliary junctions.[15] However, most of these observations come either from short-term experiments carried out on mussel gill fragments or from mussel exposure to a single dose of copper, generally higher (from 50 to 400 μg/L) than that used in this study (20 μg/L). Moreover, the molecular mechanisms involved in the inhibition of filtration in mussels by copper have not been so far elucidated.

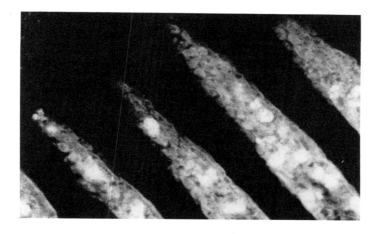

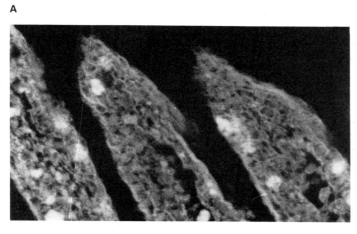

Figure 3. Gill tissue of mussels. Indirect immunofluorescence reaction with anti-α-actinin primary antibodies and FITC-conjugated secondary antibodies (400×). (A) Control mussels; (B) mussels exposed to copper for 3 days.

Table 2. Glutathione and Malondialdehyde Content in the Gills of Control and Copper-Exposed Mussels

Treatment	Glutathione	MDA
	(nmol/g wet weight)	
Control	123.89 ± 27.77	44.22 ± 2.65
Cu 1 day	89.69 ± 22.50 n.s.	45.03 ± 1.92 n.s.
Cu 3 days	59.46 ± 13.65[a]	50.72 ± 5.44 n.s.
Cu 6 days	63.22 ± 20.24[a]	54.88 ± 3.22[a]

Note: Mussels were exposed to Cu^{2+} (20 μg/L/animal) for 1, 3, and 6 days. Data represent the mean ± SD of 5 experiments each involving 12 animals. n.s., Not significant.
[a] $p < 0.01$ (Mann–Whitney U-test).

The results presented here demonstrate that copper is rapidly accumulated in mussel gill cells and that the excess amount of metal is responsible for a dramatic alteration of their structure. In addition, our data show a correlation between the histological and biochemical changes induced by the metal in the tissue. In fact, immunohistochemical data indicate that the effects of copper on mussel gills are related to disruption of ciliary microtubule organization by tubulin disassembling. Moreover, biochemical data suggest that the large amount of copper ions may affect the microtubule structure of the cilia either by direct binding to the SH groups of tubulin or, indirectly, by inducing alteration of the redox balance and oxidative stress conditions in the tissue, as evidenced by the decrease in the glutathione content and the increase in the MDA level.

Data on the effects of heavy metals on cytoskeletal structures indicate that cadmium chloride and methylmercury can cause microtubule disassembly.[37,38] In light of our results, one of the main targets of an excess of copper, an essential metal, in mussel gill cells may be represented by tubulin, the main component of the cilia structure.

Tubulin assembly is known to be affected by various factors, such as the presence of calcium ions[39] and the state of oxidation of SH groups in the protein.[40] There are, in fact, between 7 and 10 cysteine residues per tubulin subunit,[41-43] and blockade of at least one of these SH groups prevents polymerization and causes disruption of preformed microtubules.[40] With regard to this, it has been demonstrated that the maintenance of the cellular GSH content is essential for microtubule assembly *in vivo*;[44] moreover, oxidative stress conditions can alter the *in vitro* ability of tubulin to polymerize.[43] In mammalian cells, oxidative stress leads both to cytoskeletal alterations, due to oxidation of critical thiol groups of tubulin and actin,[45] and to impairment of calcium homeostasis, through inhibition of Ca-ATPases, the SH enzymes involved in calcium transport across cell membranes,[46] and consequent increase in cytosolic Ca^{2+} levels.[47]

In *Mytilus*, Ca^{2+} flux through the plasma membrane of gill cells is the controlling factor in ciliary motility, the effect depending on the type of cilia. In particular, lateral cilia, which beat continuously to produce the feeding current, arrest as a consequence of an inward flux of Ca^{2+} ions, raising the intracellular concentration of this cation up to the micromolar range (1–2 μM) and acting directly on the axoneme.[48-51] The intracellular Ca^{2+} would be then reduced to the normal level ($= 10^{-7}$ M) by the activity of the membrane Ca-ATPases.

With regard to this, recent data indicate that copper ions strongly inhibit the activity of the Ca-ATPase present in mussel gill cell plasmamembranes.[52] Moreover, it has been demonstrated that n-μM copper concentrations cause a sustained increase of the cytosolic Ca^{2+} content in mussel hemocytes.[53] In this review, it is tempting to speculate that an impairment of calcium homeostasis in mussel gill cells by copper ions may be an additional factor contributing to microtubule disassembling and alteration of cilia structure and functions.

REFERENCES

1. **Bernhard, M. and Zattera, A.,** Major pollutants in the marine environment, in *Marine Pollution and Marine Waste Disposal,* Pearson, J. and Frangipane, S., Eds., Pergamon Press, Oxford, 1975, 195.
2. **Bryan, G. W.,** Heavy metal contamination in the sea, in *Marine Pollution,* Johnston, R. Ed., Academic Press, London, 1976, 185.
3. **Viarengo, A.,** Biochemical effects of trace metals, *Mar. Pollut. Bull.,* 16, 153, 1985.
4. **Viarengo, A.,** Heavy metals in marine invertebrates: mechanisms of regulation and toxicity at the cellular level., *CRC Rev. Aquat. Sci.,* 1, 295, 1989.
5. **Baker, J. T. P.,** Histological and electron microscopical observations on copper poisoning in the winter flounder (*Pseudopleuronectes americanus*), *J. Fish Res. Bd. Can.,* 26, 2785, 1969.
6. **Gardner, R. G. and Yevich, P. P.,** Histological and hematological responses of an estuarine teleost to cadmium, *J. Fish Res. Bd. Can.,* 27, 2185, 1970.
7. **Brown, B. E., and Newell, R. C.,** The effects of copper and zinc on the metabolism of the mussel *Mytilus edulis, Mar. Biol.,* 16, 108, 1072.
8. **Scott, D. M. and Major, C. W.,** The effects of copper (II) on the survival, respiration and heart rate in the common blue mussel, *Mytilus edulis, Biol. Bull.,* 143, 679, 1972.
9. **Thurberg, F. P., Dawson, M. A., and Collier, R. S.,** Effects of copper and cadmium on osmoregulation and oxygen consumption in two species of estuarine crabs, *Mar. Biol.,* 23, 171, 1973.

10. **Betzer, S. B. and Yevich, P. P.**, Copper toxicity in *Busycon canalicolatum* L., *Biol. Bull.*, 148, 16, 1975.
11. **Viarengo, A., Pertica, M., Mancinelli, G., Capelli, R. and Orunesu, M.**, Effects of copper on the uptake of amino acids, protein synthesis and on ATP content in different tissues of *Mytilus galloprovincialis* lam., *Mar. Environ. Res.*, 4, 145, 1980.
12. **Young, J. S.**, Toxic and adaptive responses to copper by the sabellid polychaete, *Eudistylia vancouver*, in *Physiological Mechanisms of Marine Pollutant Toxicity*, Vernberg, W. B., Calabrese, A., Thurberg, F. P. and Vernberg, F. J., Eds., Academic Press, New York, 1982, 53.
13. **Moore, M. N., Widdows, J., Cleary, J. J., Pipe, R., Salkeld, R. N., Donkin, P., Farrar, S. V., Evans, S. V., and Thomson, P. E.**, Responses of the mussels *Mytilus edulis* to copper and phenanthrene: interactive effects, *Mar. Environ. Res.*, 14, 167, 1984.
14. **Papathanassiou, E.**, Effects of cadmium ions on the ultrastructure of the gill cells of the brown shrimp *Crangon crangon* (L.) (Decapoda, Caridea), *Crustaceana*, 48, 6, 1985.
15. **Sunila, I.**, Toxicity of copper and cadmium to *Mytilus edulis* L. (Bivalvia) in brackish water, *Ann. Zool. Fenn.*, 18, 231, 1981.
16. **Grace, A. L. and Gainey Jr., L. F.**, The effects of copper on the heart rate and filtration rate of *Mytilus edulis*, *Mar. Pollut. Bull.*, 18, 87, 1987.
17. **Widdows, J. and Johnson, D.**, Physiological energetics of *Mytilus edulis*: scope for growth, *Mar. Ecol. Progr. Ser.*, 46, 113, 1988.
18. **Freeman, H. C.**, Metal complexes of amino acids and peptides, in *Inorganic Biochemistry*, Eichorn, G. L., Ed., Elsevier, Amsterdam, 1973, 121.
19. **Halliwell, B. and Gutteridge, M. C.**, Oxygen toxicity, oxygen radicals, transition metals and disease, *Biochem., J.*, 219, 1, 1984.
20. **Kurijama, R. and Sakai, H.**, Role of tubulin SH groups in polymerization to microtubules, *J. Biochem.*, 76, 651, 1974.
21. **Luduena, R. F., Roach, M. C., Jordan, M. A., and Murphy, D. B.**, Different reactivities of brain and erythrocyte tubulins towards sulphydryl groups directed reagent that inhibits microtubule assembly, *J. Biol. Chem.*, 260, 1257, 1985.
22. **Meister, A. and Anderson, M. E.**, Glutathione, *A. Rev. Biochem.*, 52, 711, 1983.
23. **Esterbauer, H.**, Lipid peroxidation products: formation, chemical properties and biological activities, in *Free Radicals in Liver Injury*, Poli, G. Cheeseman, K. H., Dianzani, M. U., and Slater, T. L., Eds., IRL Press, Oxford, 1985, 29.
24. **La Roche, G., Eisler, G., and Tarzwell, C.**, Bioassay procedure for oil and oil dispersant toxicity evaluation, *J. Water Pollut. Cont. Fed.*, 42, 1982, 1970.
25. **Arena, N., Bodo, M., Baroni, T., Alia, F. A., Gaspa, L., and Becchetti, E.**, Effects of lectins on cytoskeleton and morphology of cultured chick embryo fibroblasts, *Cell Mol. Biol.*, 36, 317, 1990.
26. **Bodo, M., Becchetti, E., Pezzetti, E., Baroni, T., Calvitti, M., Alia, F. A., and Arena, N.**, Cytoskeletal and DNA synthesis modification by concavalin A in embryonic fibroblasts maintained in serum-free and serum-added medium, *Cell Mol. Biol.*, 36, 637, 1990.
27. **Gozes, I. and Barnestable, C. J.**, Monoclonal antibodies that recognize discrete forms of tubulin, *Proc. Natl. Acad. U.S.A.*, 79, 2579, 1982.
28. **Alia, E. E. and Arena, N.**, On the presence of a 100K protein in rabbit skeletal muscle: the brush border of small intestive, cultured fibroblasts and red blood cells, in *Contractile proteins in muscle and non muscle cell systems*, Alia, E., Arena, N., and Russo, M., Eds., Praeger Publishers, New York, 1985, 181.
29. **Arena, N., Barni, S., Sciola, L., Gaspa, L., Tilloca, G., Meloni, M., and Pippia, P.**, Morphological and functional change of the cytoskeleton in neoplastic and syngeneic normal cells treated with anti-microtubular substances, *Acta Oncologica*, 12, 101, 1991.
30. **Akerboom, T. P. M. and Sies, H.**, Assay of glutathione, glutathione disulphides and glutathione mixed disulphides in biological samples, *Meth. Enzym.*, 71, 373, 1981.
31. **Esterbauer, H., Lang, S., Zadravec, S., and Slater, T.**, Detection of malonaldheyde by high performance liquid chromatography, *Meth. Enzym.*, 105, 319, 1984.
32. **Viarengo, A., Mancinelli, G., Martino, G., Pertica, M., Canesi, L., and Mazzucotelli, A.**, Integrated stress indices in trace metal contamination: a critical evaluation in a field study, *MEPS Special— Biological Effects of Pollutants: Results of a Practical Workshop, Mar. Ecol. Progr. Ser.*, 46, 65, 1988.

33. **Doughtie, D. G., and Ranga Rao, K.,** Histopathological and ultrastructural changes in the antennal gland, midgut, hepatopancreas and gills of grass shrimp following exposure to hexavalent chromium, *J. Invert. Pathol.,* 43, 89, 1984.
34. **Manley, A. R.,** The effects of copper on behaviour, respiration, foltration and ventilation activity of *Mytilus edulis, J. Mar., Biol. Assoc., U.K.,* 63, 205, 1983.
35. **Redpath, K. J.,** Growth inhibition and recovery in mussesl (*Mytilus edulis*) exposed to low copper concentrations, *J. Mar. Biol. Assoc. U.K.,* 65, 412, 1985.
36. **Sunila, I. and Lindstrom, R.,** The structure of interfilamentar junction of the mussel (*Mytilus edulis L.*) gill and its uncoupling by copper and cadmium exposures, *Comp. Biochem. Physiol.,* 81C, 267, 1985.
37. **Sager, R. P. and Syversen, T. M. L.,** Disruption of microtubules by methylmercury, in *The Cytoskeleton: A Target for Toxic Agents,* Clarkson, T. W., Sager, R. P., and Syversen, T. L. M., Eds., Plenum Press, New York, 1984, 87.
38. **Perrino, A. B. and Chou, I. N.,** Role of calmodulin in cadmium-induced microtubule disassembly, *Cell Biol. Int. Rep.,* 10, 565, 1986.
39. **Weisemberg, R. C.,** Microtubule formation in vitro in solutions containing low calcium concentrations, *Science,* 1977, 1104, 1972.
40. **Nishida, E. and Kobayashi, T.,** Relationship between tubulin SH groups and bound guanine nucleotides, *J. Biochem.,* 81, 343, 1977.
41. **Lee, J. C., Frigon, R. P., and Timasheff, S. N.,** The chemical characterization of microtubule protein subunits, *J. Biol. Chem.,* 218, 7253, 1973.
42. **Mellon, M. G. and Rebhun, L. I.,** Sulphydryls and the *in vitro* polymerization of tubulin. *J. Cell Biol.,* 70, 226, 1979.
43. **Davison, A. J., Legault, A. N., and Steele, D.,** Effects of 6-hydroxydopamine on polymerization of tubulin, *Biochem. Pharmacol.,* 35, 1411, 1986.
44. **Oliver, J. M., Albertini, D. F., and Berlin, R. D.,** Effects of glutathione-oxidizing agents on microtubule assembly and microtubule-dependent surface properties of human neutrophils, *J. Cell Biol.,* 71, 921, 1976.
45. **Bellomo, G., Mirabelli, F., Vairetti, M., Iosi, F., and Maiorini, W.,** Cytoskeleton as a target in menadione-induced oxidative stress in cultured mammalian cells. I. Biochemical and Immunocytochemical features, *J. Cell. Physiol.,* 143, 118, 1990.
46. **Bellomo, G. and Orrenius, S.,** Altered thiol and calcium homeostasis in oxidative hepatocellular injury, *Hepatology,* 5, 876, 1985.
47. **Nicotera, P., McConkey, D. J., Svensson, S. S., Bellomo, G., and Orrenius, S.,** Correlation between cytosolic Ca concentration and cytotoxicity in hepatocytes exposed to oxidative stress, *Toxocology,* 52, 55, 1988.
48. **Tsuchija, T.,** Effects of calcium ions on triton extracted lamellibranch cilia, *Comp. Biochem. Physiol.,* 56A, 353, 1977.
49. **Stommel, E. W. and Stevens, R. E.,** Calcium dependent phosphatidylinositol phosphorylation in lamellibranch gill lateral cilia, *J. Comp. Physiol.,* 157, 441, 1985a.
50. **Stommel, E. W. and Stevens, R. E.,** Cyclic AMP and calcium in the differential control of *Mytilus* gill cilia, *J. Comp. Physiol.,* 157, 4512, 1985b.
51. **Stommel, E. W. and Stevens, R. E.,** EGTA induces prolonged summed depolarizations in *Mytilus* gill coupled ciliated epithelical cells: implications for the control of ciliary motility, *Cell Motil. Cytoskel.,* 10, 464, 1988.
52. **Viarengo, A. and Nicotera, P.,** Possible role of Ca^{2+} in heavy metal cytotoxicity, *Comp. Biochem. Physiol.,* 100C, 81, 1991.
53. **Viarengo, A.,** Heavy metal effects on Ca^{2+} homeostasis and on signal transduction pathways, 13th Conference of the European Society for Comparative Biochemistry and Physiology, Research for Aquaculture: Fundamental and Applied Aspects, *Antibes,* 245, 1991 (Abstr.).

CHAPTER 15

Alternative Methods in Ecotoxicological Research and Testing

Julia Fentem and Michael Balls

INTRODUCTION

Many thousands of agricultural and industrial chemicals, and other pollutants produced as a result of various domestic and urban activities, are present in our environment. These pose a potential hazard to both human and animal health, and to the maintenance of homeostasis in a wide variety of ecosystems. However, for only a relatively few individual chemicals and products (e.g., some pesticides) is detailed information available on their toxicological properties and possible adverse effects on the environment.

Increasing awareness of our need to protect the environment from chemical pollutants has resulted in greater demands for rigorous environmental hazard and risk assessment, which currently involves undertaking extensive toxicity testing in a wide range of species. In 1991, more than 3.2 million scientific procedures were carried out on living animals in Great Britain alone.[1] Of these, 8% were concerned with the safety evaluation of nonmedical/veterinary products, including substances used in industry (87,217 procedures), substances used in agriculture (77,657 procedures), and environmental pollutants (58,224 procedures). There has been a marked increase during the past 5 years in the number of animal procedures undertaken in the interests of protecting man, animals, or the environment from the adverse effects of environmental pollution, with 30,000 more procedures undertaken in this respect in 1991 than in 1987.

Ironically, in spite of increasing public concern about the use of animals in toxicity (safety) testing, the trend is toward a need for more animal testing in the 1990s. This is because of concomitant demands for greater protection of the environment and for greater safety at work, and the need to test existing chemicals (chemicals in use before the present regulations came into effect) and retest certain kinds of products (such as pesticides) to meet current standards. In addition to the main mammalian toxicology studies routinely undertaken to provide information on the potential human hazard of new pharmaceuticals, agrochemicals, food additives, etc. (i.e., tests for acute, subacute, and chronic toxicity, eye and skin irritancy, carcinogenicity and mutagenicity, reproductive toxicity, and sensitization), environmental protection legislation requires that other types of toxicity tests are also undertaken for certain chemicals (e.g., industrial chemicals and pesticides), as a basis for assessing their potential ecological effects.

In recent years, efforts have been directed toward the development of nonanimal methods to replace the animal toxicity tests currently required by law for evaluating the potential human hazard of chemicals, but ecotoxicity testing has received less attention with regard to the possibility of developing alternative testing procedures. Therefore, in this chapter we will concentrate, in particular, on the potential use of alternative methods in ecotoxicological research and testing.

ALTERNATIVES TO ANIMAL EXPERIMENTS

The term "alternatives" has come to have a special meaning in the context of animal experimentation, and includes all procedures that can completely replace the need for animal experiments, reduce the numbers of animals required, or diminish the amount of pain or distress suffered.[2,3] It embraces all the "Three Rs" proposed by Russell and Burch in 1959,[4] and we can think in terms of *replacement alternatives, reduction alternatives,* and *refinement alternatives.* The Three Rs concept of alternatives is now widely recognized, and legislation such as the UK *Animals (Scientific Procedures) Act, 1986* and EEC *Directive 86/609* represent statutory requirements that researchers must use alternatives to animal experiments wherever possible. For example, *Directive 86/609/EEC* stipulates that

> An experiment shall not be performed if another scientifically satisfactory method of obtaining the result sought, not entailing the use of an animal, is reasonably and practicably available.

That

> When an experiment has to be performed, the choice of species shall be carefully considered and, where necessary, explained to the authority. In a choice between experiments, those which use the minimum number of animals, involve animals with the lowest degree of neurophysiological sensitivity, cause the least pain, suffering, distress or lasting harm and which are most likely to provide satisfactory results, shall be selected.

And that

> All experiments shall be designed to avoid distress and unnecessary pain and suffering to the experimental animals.

These laws require that, notwithstanding other laws that, quite properly, require that workers, consumers, and the environment be given the greatest possible protection from risk of damage resulting from exposure to chemicals and products of various kinds, the Three Rs principle must be adhered to, and the need to use animals in toxicity tests must be specifically justified in terms of its relevance and necessity in particular circumstances.

The Need for Alternatives

Alternatives are needed for scientific, humanitarian, and logistical reasons. There is increasing criticism of the current dependence of toxicity testing on animal procedures (most of which have never been formally validated), and, in particular because of the known differences in chemical-induced toxicity between species, which may be manifested as differences in tissue-specific toxicity or differences in the magnitude of response. The need for replacement alternatives on ethical grounds is obvious since, as practiced, much toxicity testing requires that adverse effects are produced in laboratory animals, often leading to considerable pain and distress. In addition, from a practical perspective, few of the many thousands of chemicals currently in use have been adequately tested, and it would be economically and logistically impossible to test all of them by the full set of current animal procedures. Faced with the enormity of this task of adequate safety evaluation, it would seem that we really have no choice but to develop suitable alternative test procedures and testing strategies, which are less expensive, less time-consuming, and less labor-intensive than the traditional animal tests.

Types of Alternatives

Refinement

Refinement of animal experiments refers to improvements in the techniques and procedures employed, so that they cause less pain, suffering, and distress to any animals necessarily used.[5] Examples include the more widespread provision of anesthesia and/or analgesia, where appropriate, and the use of more-humane endpoints for some scientific procedures, such as the use of early clinical signs of toxicity rather than death as an endpoint measurement in acute toxic potency testing. Adverse effects may not only be caused by the scientific procedures themselves, but also by housing and husbandry conditions. Thus, refinement also includes enrichment of the environment in which the animals are kept, the housing of animals in groups rather than singly, and the feeding of better diets.

Reduction

The greater the number of animals used in an experiment, the greater will be the overall animal suffering. It is therefore important that, in any given experiment, the number of animals used is the minimum that is consistent with the aims of that experiment.[5] Careful experimental design and appropriate statistical analysis of the results obtained may make it possible to obtain data of comparable accuracy while using fewer animals. Proper advance planning of experiments should prevent the use of more animals than is statistically necessary, and can also help to ensure that experiments are not invalid (and animal lives wasted) because too few animals were used, so the experiment has to be repeated with larger numbers.

The need to repeat studies can sometimes be avoided, if existing information on the biological effects of chemicals is made available. Computerized databases now make such information more readily accessible and may contribute to a reduction in the number of animal experiments undertaken.

The modifications to the classical LD_{50} test that have been proposed in recent years are good examples of alternatives involving both refinement and reduction. Whereas a classical LD_{50} test requires between 60 and 80 animals, a validated alternative, the "Fixed Dose Procedure,"[6] requires only up to 20 animals per test and does not use lethality as the endpoint. Similarly, the so-called "step-wise" and "up-and-down" methods limit the maximum dose of a chemical administered to the animals and also require the use of fewer animals.

Replacement

The range of replacement alternatives includes the following:[7,8]

1. *Physicochemical methods*—the use of physical and chemical techniques, and predictions based on the physicochemical properties of molecules.
2. *Mathematical and computer models*—including the modeling of quantitative structure–activity relationships (QSAR), molecular modeling and the use of computer graphics in the design of chemicals for specific purposes, and the modeling of biochemical, physiological, pharmacological, and toxicological processes.

3. *Lower organisms*—the use of species, such as bacteria, fungi, higher plants, and invertebrates, which are not protected by legislation controlling animal experiments. For example, the use of bacteria in genotoxicity tests.
4. *Vertebrates at early stages of development*—the current British law protects animals only after they have reached a certain stage of development, and several alternative tests have been proposed that involve using the early developmental stages of vertebrates, before half-way through gestation (mammals) or through incubation (birds and reptiles), or when independent feeding occurs (amphibians and fish). For example, the use of the chorioallantoic membrane (CAM) of hen's eggs in testing for irritancy.[9]
5. *In vitro methods*—including the short-term maintenance of cell fractions, cell suspensions, tissue slices, and perfused organs, and long-term cell culture. For example, general cell culture tests for measuring the effects of chemicals on cell viability and cell proliferation, and more complex cell and tissue culture methods for studying mechanisms of toxicity, target organ specificity, and species differences in toxicity.[10]
6. *Human studies*—including clinical investigations, the properly controlled use of human volunteers, postmarketing surveillance, and epidemiological studies.

Advantages and Limitations of Replacement Alternatives

Replacement alternative methods offer scientific, humanitarian, and economic advantages to the use of animal procedures. For example, *in vitro* systems can be used to investigate the molecular and cellular mechanisms of chemical-induced toxicity (which cannot readily be investigated *in vivo*) and to study target-organ and target-species toxicity.[10] Tissues can be taken from a wide range of possible target species, which is particularly relevant when investigating the potential toxic effects of, for example, pesticides and environmental pollutants. However, the complexity of factors to be considered in assessing the value of nonanimal toxicity tests and testing strategies (Table 1) must always be borne in mind.

The main justification for developing replacement alternative methods for toxicity testing is that they will make toxicology a more scientifically based practice. Understanding the mechanisms by which chemicals cause cell and tissue damage, and the reasons for the increased susceptibility of certain species, populations, individuals, or tissues to particular chemicals, will markedly improve our ability to undertake risk assessments.

Nonanimal tests are more flexible than animal procedures, so the experimental design can be altered more readily, and the results they provide are usually more quantitative, less subjective, and more reproducible. The development of replacement alternatives has the potential not only to reduce the number of animals used, but also to contribute to the design of any subsequent animal procedures still considered necessary, so that any pain and suffering they may cause can be minimized (i.e., refinement). The nonanimal methods are usually less expensive to undertake than animal experiments, so they can help reduce the costs involved in developing new chemicals and products.

However, the limitations of replacement alternative methods must also be recognized. Most of them are still at relatively early stages of development, and the relevance of the results they provide has yet to be established. It is obvious that cell cultures, which are relatively static systems, cannot mimic the complex interactions of all the cell and tissue types in the body, and that the cells and tissues are maintained in an artificial environment in which the influences of the circulatory, nervous, and other systems of the intact organism are absent. Nor do nonanimal methods always guarantee an escape from ethical dilemmas, not least of which are concerns about the source of the animal or human tissue used in *in vitro* procedures.

The Strategic Use of Replacement Alternatives

In certain areas of toxicity testing, replacement alternative methods are currently being used as initial screens, to assess the biological effects of large numbers of new chemicals, and as

Table 1. Factors to be Taken into Account in Assessing the Value of Nonanimal Toxicity Tests and Testing Strategies

1. **Level of assessment of toxicity**
   Potential (inherent toxicity)
   Potency (relative toxicity)
   Hazard (toxicity under defined conditions of exposure)
   Risk (safety in use)
2. **Type of toxicity testing**
   Screening
   Adjunct (= complementary)
   Replacement
3. **Type of toxicity**
   General toxicity
   Target organ toxicity
   Target system toxicity
   Genotoxicity/carcinogenicity
4. **Type of test material**
   Full spectrum of chemicals
   Specific types of chemicals
   Chemicals/formulations with specific uses
     Pharmaceuticals
     Pesticides
     Cosmetics
     Household chemicals
     Industrial chemicals
5. **Likely level of toxicity**
   Very high
   High
   Moderate
   Low
   Insignificant
6. **Purpose of testing**
   General classification and labelling
   Specific worker and patient protection
   Protection of the environment
7. **Main object of interest**
   Humans
     Patient
     Worker
     Consumer
     All
   Other vertebrate species
   Invertebrate species
   Ecosystems

adjuncts/complementary tests in predominantly animal-based testing programs. Their use should result in the better design of subsequent animal tests, and a reduction in the overall number and suffering of the animals used, since *in vivo* testing should be undertaken only to confirm the absence of significant toxicity. However, there is little sign that such alternative tests will be accepted as genuine replacements for animal procedures in the near future. This is partly because they have to undergo rigorous validation (the process by which the relevance, reproducibility, and scientific quality of the proposed method are assessed), to establish the credibility of the test for a particular purpose,[11,12] before they can be considered for acceptance by the regulatory authorities.[13]

It is unrealistic to expect a single alternative test to be an adequate substitute for a particular animal test, and it is widely accepted that batteries of complementary and interdependent non-animal methods will be required.[14,15] A hierarchical approach, or tiered testing strategy, can be envisaged, in which the sequential use of nonbiological tests (e.g., computer models) and *in vitro* tests plays a major role, before undertaking a few animal tests, when absolutely essential (Table 2).[16]

Table 2. An Eight-Stage Scheme for Toxicity Testing[16]

1. Informational investigations	Consultation of literature, data banks, use of experience
2. Physicochemical studies	Measurement of physicochemical properties
3. Computer analysis	QSAR studies
	Modeling of biological processes and biokinetics
4. First-order *in vitro* tests	General cytotoxicity tests
	Special cytotoxicity tests
	Genotoxicity tests
5. Biokinetics tests	Absorption/uptake tests
	Binding/breakdown tests
	Biotransformation tests
	Clearance tests
6. Second-order *in vitro* tests	Target organ tests, e.g., for
	Ocular irritancy
	Dermal irritancy
	Pulmonary toxicity
	Hepatotoxicity
	Nephrotoxicity
	Target system tests, e.g., for
	Neurotoxicity
	Immunotoxicity
	Special tests, e.g., for
	Carcinogenicity
	Reproductive toxicity
7. Essential tests in animals	Designed on basis of knowledge from stages 1 to 6
8. Tests in human volunteers	Preclinical testing of drugs
	Premarketing testing of cosmetics

ECOTOXICOLOGY

Widespread concern about the possible ecological effects of chemicals resulted in the introduction of new legislation in many industrialized countries in the 1970s, which was directed toward the protection of the environment. For example, in the United States, the *Toxic Substances Control Act* was passed in 1976, and, in Europe, *Directive 79/831/EEC* (the VIth Amendment of the EEC Directive on Dangerous Chemicals) was introduced in 1979. Ecotoxicology has emerged as a true scientific discipline only in more recent times, and it is still very much driven by the managerial and legislative requirements of environmental risk assessment procedures.[17] This is not surprising, since most ecotoxicological studies are undertaken to provide the information necessary to make risk assessments pertaining, in particular, to chemical wastes produced by industry, to pesticides, and to industrial chemicals.

Ecotoxicology has been defined as "the study of the harmful effects of chemicals upon ecosystems."[18,19] In contrast to mammalian toxicology studies, in which the data generated from a few species (typically rodents) are extrapolated to give an indication of likely human hazard, results from ecotoxicity tests undertaken with a few species are extrapolated to an entire ecosystem.[20] Classical mammalian toxicology studies are concerned with identifying the adverse effects of chemicals on single organisms, whereas ecotoxicology studies are ultimately concerned with assessing the potential hazards to numerous wildlife species and with determining the effects of chemicals on whole ecosystems.

The ecological significance of a particular environmental pollutant typically resides in its indirect impact (e.g., its effects on predators or habitats), at sublethal concentrations, on populations and communities of species, such as effects on reproduction or genetic composition, rather than on its immediate toxic effects on individual organisms. Ecotoxicology necessarily involves consideration of the movement and distribution of chemicals in air, soil, and aquatic systems. This movement of pollutants through ecosystems, including their bioaccumulation, binding, and degradation, depends on a variety of factors, such as migration, behavior, feeding ecology, and species differences in sensitivity to chemical-induced toxicity,[19] and these all need to be taken into account in the environmental risk assessment process.

Predicting the Potential Ecotoxicological Effects of Chemicals

Ecotoxicity testing poses two major problems: first, the very wide range of species that may be at risk, and, second, the difficulty of predicting levels of exposure in the field.[19] Often, it is not possible to perform tests using the species thought to be at risk, so data obtained with surrogate species are used for hazard assessment. In practice, tests are performed on a small number of surrogate ("indicator") species and a safety factor (of at least one order of magnitude) is then usually employed in making decisions about environmental safety.

Predicting the ecotoxicological effects of a particular chemical with confidence is a very difficult, if not impossible, task for several reasons. These include the following:[18]

1. Usually more than one chemical is involved. The original product may be a mixture of chemicals, it may contain biologically active impurities, or its metabolites or degradation products may have biological effects. In addition, there is the distinct possibility that it may interact with other chemicals already present in the environment; in nearly all cases, the stress of pollution on natural ecosystems is attributable to the combined effects of different chemicals.[21]
2. It is often difficult to determine exposure with any degree of precision, since the distribution of a chemical, which affects exposure, is not uniform.
3. The relationships between exposure, the amount of pollutant within an organism, and the effects on that particular individual are complex.
4. Different species, and even individuals within a species, can react differently to the same exposure, for both genetic and environmental reasons.
5. The consequences of interactions between individuals within a population, and between species within a community, are also complex and are poorly understood.

The need to be able to predict the potential ecotoxicological effects of new chemicals before environmental contamination occurs is obvious. However, while ecotoxicological research efforts are directed toward trying to define the relationships between exposure and the amount of pollutant present within organisms, to determining the biological effects of these amounts of pollutant on individual organisms, and to assessing the overall effects on ecosystems, current regulatory ecotoxicity testing, according to a Royal Society Study Group in 1978, is "swamped by routine tests of limited value and governed by regulations rather than rational thought."[18]

Ecotoxicity Testing to Meet Regulatory Requirements

The ecotoxicological testing requirements for industrial chemicals and pesticides have been outlined by Moriarty[18] and Walker et al.[19] The approach to toxicity testing typically differs between the general screening procedure employed for industrial chemicals and the more-detailed information needed for evaluating the environmental safety of new pesticides. In general, tiered testing schemes are followed, in which the numbers of individual tests carried out, and the numbers and types of species involved, depend on the proposed uses and levels of production of new chemicals or products.

A "base-set" of data is required for all new industrial chemicals, which incorporates tests for acute toxicity in fish and *Daphnia*, and determination of the rate of biotic and abiotic degradation. Aquatic species are used in these ecotoxicity tests, because release into the environment via sewage discharged into rivers is a major route for many chemicals.[18] Further tests are required as the level of production of the chemical increases, and additional tests on terrestrial organisms are required when species in other, nonaquatic, habitats are thought to be at risk.

In the case of pesticides, there are standard requirements for tests for mammalian toxicity. In addition, the assessment of environmental safety is based on toxicity data for two species of birds (usually the mallard and the Japanese quail) and for one or more species of fish, as well as for

Daphnia, honey bees, and algae. Further tests may be required where particular risks are suspected.[19]

Limitations of the Current Procedures for Ecotoxicity Testing

Choice of Species

Although ecotoxicology is ultimately concerned with effects at the levels of population and community, most of the testing to meet legislative requirements is undertaken in the laboratory.[22] Assessments of hazard to fish and wildlife species are typically based on extrapolations from laboratory data obtained using a few surrogate species. Scientific, practical, and ethical considerations influence the selection of appropriate species for toxicity testing. Although the uncertainty inherent in extrapolating data from one species to another could be minimized by using different, ecologically representative species, in practice there are only a limited range of species available for testing purposes. This is mainly due to difficulties associated with keeping some wild species in captivity, and because of the lack of background data available on them with regard to their sensitivities to a range of chemicals, which is invaluable when trying to interpret toxicity data obtained with a new chemical or product.

Measures of Toxicity

Tests for acute toxicity and, in particular, lethality (i.e., the determination of LD_{50} and LC_{50} values) are the most common regulatory ecotoxicity testing requirements. However, the relevance of determining the lethal concentrations of a chemical in a few surrogate species (usually fish and *Daphnia*) to an assessment of its possible ecological effects is questionable. There is considerable interspecies variation in LD_{50} values (for example, the variation between mammalian species for TCDD is 200-fold[23]), which makes extrapolation of LD_{50} and LC_{50} data between invertebrates, fish, birds, and other wildlife species difficult to justify scientifically.

A major concern in ecotoxicology is the frequency and extent to which particular species may survive the initial direct impact of pollutants, but subsequently function less effectively.[18] Such sublethal effects, especially on reproductive potential, are often of much greater ecological significance than lethality.[19] Thus, no-observed effect levels (NOELs) are more relevant than are LD_{50} or LC_{50} values for assessing environmental hazards.

Other Limitations

The problems of species differences and appropriate measures of toxicity are by no means unique to ecotoxicity testing. Similarly, the other limitations evident in undertaking ecotoxicological tests according to the rigid protocols required by the regulatory authorities also apply to the majority of mammalian toxicology studies conducted for legislative purposes. These include[16]

1. dosimetry considerations, i.e., the relevance of the doses administered to the test organisms to probable levels of exposure in the field;
2. extrapolation from high dose to low dose effects;
3. extrapolation from controlled experimental conditions to variable environmental situations;
4. inadequate consideration of the effects of exposures to mixtures of chemicals; and
5. lack of adequate mechanistic understanding of toxicological phenomena.

It is the intelligent use of appropriate and relevant replacement alternative methods in ecotoxicological research and testing that offers a means of overcoming many of these limitations, since nonanimal procedures undoubtedly have the potential to improve the scientific basis of ecotoxicity testing and environmental risk assessment.

POTENTIAL USE OF ALTERNATIVES IN ECOTOXICOLOGY

Ecotoxicological Research

Fundamental and applied research in ecotoxicology, as in many other areas of science, is increasingly being conducted at the cellular and molecular levels by using nonanimal (i.e., alternative) methods. There is growing interest in the possible use of chemical-induced changes at the biochemical level for measuring exposure and sublethal effects under field conditions, and the development of biochemical assay systems that can be employed to measure early sublethal effects in the field (e.g., ELISA assays for use following nondestructive sampling) is currently an important area of ecotoxicological research.[18,19,23] Such studies should eventually lead to both the refinement of the animal procedures undertaken and a reduction in the numbers of living vertebrates used in ecotoxicity testing. Similarly, the application of new analytical methods (e.g., nuclear magnetic resonance, mass spectrometry), and other noninvasive and nondestructive imaging techniques, should result in refinement of the current approaches to environmental risk assessment.

Ecotoxicological research being conducted in five main areas, in particular, is of great importance with regard to the possible replacement of some animal tests in the future: the development of QSARs for predicting the possible ecotoxicological effects of chemicals, the development and evaluation of tests using lower organisms (e.g., bacteria and invertebrates), studies of the effects of pollutants on embryos, especially those of nonmammalian vertebrates, investigations using a wide range of *in vitro* techniques, and the use of biomarkers to try to provide a link between the exposure of an organism to a particular xenobiotic and its subsequent biological effects.

QSARs

A QSAR is a mathematical equation that relates the biological response of a chemical (e.g., toxicity) to one or more descriptors of its molecular structure and/or its physicochemical properties (e.g., hydrophobic, steric, and electronic descriptors).[24,25] In recent years, a lot of effort has been invested in QSAR studies of toxicity and, in particular, in the development of QSARs for predicting the potential ecotoxicological effects of industrial organic chemicals.[26] Bioaccumulation, biodegradability, and acute toxicity have all been subjected to QSAR analysis, and data obtained from such studies have helped in the prediction of toxic effects and in the setting of chemical testing priorities.

Hydrophobicity, which is usually modeled by log P (where P is the octanol–water partition coefficient), is a major determinant of movement into and within an organism,[24,27] and, in general, bioaccumulation increases rectilinearly with log P.[28] Biodegradability has been shown to be a function of the difference in atomic charge across key bonds in a molecule.[29] Various QSAR equations have been described for predicting fish LC_{50} values for chemicals belonging to certain classes.[25,30]

Up to now, QSAR studies have mainly demonstrated the importance of hydrophobicity in determining the ecotoxic effects of chemicals. QSARs have been developed for several groups of chemicals and, within these classes, can predict the toxicities of the substances concerned, and those of related chemicals, reasonably well. However, the QSAR approach needs to be extended to many more chemical types and to incorporate descriptors pertaining to possible mechanisms of toxicity, before its potential value in predictive ecotoxicity testing can be fully realised. A critical assessment of the use of QSARs in toxicology and ecotoxicology has been undertaken by a task force established by ECETOC (European Chemical Industry Ecology and Toxicology Centre).[30]

Lower Organisms

During the past 10 years, the urgent need for faster and less expensive ecotoxicity testing procedures, particularly for screening purposes, has been recognized, and, in response to this, a variety of rapid and relatively simple bioassays are being developed, involving the use of lower organisms. They include tests employing bacteria, yeasts, protozoans, algae, invertebrates, insects, and various plant species.[31-35] Several such methods are currently being evaluated with regard to their ability to predict the acute toxicity of aquatic pollutants to fish, while others, the so-called "microbiotests,"[36] are being used by environmental scientists for screening the toxic potentials and potencies of complex chemical wastes and sewage effluents, and for monitoring aquatic samples. Microbiotests include short-term tests for determining the effects of pollutants on enzyme activities or enzyme biosynthesis (e.g., the ATP-TOX[37] and Toxi-Chromotest[38] assays), on the bioluminescence of *Photobacterium phosphoreum* (the Microtox assay[39]), and on bacterial growth and viability.[40] Several commercial test kits are now available for monitoring the toxicity of environmental samples,[40] but these have not been extensively evaluated, and their role in ecotoxicity testing would appear to be very limited.

In contrast, research currently being undertaken to investigate the effects of selected chemicals on various invertebrate species, both aquatic and terrestrial,[33,34,41] and attempts to develop ecotoxicity testing procedures that employ multispecies assemblages,[42] should eventually prove of value for predicting the potential effects of industrial chemicals, pesticides, and other environmental pollutants on entire ecosystems.

Embryos

The UK *Animals (Scientific Procedures) Act, 1986* protects animals only after they have reached a certain stage of development. Research has therefore been directed toward the development of tests that employ early developmental stages of vertebrates and, in particular, those that use embryos. The embryos of birds, fish, amphibians, and reptiles are readily obtained and easily examined, since they typically develop externally to their parents. In addition, relatively large numbers of eggs are often produced, which is advantageous for toxicological studies, and, with regard to conservation, the impact on a population of removing eggs is minimal.[23] For example, even in the case of birds, eggs that are taken early in the breeding cycle are usually replaced.

Studies undertaken using embryos are mainly concerned with determining the acute toxicity (lethality), embryotoxicity, and teratogenicity of chemicals, although there have been a few investigations on the effects of environmental pollutants on certain enzyme activities (e.g., acetylcholinesterase and aryl hydrocarbon hydroxylase activities[23]) and other biochemical parameters. Further details on the use of embryos for toxicological studies are given by Cooke,[43] Hoffman and Albers,[44] and Weis and Weis.[45]

In Vitro Methods

Techniques are now available that permit the culture of a wide variety of cell types from many different tissues and species. *In vitro* test systems have found widespread application for studying species differences in the metabolism and toxicity of chemicals, and for investigations aimed at elucidating the mechanisms of toxicity of certain chemicals.[10,23] The type of *in vitro* method employed depends on the specific question being asked. For example, studies of biotransformation, including the measurement of kinetic parameters,[46] are typically undertaken using isolated enzyme preparations, microsomal systems,[47] and/or primary cell cultures (e.g., hepatocytes), whereas continuous cell lines are often used in assessing the toxic potentials and relative toxicities (i.e., potencies) of chemicals.[16] Attempts to elucidate mechanisms of chemical-induced toxicity generally require complementary information derived from subcellular preparations, cell cultures, and tissue slices, etc., used in combination, in addition to information obtained from *in vivo* studies.

Data derived from *in vitro* studies generally give a good indication of the likely *in vivo* metabolism and toxicity of a particular chemical across a range of species, although they must be

Table 3. Ecotoxicity Testing: Recommendations of the Working Party of the FRAME Toxicity Committee[22]

1. Ecotoxicity tests with surrogate species should be carefully designed to use the minimum number of animals and to obtain the maximum amount of information from them.
2. Tests for lethal effect (LD_{50}, LC_{50}) should be replaced by flexible sequential tests starting from low doses or concentrations: in some cases, a no-effect level is all that is necessary.
3. A better understanding of mechanisms of detoxication and toxic action in different species should lead to the improvement of testing procedures by predicting sensitivity, and by giving guidance on extrapolation from surrogates to species deemed to be at risk.
4. More work should be done on the development of *in vitro* systems and on the mathematical modeling of empirical relationships to predict the toxicities of chemicals to wild species.
5. More efficient risk assessment procedures are required, incorporating estimates of exposure in the field, as well as toxicity data obtained in the laboratory.
6. Greater use should be made of semifield testing of pesticides, which could resolve some of the uncertainties about exposure in the field.
7. Field testing procedures for pesticides could be considerably improved in a number of ways (e.g., better design of experiments and use of statistical procedures and more use of biochemical tests for toxic effects and of behavioral tests).
8. The operation of postapproval monitoring schemes for pesticides helps to give confidence in assessment of environmental safety, and should be encouraged.
9. A more flexible approach to the testing of industrial chemicals for fish toxicity would lead to a significant reduction in the numbers of fish used.

interpreted with care. Such an approach can provide information on a large number of species that cannot readily be used for *in vivo* studies, can substantially reduce the number of animals needed,[23] and can also markedly improve the selection of appropriate surrogate species for use in any subsequent *in vivo* toxicity tests considered necessary.[19]

The results of comparative biochemistry and toxicity studies, and mechanistic investigations, should eventually make it easier to predict which species will be susceptible to particular chemicals. However, at the present time, only limited predictions of *in vivo* toxicity can be made from *in vitro* test data,[19,48] and, as a Working Party of the FRAME Toxicity Committee recommended (Table 3),[22] more research is required to develop *in vitro* systems that can predict the toxicities of chemicals to wild species.

Biomarkers

Evaluation of the ecotoxicological effects of chemicals traditionally involves laboratory and/or field studies using living animals, and, although both types of approach provide useful information on the initial exposure of an organism to a chemical, and the final residue levels occurring in various tissues of that organism, there is still the problem of trying to extrapolate the effects caused by the chemical in the test species to its likely effects at the population and community levels. The use of biological markers ("biomarkers") is an exciting development in ecotoxicology, which may provide a solution to this problem.

According to the definition proposed by the U.S. National Academy of Sciences, a biomarker is *a xenobiotically-induced variation in cellular or biochemical components or processes, structures, or functions that is measurable in a biological system or sample.*[23] Biomarkers can provide sensitive and specific measures of exposure, and sometimes of toxic effects, using samples obtained from the field by nondestructive procedures. The ability to measure specific biochemical responses induced by an environmental pollutant can provide the data necessary for establishing causality, and thereby provide a link between exposure and effect.[49] It is also possible to relate such biochemical responses to consequent effects at the population level.

Biomarkers are typically changes in the activities of particular enzymes (e.g., inhibition of acetylcholinesterase is a predictive indicator of the long-term adverse effects of organophosphates[23]) or in the levels of specific biogenic compounds (e.g., changes in retinol levels are indicative of exposure to polycyclic halogenated aromatic hydrocarbons[23]). The formation of DNA and hemoglobin adducts, induction of the mixed-function oxidase system, and changes in the levels of specific serum proteins have also been investigated as potential biomarkers for certain classes of pollutants. In many cases, the biomarker can be assayed in blood, serum, or plasma. Sampling is therefore nondestructive and can be carried out in the field, thereby reducing reliance on unrepresentative captive animals. A series of samples can be taken from the same individual, which thus acts as its own control, and this also reduces the variation inherent in sampling.

Research into the use of biomarkers as indicators of the ecotoxicological effects of chemicals is still at a very early stage, and the validity of many of them as measures of toxic effect remains to be established. While they do not represent potential replacement alternatives for experiments using living animals, the use of appropriate biomarkers should lead to both reduction and refinement of the animal procedures undertaken in ecotoxicology, in addition to markedly improving the scientific basis of environmental risk assessment. The role of biomarkers in ecotoxicology has been discussed at length by Peakall.[23]

Ecotoxicity Testing

Currently, the most commonly used ecotoxicity test for screening for aquatic pollutants is the short-term fish lethality (LC_{50}) test, which has been criticized on economic, logistical, and ethical grounds. This particular test is required for the regulatory approval of all new industrial chemicals and pesticides.[18,19] In 1991, 132,005 scientific procedures were undertaken using fish in Great Britain alone, an increase of 70% on the figure for 1989.[1] Many of these were LC_{50} tests on agricultural, industrial, and other environmental chemicals, conducted for legislative purposes.

Strategies for reducing the number of fish used in acute toxicity tests,[50] and for refining the procedure employed for determining LC_{50} values, so that for compounds of low fish toxicity, a more limited test would be undertaken where the LC_{50} was greater than 100 m gL^{-1} (the concentration below which a substance is classified as "ecotoxic"),[19] have been proposed. Walker et al.[19] recently questioned whether fish toxicity data should be mandatory for all new industrial chemicals, since they argue that fish are not necessarily an appropriate choice for base-set tests. It is certainly difficult to justify using such large numbers of fish solely for screening purposes, and it would appear that, of all the ecotoxicity tests currently required by the regulatory authorities, the fish LC_{50} test could most readily be replaced, for the majority of industrial chemicals, by alternative methods using bacteria, lower organisms, or cultures of fish cells.[51]

Possible Replacement of the Fish Acute Lethality Test

Several replacement alternative methods have been evaluated, to varying extents, with regard to their ability to predict the acute toxicities of aquatic pollutants to fish. Of these, the luminescent bacteria toxicity bioassay (Microtox),[39] the *Daphnia* acute toxicity test, and basal cytotoxicity tests using fish cells in culture,[52] appear to have been the most extensively studied, in addition to recent work on QSARs. However, it is highly improbable that the bacterial luminescence bioassay could replace fish acute toxicity tests, and further studies are required before the replacement of fish LC_{50} tests by cytotoxicity tests with fish cell cultures, for certain classes of chemicals, can be considered.[51]

The introduction of a more flexible approach to ecotoxicity testing would certainly reduce the number of fish required. Walker et al.[19] have recommended that the need for fish toxicity tests should be based on a review of the physicochemical properties of the chemical, predictions from QSAR studies, and the results of acute toxicity tests with *Daphnia*. *Daphnia* are sensitive to a wide range of structurally diverse chemicals, and the results of acute toxicity tests with *Daphnia* are typically in close agreement with those of fish acute lethality tests. Thus, *Daphnia* could possibly be used as the single species for screening for the acute toxic effects of potential aquatic pollutants.[51] If data for another species were considered necessary, then an additional invertebrate species, rather than a vertebrate species, should be used (e.g., the mayfly larva, *Hexagenia limbata*).

Alternatively, replacement of the fish LC_{50} test by a battery of nonanimal tests can be envisaged. Such a battery would consist of a set of complementary tests, which could be used in combination to predict the potential toxicities of chemicals. The Microtox test, cytotoxicity tests with fish cell cultures, and the TOXKITS (cyst-based aquatic invertebrate tests), currently being evaluated,[53] could possibly be included in a test battery, in addition to QSAR predictions and acute toxicity tests with *Daphnia*.[51] However, further studies need to be undertaken to assess the suitability of some of these tests for inclusion in a battery approach to the ecotoxicity testing of industrial chemicals.

CONCLUSIONS

The recommendations contained in the Second Report of the FRAME Toxicity Committee (Table 3)[22] make it clear that greater consideration must be given to the application of the Three Rs in ecotoxicology, as is required by the current animal protection legislation in various countries and in the EEC. There is certainly scope for reducing the number of animals used and for refining the procedures undertaken. There are also prospects for replacing some procedures involving living vertebrates in ecotoxicity testing, most notably the fish acute lethality test for screening industrial chemicals, with alternative tests or testing strategies.

Ecotoxicological research and ecotoxicity testing to meet regulatory requirements must not be viewed as distinct entities, but rather as complementary approaches. Scientists and regulators must consider incorporating the most recent research developments into their testing procedures as soon as possible, in order to continually reduce our dependence on toxicity tests using large numbers of animals as a means of protecting the environment.

It is also important that the amount of information obtained from individual animals is maximized,[54] and more emphasis must be placed on the development and use of methods likely to improve our understanding of the mechanisms underlying the ecotoxicological effects of chemicals. Information from *in vitro* studies will help to increase our knowledge of mechanisms, and thereby improve our ability to extrapolate data obtained using an "indicator" species to other species that may be exposed to the chemical. In addition, the use of appropriate biomarkers should help the extrapolation of the effects observed in a test species to the potential effects of a particular chemical on whole ecosystems. *In vitro* tests for investigating biokinetics, and for studying species differences in the effects of certain chemicals and formulations (e.g., pesticides, which are designed to be highly, but selectively, toxic) on specific biological target systems, need to be developed, to provide a more scientific basis for environmental risk assessment.

We are hopeful that, in the future, the intelligent use of appropriate *in vitro* and lower organism test systems, in conjunction with predictive computer modeling, will not only result in the replacement of many toxicity tests using protected animals, but will also markedly improve current hazard identification and risk assessment practices in ecotoxicology.

REFERENCES

1. **Anon.,** *Statistics of Scientific Procedures on Living Animals, Great Britain 1991, Cm 2023,* HMSO, London, 1992.
2. **Smyth, D. H.,** *Alternatives to Animal Experiments,* Scolar Press, London, 1978.
3. **Anon.,** *Alternatives to Animal Use in Research, Testing, and Education,* U.S. Congress Office of Technology Assessment, Washington, D.C., 1986.
4. **Russell, W. M. S., and Burch, R. L.,** *The Principles of Humane Experimental Technique,* Methuen, London, 1959.
5. **Smith, J. A., and Boyd, K. M., Eds.,** *Lives in the Balance. The Ethics of Using Animals in Biomedical Research,* Oxford University Press, Oxford, 1991, chap. 5.

6. **van den Heuvel, M. J., Clark, D. G., Fielder, R. J., Koundakjian, P. P., Oliver, G. J. A., Pelling, D., Tomlinson, N. J., and Walker, A. P.,** The international validation of a fixed-dose procedure as an alternative to the classical LD_{50} test, *Food Chem. Toxicol.,* 28, 469, 1990.
7. **Balls, M.,** Alternatives to animal experimentation, *ATLA,* 11, 56, 1983.
8. **Fentem, J., and Balls, M.,** *In vitro* alternatives to toxicity testing in animals, *Chem. Indust.,* 6, 207, 1992.
9. **Luepke, N. P.,** Hen's egg chorioallantoic membrane test for irritation potential, *Food Chem. Toxicol.,* 23, 287, 1985.
10. **Jolles, G., and Cordier, A., Eds.,** *In Vitro Methods in Toxicology,* Academic Press, London, 1992.
11. **Balls, M., Blaauboer, B., Brusick, D., Frazier, J., Lamb, D., Pemberton, M., Reinhardt, C., Roberfroid, M., Rosenkranz, H., Schmid, B., Spielmann, H., Stammati, A.-L., and Walum, E.,** Report and recommendations of the CAAT/ERGATT workshop on the validation of toxicity test procedures, *ATLA,* 18, 313, 1990.
12. **Balls, M.,** The validation and acceptance of *in vitro* toxicity tests, in *In Vitro Methods in Toxicology,* Jolles, G., and Cordier, A., Eds., Academic Press, London, 1992, chap. 25.
13. **Balls, M., Botham, P., Cordier, A., Fumero, S., Kayser, D., Koëter, H., Koundakjian, P., Gunnar Lindquist, N., Meyer, O., Pioda, L., Reinhardt, C., Rozemond, H., Smyrniotis, T., Spielmann, H., Van Looy, H., van der Venne, M.-T., and Walum, E.,** Report and recommendations of an international workshop on promotion of the regulatory acceptance of validated non-animal toxicity test procedures, *ATLA,* 18, 339, 1990.
14. **Balls, M.,** Replacing experiments on laboratory animals, *Trends Biochem. Sci.,* 11, 236, 1986.
15. **Balls, M., Atkinson, K. A., and Gordon, V. C.,** Complementation in the development, validation and use of non-animal test batteries, with particular reference to their ocular irritancy, *ATLA,* 19, 429, 1991.
16. **Balls, M., and Fentem, J. H.,** The use of basal cytotoxicity and target organ toxicity tests in hazard identification and risk assessment, *ATLA,* 20, 368, 1992.
17. **Depledge, M.,** Series foreward, in *Animal Biomarkers as Pollution Indicators,* Peakall, D., Chapman & Hall, London, 1992, xi.
18. **Moriarty, F.,** *Ecotoxicology. The Study of Pollutants in Ecosystems,* 2nd ed., Academic Press, London, 1988.
19. **Walker, C. H., Greig-Smith, P. W., Crossland, N. O., and Brown, R.,** Ecotoxicology, in *Animals and Alternatives in Toxicology. Present Status and Future Prospects,* Balls, M., Bridges, J., and Southee, J., Eds., Macmillan Press, Basingstoke, England, 1991, chap. 9.
20. **Calamari, D.,** The role of ecotoxicology in the assessment of human exposure to chemical substances, *Human Exp. Toxicol.,* 11, 307, 1992.
21. **Peterson, P. J., Batt, S., and Burton, M. A. S.,** The significance of ecotoxicology, in *Proceedings of the Fifth International Congress of Toxicology. Basic Science in Toxicology,* Volans, G. N., Sims, J., Sullivan, F. M., and Turner, P., Eds., Taylor & Francis, London, 1990, 155.
22. **Anon.,** Animals and alternatives in toxicology: Present status and future prospects. The Second Report of the FRAME Toxicity Committee, *ATLA,* 19, 116, 1991.
23. **Peakall, D.,** *Animal Biomarkers as Pollution Indicators,* Chapman & Hall, London, 1992.
24. **Basketter, D. A.,** Quantitative structure-activity relationships, *Toxic. In Vitro,* 3, 351, 1989.
25. **Bawden, D., Tute, M. S. and Dearden, J. C.,** Computer modelling and information technology, in *Animals and Alternatives in Toxicology. Present Status and Future Prospects,* Balls, M., Bridges, J., and Southee, J., Eds., Macmillan Press, Basingstoke, England, 1991, chap. 10.
26. **Blum, D. J. W., and Speece, R. E.,** Determining chemical toxicity to aquatic species. The use of QSARs and surrogate organisms, *Environ. Sci. Technol.,* 24, 284, 1990.
27. **Calamari, D., and Vighi, M.,** The role of ecotoxicology in environmental protection, in *Proceedings of the Fifth International Congress of Toxicology. Basic Science in Toxicology,* Volans, G. N., Sims, J., Sullivan, F. M., and Turner, P., Eds., Taylor & Francis, London, 1990, 193.
28. **Esser, H. O.,** A review of the correlation between physicochemical properties and bioaccumulation, *Pesticide Sci.,* 17, 265, 1986.
29. **Dearden, J. C., and Nicholson, R. M.,** The prediction of biodegradability by the use of quantitative structure-activity relationships: Correlation of biological oxygen demand with atomic charge difference, *Pesticide Sci.,* 17, 305, 1986.
30. **Turner, L., Choplin, F., Dugard, P., Hermens, J., Jaeckh, R., Marsmann, M., and Roberts, D.,** Structure-activity relationships in toxicology: An assessment, *Toxic. In Vitro,* 1, 143, 1987.

31. **Ribo, J. M., and Kaiser, K. L. E.,** Effect of chemicals on photoluminescent bacteria and their correlations with acute and sublethal effects on other organisms, *Chemosphere*, 12, 1421, 1983.
32. **Bitton, G., and Dutka, B. J., Eds.,** *Toxicity Testing Using Microorganisms*, Vol. 1, CRC Press, Boca Raton, FL, 1986.
33. **Diamond, J. M., Winchester, E. L., Mackler, D. G., and Gruber, D.,** Use of the mayfly *Stenonema modestum* (Heptageniidae) in subacute toxicity assessments, *Environ. Toxicol. Chem.*, 11, 415, 1992.
34. **Hickey, C. W., and Vickers, M. L.,** Comparison of the sensitivity to heavy metals and pentachlorophenol of the mayflies *Deleatidium* spp. and the cladoceran *Daphnia magna*, *Freshwater Res.*, 26, 87, 1992.
35. **Wang, W.,** Literature review on higher plants for toxicity testing, *Water, Air Soil Pollut.*, 59, 381, 1991.
36. **Blaise, C.,** Microbiotests in aquatic ecotoxicology: Characteristics, utility, and prospects, *Environ. Toxicol. Water Qual.*, 6, 145, 1991.
37. **Xu, H., and Dutka, B. J.,** ATP-TOX system: A new rapid sensitive bacterial toxicity screening system based on the determination of ATP, *Tox. Assess.*, 2, 149, 1987.
38. **Reinhartz, A., Lampert, I., Herzberg, M., and Fish, F.,** A new short-term, sensitive bacterial assay kit for the detection of toxicants, *Tox. Assess.*, 2, 193, 1987.
39. **Bulich, A. A.,** Use of luminescent bacteria for determining toxicity in aquatic environments, in *Aquatic Toxicology*, Marking, L. L., and Kimerle, R. A., Eds., American Society for Testing Materials, Philadelphia, 1979, 98.
40. **Bitton, G., and Koopman, B.,** Bacterial and enzymatic bioassays for toxicity testing in the environment, *Rev. Environ. Contam. Toxicol.*, 125, 1, 1992.
41. **Kohler, H. R., Triebskorn, R., Stocker, W., Kloetzel, P. M. and Alberti, G.,** The 70kD heat shock protein (hsp 70) in soil invertebrates: A possible tool for monitoring environmental toxicants, *Arch. Environ. Contam. Toxicol.*, 22, 334, 1992.
42. **Mothes-Wagner, U., Reitze, H. K., and Seitz, K. A.,** Terrestrial multispecies toxicity testing. I. Description of the multispecies assemblage, *Chemosphere*, 24, 1653, 1992.
43. **Cooke, A. S.,** Tadpoles as indicators of harmful levels of pollution in the field, *Environ. Pollut.*, 25A, 123, 1981.
44. **Hoffman, D. J., and Albers, P. H.,** Evaluation of potential embryotoxicity and teratogenicity of 42 herbicides, insecticides, and petroleum contaminants to Mallard eggs, *Arch. Environ. Contam. Toxicol.*, 13, 15, 1984.
45. **Weis, J. S., and Weis, P.,** Pollutants as development toxicants in aquatic organisms, *Environ. Health Perspect.*, 71, 77, 1987.
46. **Walker, C. H.,** Kinetic models to predict bioaccumulation of pollutants, *Funct. Ecol.*, 4, 295, 1990.
47. **Walker, C. H., Chipman, J. K., and Kurukgy, M.,** Microsomal systems as models for *in vivo* metabolism, *Ecotoxicol. Environ. Safety*, 3, 39, 1979.
48. **Walker, C. H.,** The correlation between *in vivo* and *in vitro* metabolism of pesticides in vertebrates, in *Progress in Pesticide Biochemistry*, Vol. 1, Hutson, D. H., and Roberts, T. R., Eds., John Wiley & Sons, Chichester, England, 1981, 247.
49. **Walker, C.,** Foreword, in *Animal Biomarkers as Pollution Indicators*, Peakall, D., Chapman & Hall, London, 1992, xv.
50. **Douglas, M. T., Chanter, D. O., Pell, J. B., and Burney, G. M.,** A proposal for the reduction of animal numbers required for the acute toxicity to fish test (LC_{50} determination), *Aquat. Toxicol.*, 8, 243, 1986.
51. **Fentem, J., and Balls, M.,** Replacement of fish in ecotoxicology testing: Use of bacteria, other lower organisms and fish cells *in vitro*, in *Ecotoxicology Monitoring*, Richardson, M., Ed., VCH Publishers, Weinheim, Germany, 1993, 71.
52. **Babich, H., and Borenfreund, E.,** Cytotoxicity and genotoxicity assays with cultured fish cells: A review, *Toxic. In Vitro*, 5, 91, 1991.
53. **Persoone, G.,** Cyst-based toxicity tests. I. A promising new tool for rapid and cost-effective toxicity screening of chemicals and effluents, *Zeit. angewandte Zool.*, 78, 235, 1991.
54. **Osborn, D., and French, M. C.,** Ecotoxicology, in *Animals and Alternatives in Toxicology. Present Status and Future Prospects*, Balls, M., Bridges, J., and Southee, J., Eds., Macmillan Press, Basingstoke, England, 1991, 242.

SECTION IV

Contaminants And Risks For Human Health

CHAPTER 16

Cancer Risks from Arsenic in Drinking Water

Allan H. Smith, Claudia Hopenhayn-Rich, Michael N. Bates, Helen M. Goeden, Hertz-Picciotto, Heather M. Duggan, Rose Wood, Michael J. Kosnett, and Martyn T. Smith

Ingestion of arsenic, both from water supplies and medicinal preparations, is known to cause skin cancer. The evidence assessed here indicates that arsenic can also cause liver, lung, kidney, and bladder cancer and that the population cancer risks due to arsenic in U.S. water supplies may be comparable to those from environmental tobacco smoke and radon in homes. Large population studies in an area of Taiwan with high arsenic levels in well water (170 800 μg/L) were used to establish dose response relationships between cancer risks and the concentration of inorganic arsenic naturally present in water supplies. It was estimated that at the current EPA standard of 50 μg/L, the lifetime risks of dying from cancer of the liver, lung, kidney, or bladder from drinking 1 L/day of water could be as high as 13 per 1000 persons. It has been estimated that more than 350,000 people in the United States may be supplied with water containing more than 50 μg/L. For average arsenic levels and water consumption patterns in the United States, the risks estimate was around 1/1000. Although further research is needed to validate these findings, measures to reduce arsenic levels in water supplies should be considered.

INTRODUCTION

Arsenic is a ubiquitous element present in various compounds throughout the earth's crust. It was identified in ancient times; the Greek alchemist Olympiodorus reportedly obtained metallic arsenic by roasting one of its sulfides. The use of arsenical compounds increased greatly during the 18th and 19th centuries, including use in pigments and dyes, in preservatives of animal hides, in glass manufacture, agricultural pesticides, and various pharmaceutical substances.

The first described health effect, reported by Agricola in De Re Metallica in 1556,[1] involved arsenical cobalt, which ate away the skin of the hands of workmen. In 1888 Hutchison first described carcinoma of the skin in patients treated with arsenical mixtures for psoriasis and other skin conditions.[2] Subsequent investigations have confirmed that ingestion of inorganic arsenic can cause skin cancer and that inhalation of inorganic arsenic can cause skin cancer and that inhalation of inorganic arsenic can cause lung cancer.[3]

Both organic and inorganic arsenic are present in varying amounts in food. Fish, for example, contain relatively high concentrations of organic arsenic. However, inorganic forms of arsenic can be present as either arsenate [As(V)] or arsenite [As(III)]. Although As(III) is more toxic, human metabolism of As(V) involves reduction to As(III) before undergoing detoxification by methylation.[4]

Arsenic is present in soil at levels ranging from 0.2 to 40 µg/g (rarely more than 10 µg/g) and in urban air at levels around 0.02 µg/m^3, but for the general population the main exposure to inorganic arsenic is through ingestion. Although most major U.S. drinking water supplies contain levels lower than 5 µg/L, it has been estimated that about 350,000 people might drink water containing more than 50 µg/L,[5] the standard for arsenic set by the U.S. Environmental Protection Agency (EPA).

Cancer risk estimates attributed to ingested arsenic have been based on skin cancer risks alone. There is now sufficient evidence to consider other internal and more fatal cancers caused by ingested arsenic as well. The purpose of this chapter is to present the findings of a cancer risk assessment of ingestion of inorganic arsenic in drinking water based on mortality from internal cancers.

Cancer risk assessments such as this one can be divided into four steps: hazard identification, dose–response analysis, exposure assessment, and risk characterization.[6] The first section of this chapter deals with hazard identification and presents evidence indicating that, in addition to the well-known association with skin cancer, ingestion of arsenic may also cause liver, lung, bladder, and kidney cancer. A dose–response analysis and risk extrapolation based on epidemiological studies of populations exposed to elevated levels of inorganic arsenic in their drinking water follow, along with an assessment of the possibility of a threshold (i.e., a level of exposure below which there would be no increase in population cancer risks). Exposure data are then described concerning the levels of arsenic in U.S. water supplies.

Finally, results of risk characterization are presented, and the estimates of cancer risks from ingestion of arsenic in U.S. drinking water are compared with those from two other high-risk environmental exposures: environmental tobacco smoke and radon.

HAZARD IDENTIFICATION

The main sources of evidence for the carcinogenicity of ingested inorganic arsenic come from human studies, with some limited evidence from animal studies. Results of human investigations are described first, followed by those of animal experiments.

Human Studies

Inhalation of inorganic arsenic has been shown to cause lung cancer in studies of smelter workers.[3] However, evidence considered below suggests that systemic absorption by these workers was insufficient to identify significantly increased risks of cancers at other sites. Much higher systemic exposures have occurred among populations ingesting water with high arsenic concentrations, and studies of such populations have shown increased risks of skin, liver, lung, bladder, and kidney cancers.

The plausibility of a casual association is supported by exposure studies that found arsenic concentrations in skin, liver, lung, and kidney tissues.[7] Animal studies generally show arsenic to accumulate at these site.[8–12] Studies of arsenic in human exposed to background levels have found varying concentrations in different organs, including skin, lung, liver, and kidney.[13,14] This section presents a brief overview of the skin cancer evidence, and a more detailed analysis of the less publicized evidence linking ingestion of inorganic arsenic with cancer at the other more fatal sites.

Skin Cancer. Substantial evidence led the International Agency for Research on Cancer (IARC) to conclude that ingestion of inorganic arsenic can cause skin cancer.[3] Populations in countries

Table 1. Estimated Mortality Risk Ratios for Liver, Lung, Bladder, and Kidney Cancer by Arsenic Levels in Drinking Water in Southwestern Taiwan, Using Cancer Mortality Rates of the General Taiwanese Population as Reference[30]

Cancer site	Sex	Water levels, µg/L				p-Value for linear trend
		Background	170	470	800	
Liver	M	1.0	1.2	1.5	2.5	<0.001
	F	1.0	1.6	2.1	3.6	<0.001
Lung	M	1.0	1.8	3.3	4.5	<0.001
	F	1.0	2.8	4.3	8.8	<0.001
Bladder	M	1.0	5.1	12.1	28.7	<0.001
	F	1.0	11.9	25.1	65.4	<0.001
Kidney	M	1.0	4.9	11.9	19.6	<0.001
	F	1.0	4.0	13.9	37.0	<0.001

such as Taiwan, Mexico, India, and Chile who consumed drinking water with high levels of arsenic had high rates of skin cancer.[15–18] In Taiwan, the prevalence of skin cancer among highly exposed males aged 60 years and older reached 25%.[15] There are also many corroborating reports of skin cancer cases resulting from the use of orally administered arsenical medications, particularly Fowler's solution, which was widely used for the treatment of a variety of conditions such as asthma and psoriasis.[19–22]

Liver Cancer. Angiosarcoma of the liver is a very rare tumor, often associated with exposure to vinyl chloride or thorotrast. It is estimated that only about 25 cases occur in the United States each year.[23] In light of its rarity, even a small number of angiosarcoma cases associated with exposure to arsenic must be considered meaningful.

In 1957, Roth reported three cases of liver angiosarcoma in a series of 27 autopsies performed between 1950 and 1956 among arsenic-poisoned German vintners.[24] A study in Chile found that among a group of 16 male cancer patients exposed to high arsenic levels through the water supply (200–2000 µg/L), 15 had skin carcinomas and one had a liver angiosarcoma in addition to chronic arsenical dermatosis.[25]

Falk et al. identified 168 cases of liver angiosarcoma in the United States between 1964 and 1974,[26] of whom 7 had used Fowler's solution for 6–17 years. Other individual case reports of liver angiosarcoma associated with medicinal ingestion of arsenic have also been published.[23,27,28]

Increased mortality from primary liver cancer has also been associated with arsenic ingestion in several studies. Luchtrath compared the autopsy findings of 163 German winegrowers diagnosed as having chronic arsenic poisoning with those of a control group of 163 men of similar age.[29] The winegrowers had been heavily exposed to arsenic through drinking Haustrunk, a wine substitute made from an aqueous infusion of grapes that had a high arsenic content. Liver cancers were found in five of the winegrowers but in none of the control group. These findings are consistent with an arsenic effect but are difficult to interpret due to possible confounding with alcohol intake.

Several epidemiological studies based on data from an area of southwestern Taiwan known to have high levels of inorganic arsenic in the artesian well water supply have found elevated rates of liver cancer deaths. The most striking findings come from a study in which the population was classified into three groups according to the arsenic level in their drinking water (300, 300–600, and >600 µg/L).[30,31] Using the number of wells in each category and their arsenic concentrations, the EPA calculated the weighted averages for each of the three groups to be 170, 470, and 800 µg/L.[32] From the data given by Chen et al.,[30] an increasing mortality rate ratio for liver cancer can be calculated with increasing arsenic concentration: 1.2, 1.5, and 2.5 for males ($p < 0.001$) and 1.6, 2.1, and 3.6 for females ($p < 0.001$) (Table 1).

The similarities between residents of neighboring villages with respect to diets, sociodemographic characteristics, and lifestyle make it unlikely that confounding could explain the association between water arsenic concentrations and cancer rates.

These results are supported by (1) case–control study in the same area that reported a strong relationship between years of consumption of well water and liver cancer,[33] (2) a comparison of site-specific cancer mortality rates in the population of the high arsenic area with those of the general Taiwanese population,[34] (3) an investigation of cancer mortality of patients who suffered from blackfoot disease (BFD), a vascular disorder endemic to southwestern Taiwan and associated with the use of arsenic-rich artesian well water,[35] and (4) an ecological study of cancer mortality rates and arsenic levels in the drinking water of 314 townships in Taiwan.[36]

In summary, a causal association between ingested arsenic and liver cancer is supported by a series of case reports concerning angiosarcoma of the liver, and investigation of arsenic-poisoned winegrowers in Germany, and several studies in southwestern Taiwan where drinking water has a naturally high arsenic content.

Lung Cancer. Although inhaled arsenic is a well-known lung carcinogen,[3] little attention has been given to the evidence relating lung cancer to arsenic ingestion. A number of case reports link lung cancer to cutaneous signs of arsenicism resulting from ingestion.[28, 37–39] For example, Robson and Jelliffe described six patients with both lung tumors and skin disease characteristic of arsenic exposure, all of whom had used medications containing arsenic.[39] Case reports of a common cancer such as lung cancer are not convincing by themselves, but the following epidemiological studies provide strong additional evidence.

Lung cancer was found in 108 of the 163 arsenic-poisoned winegrowers in Germany compared to only 14 among the controls.[29] Mortality data from the trade association listed 417 deaths of winegrowers, 242 of whom had lung carcinomas. Although the winegrowers had heavy exposure to arsenic by ingestion, the possibility that the exposures responsible for the lung cancers were at least partly from inhalation cannot be ruled out. However, it should be noted that 30 of the winegrowers in the case series had skin cancer, which has been clearly linked to high levels of ingested arsenic, but has not been found in studies of smelter workers who have high lung cancer risks from arsenic inhalation.

Because data on smoking were not available, one could argue that confounding by smoking might explain the results. However, a high mortality odds ratio[40] of 14.7 can be calculated from the study results, making it unlikely that smoking alone could explain the findings. The main weakness of Luchtrach's study[29] relates to the lack of information on the choice of study subjects so that selection bias cannot be ruled out.

A study in the provincie of Corboda, Argentina, examined mortality records for all deaths occurring between 1949 and 1959 in areas with high arsenic in drinking water (average 600 µg/L) and compared the cause-specific mortality rates to those of the entire provincie.[41] Mortality from all cancers combined was found to be considerably higher than in the provincie as a whole (24% of all deaths compared to 15%). Of the 556 deaths attributed to cancer, 35% were found to be of the respiratory organs. In addition, several other published reports have mentioned elevated rates of lung cancer among patients with arsenic-related skin disorders in several areas of Argentina where levels of arsenic in drinking water are known to be high.[42,43] However, these findings can only be considered suggestive evidence because they are based on observations from dermatological practices, with little or no information on background rates, case selection, length and completeness of follow-up, or smoking habits.

The studies in southwestern Taiwan, described above in the section on liver cancer, also show evidence of elevated lung cancer mortality rates. When the population in the arsenic-rich area was divided in three groups according to the arsenic levels in drinking water, a clear dose–response relationship was observed.[30] As shown in Table 1, increasing water arsenic concentrations (170, 470, and 800 µg/L) resulted in mortality rate ratios for lung cancer of 1.8, 3.3, and 4.5 for males, and 2.8, 4.3, and 8.8 for females, respectively, using lung cancer mortality in the general Taiwanese population for comparison.

The results of the case-control study carried out in the same area showed a linear trend between lung cancer rates and years of exposure to well water ($p < 0.01$), which persisted in a multiple regression analysis that controlled for the effects of smoking.[33] Similar findings were reported in other mortality studies in the same area[34,35] and in the recent ecological study carried out in all of Taiwan.[36] In summary, the results of epidemiological studies provide evidence that ingested inorganic arsenic increases the risk of lung cancer.

Kidney and Bladder Cancer. The Taiwanese investigation of cancer mortality described above also found a clear dose–response relationship between arsenic water levels and bladder and kidney cancer.[30] In order of increasing water arsenic concentrations (170, 470, and 800 µg/L), the corresponding mortality rate ratios for bladder cancer were 5.1, 12.1, and 28.7 for males, and 11.9, 25.1, and 65.4 for females, and for kidney cancer, 4.9, 11.9, and 19.6 for men and 4.0, 13.9, and 37.0 for women (see Table 1). As with liver and lung cancer, the findings were supported by a case-control study showing a dose–response relationship between bladder cancer and years of artesian well water consumption ($p < 0.01$) after controlling for the effects of smoking, tea consumption, and other dietary factors.[33] Results of other studies in the same area of Taiwan provide additional support for the association.[34–36,44]

The magnitude of the above mortality rate ratios for bladder and kidney cancers is such that confounding by some other risk factor is most unlikely to be the explanation. The evidence suggests a causal relationship. Similar associations have so far not been reported elsewhere in relation with arsenic ingestion, but no other populations of comparable size and exposure have been studied. The mortality rate ratios for these two cancers are sufficiently high to ask why increased risks have not been consistently detected in studies of smelter workers. Based on urinary arsenic levels of workers in a large smelter study that reported marked increases in lung cancer due to arsenic inhalation,[45] one can estimate an average cumulative absorbed dose of arsenic of approximately 3000 mg. This is less than half the cumulative dose achieved by a 60-year-old worker drinking 2 L of water/day at the lowest exposure level (170 µg/L) in the studies in Taiwan. If, on the other hand, one considers the group of workers with highest cumulative exposure, the estimate of systemic absorption becomes comparable to that of Taiwan. However, rates of these cancers are not given. Thus the failure to detect significant increases in bladder and kidney cancer among smelter workers may be due to a combination of their lower systemic exposure to arsenic and lack of examination of the highest exposed groups.

Nevertheless, the findings concerning arsenic smelters are not completely negative. A recent study of arsenic-poisoned workers and residents of a mining town in Japan found them to have significantly elevated rates of cancers of the bladder, kidney, and other urinary organs [standardized mortality ratio (SMR)= 766, 95% confidence interval (CI) 136–2795] in addition to lung cancer (SMR= 566, 95% CI 266–119).[46] Although the urinary tract cancer SMR of 766 was based on only two cases, it is significant that both were accompanied by Bowen's disease (characterized by skin lesions and associated with arsenic ingestion). Further study of kidney and bladder cancer mortality among smelter workers by duration and intensity of exposure is warranted.

Animal Studies

Arsenic is unique in being the only established human carcinogen that has not been established as a carcinogen in rodents. The IARC concluded that the results of animal studies supply only limited evidence of carcinogenicity.[47] Although most arsenic inhalation bioassays have produced negative results, two groups of investigators reported positive findings in experiments involving intratracheal administration of arsenic.[48–50] Bioassays involving oral exposure to arsenic have produced inconclusive results. Studies in mice given drinking water with arsenic levels ranging from 4.0 to 100 mg/L did not show increased cancer rates.[51–53]

The effects of arsenic have also been examined in mouse strains that have a high background incidence rate of spontaneous tumors. The results from these investigations have been inconsistent. In one study, sodium arsenite appeared to inhibit the development and growth of precancerous

Table 2. Carcinogenic Effects of Arsenic on Intact Rats[60]

Treatment group	Sacrifice time points, months			p
	10	15	24	
Liver neoplastic nodules				
Saline control	0/14[a]	0/11	0/16	
Saline + As(III)	0/22	0/22	2/7	0.08[b]
Saline + As(V)	0/14	0/16	1/5	NA
Saline + DMA	0/10	0/21	2/6	0.06[b]
DEN + control	0/14	1/8	3/6	<0.01[c]
DEN + As(III)	0/13	3/14	5/6	0.22[c]
DEN + As(V)	0/9	4/9	4/7	0.21[c]
DEN + DMA	1/14	2/11	7/8	0.11[c]
Kidney tumors				
Saline control	1/14	0/11	1/16	
Saline + As(III)	0/22	2/22	2/7	0.28[b]
Saline + As(V)	2/14	0/16	2/5	0.12[b]
Saline + DMA	0/10	0/221	1/6	0.28[b]
DEN control	0/14	0/8	2/6	0.42[b]
DEN + As(III)	3/13	4/14	2/6	0.05[c]
DEN + As(V)	3/10	4/9	4/7	<0.01[c]
DEN + DMA	1/14	0/11	3/8	0.45[c]

Abbreviations: DMA, dimethylarsonic acid; DEN, diethylnitrosamine; NA, not available (insufficient cell size to perform test).
[a]Number of animals affected/number of animals in treatment group.
[b]Pooled across sacrifice time points and compared to saline controls using Mantel–Haenszel chi-square test (p-values are for one-tailed testing).
[c]Pooled across sacrifice time points and compared to DEN controls using Mantel–Haenszel chi-square test (p-values are for one-tailed testing).

cell populations, but once tumors developed, the growth rate was faster and the incidence of multiple tumors and metastases was higher in the arsenic-treated animals.[54,55] In another study, trivalent or pentavalent arsenic given in drinking water in conjunction with urethane decreased the number and size of tumors.[56,57]

Another group of investigators reported increased kidney and liver tumors in rats treated with either trivalent or pentavalent arsenic.[58–60] In their latest study, intact male Wistar rats were injected with either saline or 30 mg/kg of diethylnitrosamine (DEN), a known carcinogen. After 1 week each group was subdivided into four treatment groups receiving 0 and 160 mg/L As(III), 160 mg/L As(V), or 80 mg/L dimethylarsonic acid (DMA) in drinking water. The animals were sacrificed after 10, 15, or 24 months of exposure.

Based on limited statistical analysis, the authors concluded that As(III) and DMA were promoters for DEN-initiated hapatocellular carcinomas and that As(III) and As(V) were promoters for DEN-initiated renal tumors. Using data provided in the report, we have conducted additional analyses using the Mantel–Haenszel chi-square test.[61] In view of the relatively small number of animals per group (Table 2), the treatment groups were pooled across sacrifice time points first and then across treatments. The results of this analysis show an increase in tumor incidence, both with and without DEN, for each form of arsenic. The overall effect of arsenic on liver neoplastic nodules and kidney tumors was not likely to be attributable to chance (p = 0.012 and 0.013, respectively). It is noteworthy that the increased tumor incidence observed in rats occurred in liver and kidney, two of the target organs observed in humans. However, the fact that treated rats had significantly lower weights than control rats weakens the evidence from this study because it raises the possibility that increased cancer rates were an indirect effect of exposure affecting nutritional status.

Table 3. Regression Analysis of Arsenic in Drinking Water and Cancer Mortality Rates in Taiwan

Cancer site	Sex	Background mortality[a]	Slope estimate[b]	Standard error of the slope	p-Value for linear trend
Liver	M	28.0	0.041	0.008	<0.001
	F	8.9	0.026	0.005	<0.001
Lung	M	19.4	0.091	0.007	<0.001
	F	9.5	0.083	0.004	<0.001
Bladder	M	3.1	0.083	0.003	<0.001
	F	1.4	0.091	0.002	<0.001
Kidney	M	1.1	0.026	0.002	<0.001
	F	0.9	0.033	0.002	<0.001

[a]Age-adjusted mortality per 100,000.
[b]The slope represents the increase in cancer mortality rate (per 100,000) per microgram increase of arsenic in drinking water, based on data from Chen et al.[30] and Wu et al.[31]

The above data analysis provides some evidence that high doses of ingested arsenic may result in carcinogenic activity in rats. Although this effect was not seen in mice, the levels of arsenic in their drinking water were lower, and mice may be protected by methylating arsenic more rapidly than rats.[62]

DOSE–RESPONSE ANALYSIS

Skin Cancer

Skin cancer risk estimates have been derived based on the study Tseng et al., which involves a large population in the endemic area Taiwan.[15] Using skin cancer prevalence rates from populations having different arsenic levels in their drinking water, a clear dose–response relationship was observed. Brown et al.[63] calculated the lifetime risks of skin cancer to be 1.3/1000 for males and 0.6/1000 for females per microgram of arsenic per day. In the section that follows, we examine the dose–response relationships between arsenic levels in water and cancers of the lung, liver, bladder, and kidney and extrapolate the risk to the current U.S. drinking water standard of 50 g/L. We focus on these sites because mortality from these cancers is much higher than from skin cancer.

Other Cancers

Estimates of the risks of bladder, kidney, lung, and liver cancer were based on data from southwestern Taiwan, where a dose–response gradient between arsenic water levels and cancer mortality rates for these sites was observed.[30] The statistical testing consisted of a trend analysis of proportions using linear regression.[64] The well water concentrations were divided into three categories, with weighted average concentrations estimated to be 170, 470, and 800 g/L.[32] The mortality cancer rates reported by Chen et al. for the three arsenic levels were used,[30] weighted by the person-years of exposure at each dose group, as given by Wu et al.,[31] assuming that all cancer deaths for the four sites occurred in persons more than 20 years of age. The background population mortality rates for Taiwan were used as intercepts. The results of the regression analysis are shown in Table 3 and plotted in Figure 1, where it can be seen that the findings are reasonably consistent with linear dose–response relationships.

IS THERE A THRESHOLD?

Inorganic arsenic is methylated into less toxic, organic forms at various sites in the body, in particular the liver and kidney.

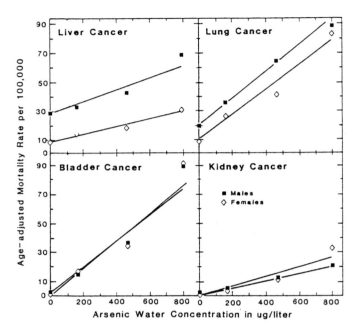

Figure 1. Age-adjusted mortality rates for liver, lung, bladder, and kidney cancer by the arsenic concentrations in drinking water and plots of linear regressions for data from studies in Taiwan.[30,31]

Urinary excretion of inorganic arsenic is thus a combination of unchanged inorganic arsenic and its methylated forms, DMA and MMA (dimethylarsonic and manomethylarsinic acids). It has been suggested that one might expect a threshold for the carcinogenic effect of arsenic ingestion if methylation activity had to be saturated before cancer risks increased.[32,65] If the carcinogenic effects of arsenic are caused only by the inorganic forms, and if ingested inorganic arsenic reaches target organs only when methylation is saturated, then one might expect a threshold for arsenic ingestion below which there would be no risk of it causing cancer. If, on the other hand, some inorganic arsenic always passes through the liver unchanged and reaches target organs, but the proportion remaining in inorganic form increased with increasing ingestion of arsenic, then one would expect a sublinear dose–response relationship between ingested arsenic and cancer risk.

A study in mice showed that increasing the dose of either As(III) or As(V) two orders of magnitude resulted in a 2- to 2.5-fold increase in the proportion excreted in the organic form.[62]

The data on methylation in humans, however, does not show evidence for such a relationship. Table 4 illustrates the distribution of urinary metabolites of inorganic arsenic excreted after different levels of exposure. Studies of individuals with background exposure found that unmethylated inorganic arsenic constituted between 15 and 32% of the urinary excretion of the metabolites of inorganic arsenic.[66–70] Occupationally exposed populations had values within the same range,[67,69,70] as well as human volunteers who ingested measured doses of inorganic arsenic.[66,68,71] It can be clearly seen in Table 4 that even at low background levels of arsenic exposure methylation is far from complete.

A recent paper[72] seems to contradict this evidence by showing that at low exposures methylation is near complete. However, it should be noted that this inference cannot be made based on the way the data were analyzed: nondetectable values were averaged in as zero. If, in fact, each inorganic arsenic species was just at the detection level of 0.5 µg/L, then the total inorganic

Table 4. Relative Distribution of Urinary Metabolites of Inorganic Arsenic (Unchanged In-As, MMA, DMA) After Exposure to Different Levels of Inorganic Arsenic

Source of exposure	Number exposed	% Excreted in urine[a]			Reference
		In-As	MMA	DMA	
Background	148	32	32	36	67
	4	18	4	78	66
	16	18	4	78	68
	6	18	16	65	69
	41	15	20	66	70
	102	23	7	70	80
Occupational					
Smelter workers	9	19	20	61	69
Smelter workers	30	20	22	66	70
Glass workers	38	23	15	62	67
Gallium arsenide workers	27	32	9	61	80
Experimental, µg/kg/L[b]					
0.000143	6	27	21	51	71
7.1	3	25	21	54	68
9[c]	1	16	34	50	66
18[c]	1	7	20	73	66
35.5[c]	1	19	21	60	66
71.5[c]	1	26	32	42	66

Abbreviations: In-As, inorganic arsenic; MMA, monomethylarsinic acid; DMA, dimethylarsonic acid.
[a]When a number exposed is >1, numbers represent group mean values.
[b]Experimental doses are via ingestion of arsenic in water, assuming a body weight of 70 kg for human studies.
[c]Dose represents cumulative dose from five consecutive days.

arsenic would be roughly around 20%. Therefore, the evidence described above does not support a threshold for inorganic arsenic carcinogenicity.

EXPOSURE ASSESSMENT

Current dietary intake of total arsenic in U.S. adults, excluding tap water, has been estimated to be around 45–50 µg/day.[32,73] Most of the arsenic derives from seafood, meat, and poultry (80%) and grains and cereals (17%). Although seafood has a high arsenic content, only about 5–10% is in the inorganic form,[74] and the organic forms (mainly arsenobetaine) are excreted unchanged.[69] The EPA estimated the average total intake of inorganic arsenic from food, water, and other beverages to be about 17 µg/day, of which 5 µg/day come from drinking water.[32] Thus, for the majority of the U.S. population, about 30% of ingested inorganic arsenic comes from drinking water. However, if inorganic arsenic is present at the current water standard (50 µg/L), drinking water could contribute almost 100 µg more to the daily intake of inorganic arsenic. This would represent about 90% of the daily intake. Even at half that concentration, water would still be by far the main source of inorganic arsenic (more than 80%).

Based on several national water surveys, the EPA estimated the national occurrence of inorganic arsenic in drinking water.[5] Although most public water supplies contain levels below 5 µg/L, it is estimated that about 350,000 people might drink water containing more than 50 µg/L, and about 2,500,000 people drink water containing more than 25 µg/L.

RISK CHARACTERIZATION

Risk Estimation for the U.S. Population

The estimated mortality rate rations for males and females are presented in Table 5. Risks for each cancer were extrapolated linearly to an arsenic concentration of 50 µg/L from the regression

Table 5. Estimation of Lifetime Risks Per 1000 of Dying from Cancer of the Kidney, Bladder, Lung, and Liver from Consumption of 1 L/day of Water with an Arsenic Concentration of 50 µg/L

Cancer site/sex	Estimated mortality rate ratio[a]	Rate ratio (RR)[b]	B[c]	$B(RR-1)$[d]
Liver				
Male	1.07	1.02	3.67	0.1
Female	1.14	1.07	2.82	0.2
Lung				
Male	1.24	1.07	76.4	5.3
Female	4.25	2.63	39.2	8.6
Bladder				
Male	2.34	1.38	6.08	2.3
Female	4.25	2.63	3.14	5.1
Kidney				
Male	2.18	1.34	5.02	1.7
Female	2.83	1.92	3.64	3.4
Total risk per daily consumption of 1 L of water				
Male				9.4
Female				17.3
Average of male and female risks				13.4

[a]Based on data from southwestern Taiwan.[30,31]
[b]RR = estimated mortality rate ratio calculated for water consumption of 1 L/day.
[c]B = U.S. background risk per 1000.
[d]$B(RR-1)$ = lifetime risk per 1000 from water consumption of 1 L/day.

lines shown in Figure 1. Based on the EPA's assumption of daily water consumption rates of 3.5 L for men and 2.0 L for women in the hot climate of Taiwan,[32] mortality rate ratios (RR) were calculated for a daily water intake of 1 L/day. The excess rate ratios, RR-1, were multiplied by the U.S. background rate (B), which was estimated by dividing the number of deaths from each cancer by the total number of deaths, using figures from U.S. Vital Statistics for the year 1985.[75] Lifetime risks of death from cancers of the liver, lung, bladder, and kidney were thus derived for consumption of 1 L/day in the United States (Table 5). The estimates of dying from one of these cancers due to a lifetime consumption of water containing 50 µg of arsenic/L at a rate of 1 L/day were 9.4/1000 for males and 17.3/1000 for females, averaging around 13/1000. It is important to emphasize that contrary to most risks, these involve only a small extrapolation from clearly demonstrated effects at 170 µg/L and above.

The male/female difference in risk estimation is due in part to the lower background cancer rates for women in Taiwan, thus making the rate ratios higher for those exposed. The EPA assumption that men in Taiwan drink almost twice as much water as women also decreases the estimated carcinogenic potency of arsenic for men compared to women. Nevertheless, the male/female risk estimate differences are within the range of uncertainty of a risk assessment such as the one presented here, and there is no biological reason to consider females to be at higher risk. For these reasons the male and female risks were averaged for application to the general population.

Although there are no accurate data on the average arsenic levels in drinking water for the United States, estimates range from 2.0 to 2.5 µg/L.[7,32] Using the estimated average water intake in the United States of 1.6 L/day[76] and average arsenic water level of 2.5 µg/L, linear extrapolation yielded an estimated lifetime risk of dying from liver, lung, bladder, or kidney cancer due to arsenic in drinking water of 1/1000.

Some animal studies seem to indicate that a protein-deficient diet increases arsenic toxicity by decreasing methylation capacity.[8] It has been hypothesized that populations consuming inadequate diets and exposed to high arsenic levels may thus be more susceptible to arsenic toxicity.[32] If nutritionally inadequate diets among the Taiwanese populations exposed to arsenic made them more susceptible to the carcinogenic effects of arsenic, the extrapolated risk estimates for the average U.S. population would be too high. However, the validity of the Taiwanese findings

Table 6. Estimated Lifetime Risks of Dying from Cancer Due to Exposure to Different Environmental Carcinogens in the United States

Carcinogen	Risk
Environmental tobacco smoke (passive smoking)	
Low exposure (not married to a smoker)	4/1000
High exposure (married to a smoker)	10/1000
Radon in homes	
Average exposure	3/1000
High exposure (1–3% of homes)	20/1000
Arsenic in drinking water (1.6 L/day)	
2.5 µg/L (U.S. estimated average)	1/1000
50 µg/L (U.S. water standard)	21/1000

would remain unchanged, as the populations in Taiwan drinking water with different levels of arsenic were all similar with respect to lifestyle, education, and occupation.[33] Variation in the detoxification of arsenic by methylation, either due to nutritional or genetic factors, has not yet been investigated in humans and is an important area of research that needs to be addressed.

COMPARISON WITH CANCER RISKS FROM OTHER ENVIRONMENTAL EXPOSURES

The estimates of cancer risks associated with the current U.S. drinking water standard derived above place arsenic at the forefront of cancer risks associated with environmental exposures. There are only two known environmental exposures with comparable risks, namely, environmental tobacco smoke (ETS) and radon in homes. In the case of ETS, exposure occurs from several sources, mainly from living with a smoker or working with smokers. The National Research Council estimates the lifetime risk of lung cancer attributable to ETS to range from 4 to 10/1000.[77]

Considerable attention has been given in recent years to the increased risk of lung cancer caused by indoor concentrations of radon in U.S. homes, which poses a lifetime lung cancer risk of about 3/1000. Table 6 compares the cancer risks from ETS, radon, and arsenic in drinking water.

EVIDENCE THAT ARSENIC IS A BENEFICIAL MICRONUTRIENT

Policy decisions arising from potential cancer risks due to inorganic arsenic in drinking water need to consider the possibility that arsenic is a beneficial micronutrient. Signs of arsenic deprivation, including depressed growth and abnormal reproductive function, have been described for the chicken, goat, pig, and rat,[32] indicating the possibility that arsenic, at least in inorganic form, is an essential nutrient. Based on data from these experiments, various estimates of human nutritional requirements have been made, ranging from 10 to 30 µg/day.[32] However, the relevance of the experimental animal data to humans is unclear. No human arsenic deficiency syndrome has yet been reported, even though many water supplies contain less than 2.5 µg/L. In contrast to humans, rats sequester arsenic in red blood cells (II), and it is possible that the other animal species in the deprivation studies also differ from humans in arsenic storage and metabolism. Human requirements for inorganic arsenic warrant more investigation, but there is no evidence to suggest that deficiency effects would result from reducing levels of arsenic in water supplies containing above-average concentrations.

CONCLUSIONS

On the basis the overall consistency of results from epidemiological studies, there is persuasive evidence that inorganic arsenic is a cause of human cancer at several sites. A causal association between ingested arsenic and skin cancer has previously been established. The evidence presented here strongly supports a causal relationship between ingested arsenic and both liver and lung cancer. There is also evidence from Taiwan that arsenic causes human kidney and bladder cancer, although further studies are needed to confirm these findings. Additional research also needs to be conducted concerning arsenic methylation and its relationship to genetic, dietary, or other lifestyle factors, which may affect individual differences in susceptibility to the carcinogenic effects of arsenic.

Although the drinking water of the majority of the U.S. population has levels of arsenic considerably below the standard, it is estimated that about 350,000 people drink water with levels above the standard, and more than 2.5 million people are supplied with water containing more than 25 µg/L. The comparisons with ETS and radon, two recognized carcinogens of public health concern, serve to point out the serious environmental cancer risks posed by arsenic in drinking water. Although further research is needed to validate the findings of this risk assessment, measures should be taken to reduce arsenic levels in water supplies.

ACKNOWLEDGMENTS

This work was funded in part by a contract with the California Department of Health Services. Additional support was received from the Health Effects component of the University of California Toxic Substances Program, the Center for Occupational and Environmental Health, University of California, and NIH Grant P42-ES04705. We thank G. Alexeeff, B. Ames, C. Becker, L. Gold, E. Holly, M. Lipsett, and F. Spear for reviewing and commenting on the manuscript.

REFERENCES

1. **Dibner, B.,** *Agricola on Metals,* Burndy Library, Norwalk, CT, 1958.
2. **Hunter, D.,** *The Diseases of Occupations,* English Universities Press, London, 1957.
3. **IARC,** *IARC Monographs on the Evaluation of the Carcinogenic Risk of Chemical to Man: Some Metals and Metallic Compounds,* Vol. 23, International Agency for Research on Cancer, Lyon, 1980.
4. **Squibb, K. S., and Fowler, B. A.,** The toxicity of arsenic and its compounds, in *Biological and Environmental Effects of Arsenic,* Fowler, B. A., Ed., Elsevier, Amsterdam, 1983, 233–269.
5. **Science Applications International Corporation,** *Estimated National Occurrence and Exposure to Arsenic in Public Drinking Water Supplies.* Revised Draft, prepared for U.S. Environmental Protection Agency under Contract no. 68-01-766, EPA, Washington, D.C., 1987.
6. **National Research Council,** *Risk Assessment in the Federal Government: Managing the Process,* National Academy Press, Washington, D.C., 1983.
7. **Life Systems, Inc.,** *Toxicological Profile for Arsenic,* Report No. ATSDR/TP-88/02 prepared for Agency for Toxic Substances and Disease Registry, Atlanta, GA, 1989.
8. **Vahter, M., and Marafante, E.,** Effects of low dietary intake of methionine, choline or proteins on the biotransformation of arsenite in the rabbit, *Toxicol. Lett.,* 37, 41–46 1987.
9. **Vahter, M., and Norin, H.,** Metabolism of ^{74}As-labeled trivalent and pentavalent inorganic arsenic in mice and rats, *Environ. Res.,* 21, 446–457, 1980.
10. **Marafante, E., Bertolero, F., Edel, J., Pietra, R., and Sabbioni, E.,** Intracellular interaction and biotransformation of arsenite in rats and rabbits, *Sci. Total Environ.,* 24, 27–39, 1982.
11. **Vahter, M.,** Metabolism of arsenic, in *Biological and Environmental Effects of Arsenic,* Fowler, B. A., Ed., Elsevier, Amsterdam, 1983, 171–198.

12. **Yamauchi, H., and Yamamura, Y.,** Metabolism and excretion of orally administered arsenic trioxide in the hamster, *Toxicology,* 34, 113–121, 1985.
13. **Dang, H. S., Jaiswal, D. D., and Somasundaram, S.,** Distribution of arsenic in human tissues and milk, *Sci. Total Environ.,* 29, 171–175, 1983.
14. **Yamauchi, H., and Yamamura, Y.,** Concentration and chemical species of arsenic in human tissue, *Bull. Environ. Contam. Toxicol.,* 31, 267–277, 1983.
15. **Tseng, W. P., Chu, H. M., How, S. W., Fong, J. M., Lin C. S., and Yeh, S.,** Prevalence of skin cancer in an endemic area of chronic arsenicism in Taiwan, *J. Natl. Cancer Inst.,* 40, 453–463, 1968.
16. **Cebrian, M. E., Albores, A., Aquilar, M., and Blakely, E.,** Chronic arsenic poisoning in the North of Mexico, *Hum. Toxicol.,* 2, 121–133, 1983.
17. **Chakraborty, A. K., and Saha, K. C.,** Arsenical dermatosis from tubewell water in West Bengal, *Indian J. Med. Res.,* 85, 326–334, 1987.
18. **Zaldivar, R.,** Arsenic contamination of drinking water and food-stuffs causing endemic chronic poisoning, *Beitr. Pathol.,* 151, 384–400, 1974.
19. **Cuzick, J., Evans, S., Gillman, M., and Evans, D. A.,** Medicinal arsenic and internal malignancies, *Br. J. Cancer,* 45, 904–911, 1982.
20. **Sommers, S. C., and McManus, R. G.,** Multiple arsenical cancers of skin and internal organs, *Cancer,* 6, 347–359, 1953.
21. **Fierz, U.,** Katamnestische ontersuchungen ober die nebenwirkungen der therapie sit anorganisches arsen bei hautkrankheiten, *Dermatologica,* 131, 41–58, 1965.
22. **Tay, C. H.,** Cutaneous manifestations of arsenic poisoning due to a certain Chinese herbal medicine, *Aust. J. Dermatol.,* 15, 121–131, 1974.
23. **Roat, J. W., Wald, A., Mendelow, H., and Pataki, K. I.,** Hepatic angiosarcoma associated with short-term arsenic ingestion, *Am. J. Med.,* 73, 933–936, 1982.
24. **Roth, F.,** The sequelae of chronic arsenic poisoning in Moselle vintners, *J. Med. Monthly,* 2, 172–175, 1957.
25. **Zaldivar, R., Prumes, L., and Ghai, G. L.,** Arsenic dose in patients with cutaneous carcinomata and hepatic hemangioendothelioma after environmental and occupational exposure, *Arch. Toxicol.,* 47, 145–154, 1981.
26. **Falk, H., Caldwell, C. G., Ishak, K. G., Thomas, L. B., and Popper, H.,** Arsenic-related hepatic angiosarcoma, *Am. J. Ind. Med.,* 2, 43–50, 1981.
27. **Lander, J. J., Stanley, R. J., Summer, H. W., Boswell, D. C., and Aach, R. D.,** Angiosarcoma of the liver associated with Fowler's solution (potassium arsenite), *Environ. Mutagen.,* 68, 1582–1586, 1975.
28. **Kasper, M. L., Schoenfield, L., Strom, R. L., and Theologides, A.,** Hepatic angiosarcoma and bronchioloalveolar carcinoma induced by Fowler's solution, *J. Am. Med. Assoc.,* 252, 3407–3408, 1984.
29. **Lunchtrath, H.,** The consequences of chronic arsenic poisoning among Moselle wine growers, *J. Cancer Res. Clin. Oncol.,* 105, 173–182, 1983.
30. **Chen, C. J., Kuo, T. L., and Wu, M. M.,** Arsenic and cancers (letter), *Lancet,* i, 414–415, 1988.
31. **Wu, M. M., Kuo, T. L., Hwang, Y. H., and Chen, C. J.,** Dose–response relation between arsenic well water and mortality from cancer, *Am. J. Epidemiol.,* 130, 1123–1132, 1989.
32. **Risk Assessment Forum,** *Special Report on Ingested Arsenic: Skin Cancer; Nutritional Essentiality,* EPA/625/3–87/013, U.S. Environmental Protection Agency, Washington, D.C., 1988.
33. **Chen, C. J., Chuang, Y. C., You, S. L., and Lin, H. Y.,** A retrospective study on malignant neoplasms of bladder, lung, and liver in Blackfoot disease endemic area in Taiwan, *Br. J. Cancer,* 53, 399–405, 1985.
34. **Chen, C. J., Chuang, Y. C., Lin, T. M., and Wu, H. Y.,** Malignant neoplasms among residents of a Blackfoot disease-endemic area in Taiwan: High-arsenic artesian well water and cancers, *Cancer Res.,* 45, 5895–5899, 1985.
35. **Chen, C. J., Wu, M. M., Lee, S. S., Wang, J. D., Cheng, S. H., and Wu, H. Y.,** Atherogenicity and carcinogenicity of high-arsenic artesian well water; multiple risk factors and related malignant neoplasms of Blackfoot disease, *Arteriosclerosis,* 8, 452–460, 1988.
36. **Chen, C. J., and Wang, C. J.,** Ecological correlation between arsenic level in well water and age-adjusted mortality from malignant neoplasms, *Cancer Res.,* 50(17), 5470–5475, 1990.

37. **Heddle, R., and Bryant, G. D.,** Small cell lung carcinoma and Bowen's disease 40 years after arsenic ingestion, *Chest,* 84, 776–777, 1983.
38. **Goldman, A. L.,** Lung cancer in Bowen's disease, *Am. Rev. Respir. Dis.,* 108, 1205–1207, 1973.
39. **Robson, A. O., and Jelliffe, A. M.,** Medicinal arsenic poisoning and lung cancer, *Br. Med. J.,* 2, 207–209, 1963.
40. **Miettinen, O., and Wang J.-D.,** An alternative to the proportionated mortality ratio, *Am. J. Epidemiol.,* 114, 144–148, 1981.
41. **Bergoglio, R. M.,** Mortality from cancer in regions of arsenical waters of the province of Cordoba, Argentina, *Prensa Med. Argent.,* 51, 9954–1008, 1964.
42. **Tello, E. E.,** Hydro-arsenicisms: What is the Argentine chronic hydro-arsenicism (HACREA)?, *Arch. Argent. Dermatol.,* 36, 197–216, 1986.
43. **Biagini, R., Rivero, M., Salvador, M., and Cordoba, S.,** Chronic arsenicism and lung cancer, *Arch. Argent. Dermatol.,* 28, 151–158, 1978.
44. **Chiang, H., Hong, C., Guo, H., Lee, E., and Chen, T.,** Comparative study on the high prevalence of bladder cancer in the Blackfoot disease endemic area in Taiwan., *J. Formosan Med. Assoc.,* 87(11), 1074–1080, 1988.
45. **Enterline, P. E., and Marsh, G. M.,** Cancer among workers exposed to arsenic and other substances in a copper smelter, *Am. J. Epidemiol.,* 116, 895–911, 1982.
46. **Tsuda, T., Nagira, T., Yamamoto, M., and Kume, Y.,** An epidemiological study on cancer in certified arsenic poisoning patients in Toroku, *Ind. Health,* 28, 53–62, 1990.
47. **IARC,** IARC Monographs on the Evaluation of Carcinogenic Risks to Humans; Overall Evaluation of Carcinogenicity: An Updating of IARC Monographs Volumes 1 to 42, Supplement 7, International Agency for Research on Cancer, Lyon, 1987, 100–106.
48. **Pershagen, G., Nordberg, G., and Bjorklund, N. E.,** Carcinomas of the respiratory tract in hamsters given arsenic trioxide and/or benzo-a-pyrene by the pulmonary route, *Environ. Res.,* 34, 227–241, 1984.
49. **Pershagen, G., and Bjorklund, N. E.,** On the pulmonary tumorigenicity of arsenic trisulfide and calcium arsenate in hamsters, *Cancer Lett.,* 27, 99–104, 1985.
50. **Ishinishi, N., Yamamoto, A., Hisanaga, A., and Inamasu, T.,** Tumorigenicity of arsenic trioxide to the lung in Syrian golden hamsters by intermittent instillations, *Cancer Lett.,* 21, 141–147, 1983.
51. **Huepr, W. C., and Payne, W. W.,** Experimental studies in metal carcinogenesis, *Arch. Environ. Health,* 5, 445–462, 1962.
52. **Baroni, C., Van Esch, G. J., and Saffiotti, U.,** Carcinogenesis test of two inorganic arsenicals, *Arch. Environ. Health.,* 7, 668–674, 1963.
53. **Kanisawa, M., and Schroeder, H. A.,** Life term studies on the effect of trace elements on spontaneous tumors in mice and rats, *Cancer Res.,* 29, 892–895, 1969.
54. **Schrauzer, G. N., and Ishmael, D.,** Effects of selenium and of arsenic on the genesis of spontaneous mammary tumors in inbred C3H mice, *Ann. Clin. Lab. Sci.,* 4, 441–447, 1974.
55. **Schrauzer, G. N., White, D. A., McGinness, J. E., Schneider, C. J., and Bell, L. J.,** Arsenic and cancer effects of joint administration of arsenite and selenite on the genesis of mammary adenocarcinoma in inbred female C3H/St mice, *Bioinorgan. Chem.,* 9, 245–253, 1978.
56. **Blakley, B. R.,** Alterations in urethan-induced adenoma formation in mice exposed to selenium and arsenic, *Drug Nutr. Interact.,* 5, 97–102, 1987.
57. **Blakley, B. R.,** The effect of arsenic on urethan-induced adenoma formation in Swiss mice, *Can. J. Vet. Res.,* 51, 240–243, 1987.
58. **Shirachi, D. Y., Johansen, J. P., McGowan, J. P., and Tu, S. H.,** Tumorigenic effect of sodium arsenite in rat kidney, *Proc. West. Pharmacol. Soc.,* 26, 413–415, 1983.
59. **Shirachi, D. Y., Tu, S.-H., and McGowan, J. P.,** *Carcinogenic Potential of Arsenic Compounds in Drinking Water,* EPA/600/1-86/003, U.S. Environmental Protection Agency, Cincinnati, OH, 1986.
60. **Shirachi, D. Y., Tu, S.-H., and McGowan, J. P.,** *Carcinogenic Effects of Arsenic Compounds in Drinking Water,* EPA/600/1-87/007, U.S. Environmental Protection Agency, Cincinnati, OH, 1987.
61. **Mantel, N., and Haenszel, W.,** Statistical aspects of the analysis of data from retrospective studies of disease, *J. Natl. Cancer. Inst.,* 22, 719–748, 1959.
62. **Vahter M.,** Biotransformation of trivalent and pentavalent inorganic arsenic in mice and rats, *Environ. Res.* 25, 286–293, 1981.

63. **Brown, K. G., Boyle, K. E., Chen, C. W., and Gibb, H. J.,** A dose–response analysis of skin cancer from inorganic arsenic in drinking water, *Risk Anal.,* 9, 519–528, 1989.
64. **Rothman, K. J.,** *Modern Epidemiology,* Little, Brown, Boston, 1986.
65. **Marcus, W. L., and Rispin, A. S.,** Threshold carcinogenicity using arsenic as an example, in *Advances in Modern Environmental Toxicology: Risk Assessment and Risk Management of Industrial and Environmental Chemicals,* Cothern, C. R., and Mehlman, M. A., Eds., Princeton Publishing Company, Princeton, NJ, 1988, 133–158.
66. **Buchet, J. P., Lauwerys, R., and Roels, H.,** Comparison of the urinary excretion of arsenic metabolites after a single oral dose of sodium arsenite, monomethylarsonate, or dimethylarsinate in man, *Int. Arch. Occup. Environ. Health,* 48, 71–79, 1981.
67. **Foa, V., Colombi, A., Maroni, M., Buratti, M., and Calzaferri, G.,** The speciation of the chemical forms of arsenic in the biological monitoring of exposure to inorganic arsenic, *Sci. Total Environ.,* 34, 241–259, 1983.
68. **Buchet, J. P., Lauwerys, R., and Roels, H.,** Urinary excretion of inorganic arsenic and its metabolites after reported ingestion of sodium metaarsenite by volunteers, *Int. Arch. Occup. Environ. Health,* 48, 111–118, 1981.
69. **Vahter, M.,** Environmental and occupational exposure to inorganic arsenic, *Acta Pharmacol. Toxicol.,* 59, 31–34, 1986.
70. **Smith, T., Crecelius, E., and Reading, J.,** Airborne arsenic exposure and excretion of methylated arsenic compounds, *Environ. Health Perspect.,* 19, 89–93, 1977.
71. **Tam, G. K. H., Charbonneau, S. M., Bryce, F., Pomroy, C., and Sandi, E.,** Metabolism of inorganic arsenic in humans following oral ingestion, *Toxicol. Appl. Pharmacol.,* 50, 319–322, 1979.
72. **Farmer, J. G., and Johnson, L. R.,** Assessment of occupational exposure to inorganic arsenic based on urinary concentrations and speciation of arsenic, *Br. J. Ind. Med.,* 47, 342–348, 1990.
73. **Gunderson, E. L.,** FDA Total Diet Study, April 1982–April 1984: Dietary intakes of pesticides, selected elements, and other chemicals, *J. Assoc. Off. Anal. Chem.,* 71, 1200–1209, 1988.
74. **Pershagen, G.,** Sources of exposure and biological effects of arsenic, in *Environmental Carcinogens: Selected Methods of Analysis, Some Metals: As, Be, Cd, Cr, Ni, Pb, Se, Zn,* Fishbein, L., Ed., International Agency for Research on Cancer, Lyon, 1986, 45–61.
75. **National Center for Health Statistics,** *Vital Statistics of the United States,* Vol. 2, *Mortality,* Part A, Public Health Service, U.S. Department of Health and Human Services, Washington, D.C., 1985.
76. **Cotruvo, J. A.,** Drinking water standards and risk assessment, *Regul. Toxicol. Pharmacol.,* 8, 288–299, 1988.
77. **National Research Council,** *Environmental Tobacco Smoke: Measuring Exposures and Assessing Health Effects,* National Academy Press, Washington, D.C., 1986.
78. **National Research Council,** *Health Risks of Radon and Other Internally Deposited Alpha-Emitters,* BEIR IV, National Academy Press, Washington, D.C., 1988.
79. **Nero, A. V., Schwehr, M. B., Nazaroff, W. W., and Rezvan, K. L.,** Distribution of airborne radon-222 concentrations in U.S. homes, *Science,* 234, 992–997, 1986.
80. **Yamauchi, H., Takahashi, K., Masiko, M., and Yamamura, Y.,** Biological monitoring of arsenic exposure of gallium arsenide- and inorganic arsenic-exposed workers by determination of inorganic arsenic and its metabolites in urine and hair, *Am. Ind. Hyd. Assoc. J.,* 50(11), 606–612, 1989.

CHAPTER **17**

Serum 2,3,7,8-Tetrachlorodibenzo-*p*-dioxin Levels of New Zealand Pesticide Applicators and Their Implication for Cancer Hypotheses

Allan H. Smith, Donald G. Patterson, Jr., Marcella L. Warner, Ron Mackenzie, and Larry L. Needham

Background: The phenoxyherbicide 2,4,5-trichlorophenoxyacetic acid (2,4,5-T) has been widely used by professional pesticide applicators in New Zealand since before 1950. Epidemiologic studies of the risk of cancer and birth defects have been conducted in this group of workers, but little is known about the extent of their exposure to the 2,4,5-T contaminant 2,3,7,8-tetrachlorodibenzo-*p*-dioxin (TCDD), a potent carcinogen in animals.

Purpose: The objective of this study was to determine whether the blood serum levels of TCDD in a group of professional 2,4,5-T applicators in New Zealand were greater than those of a matched control group not involved in 2,4,5-T spraying.

Methods: Of 548 men employed as professional pesticide applicators in New Zealand from 1979 through 1982, nine were selected who had sprayed pesticides, although not necessarily 2,4,5-T, for at least 180 months. These applicators had sprayed 2,4,5-T for a range of 83–372 months. We measured the blood serum levels of polychlorinated dibenzo-*p*-dioxins and dibenzofurans, which were substituted with chlorine at the 2,3,7,8 position, in the nine pesticide applicators and in a matched group of nine control subjects.

Results: The average serum level of TCDD for applicators was almost 10 times that for the matched control subjects, while the average levels of all other congeners and isomers measured in the two groups did not differ substantially. TCDD levels in eight of the nine applicators were higher than those in the control subjects (mean difference, 47.7 parts per trillion). The variation in TCDD levels among the applicators was related to their duration of work exposure to 2,4,5-T.

Conclusion: On the basis of our findings in these subjects in New Zealand, we conclude that increased risks of cancer from brief exposure to phenoxyherbicides reported in other countries are probably not attributable to the TCDD that contaminants 2,4,5-T. We cannot determine from these results, however, whether TCDD exposure from prolonged use of 2,4, 5-T poses significant health risks (*J. Natl. Cancer Inst.* 84, 104–108, 1992).

The results of epidemiologic studies in Sweden have indicated that exposure to phenoxyherbicides and their contaminants may induce soft-tissue sarcoma[1-5] and malignant lymphoma,[6] although conflicting evidence from Sweden has also been presented.[7-9] The results of studies in

New Zealand have not shown an association between these cancers and exposure to phenoxyherbicides.[10–14] Epidemiologic studies in other countries have produced mixed results.[15–21] The main weakness in these studies has been exposure assessment involving subjects attempting to recall short periods of herbicide use many years previous to the time of the study.[22]

The phenoxyherbicide 2,4,5-trichloro-phenoxyacetic acid (2,4,5-T) has been widely used in New Zealand since before 1950. During its manufacture, 2,4,5-T becomes contaminated with 2,3,7,8-tetrachlorodibenzo-p-dioxin (TCDD), a potent carcinogen for some animal species. Since 1970, the levels of TCDD in 2,4,5-T manufactured in New Zealand have been steadily decreasing from approximately 0.005 ppm in 1985.[10] No measurements of the level of contamination were available for 2,4,5-T produced before 1971.

Professional pesticide applicators involved in ground-level spraying of 2,4,5-T in New Zealand are perhaps the group most heavily exposed to agricultural use of 2,4,5-T in the world. Many of the applicators spray for more than 6 months per year, and some have been spraying for more than 20 years. Various epidemiologic studies of cancer and birth defects and use of 2,4,5-T in New Zealand have been conducted,[10–14,23,24] but little is known about the actual extent of human exposure to TCDD.

The objective of this study was to compare blood serum levels of TCDD between a group of professional 2,4,5-T applicators in New Zealand and a matched control group not involved in 2,4,5-T spraying. To determine whether TCDD levels were elevated in the applicators, we measured, in both groups, the serum levels of all polychlorinated dibenzo-p-dioxins (PCDDs) and polychlorinated dibenzofurans (PCDFs), which were substituted with chlorine at the 2,3,7,8 position. Here, we present the findings as they relate to TCDD exposure levels and use this information to interpret earlier epidemiologic findings regarding 2,4,5-T use and the hypothesis that TCDD causes cancer in humans.

STUDY DESIGN AND METHODS

Study Subjects

From a file of 548 men actively employed as professional pesticide applicators in New Zealand at any time from 1979 through 1982,[23,24] we identified 11 men aged 65 years or less with the greatest number of years and months per years of pesticide application experience. The file was originally established for studies of reproductive outcomes and is composed of all chemical applicators registered with the Agricultural Chemicals Board in 1979 and who were able to be located.[23] Data in this file were based on exposure to pesticides in general and were not specific for the use of 2,4,5-T.

All 11 applicators identified had started spraying pesticides, although not necessarily 2,4,5-T, before 1960, were still spraying in 1984, and had sprayed pesticides for at least 180 months. A current address was located for nine of the 11 applicators. All nine applicators were invited in writing and by telephone to participate in the study, and all responded and signed a consent form agreeing to participate. Applicators were interviewed by telephone about their work histories, which included questions concerning lifetime use of 2,4,5-T in terms of years, months per year, and days per week. The applicators then donated one unit (500 mL) of blood at the blood bank nearest their homes. The blood was drawn by qualified New Zealand blood bank technicians.

The individually matched control group was composed of nine New Zealand men selected according to the following criteria: voluntary blood donation at the same blood bank where an applicator's blood had been drawn, age within 5 years of that applicator's age, no previous employment as a farm worker, no previous spraying of 2,4,5-T or any other agricultural chemical, and written permission for use of the blood drawn for research purposes, after a letter describing the study was read. These criteria ensured that the control group was similar in sex, age, and geographic distribution to the applicator group and that identical procedures were used to collect blood from both groups. Blood samples were processed and sent to the Centers for Disease Control in Atlanta, GA, where researchers in a blinded evaluation measured levels of PCDDs and PCDFs in serum.

Laboratory Analysis

Centers for Disease Control researches measured levels of PCDDs and PCDFs in serum, using the Centers for Disease Control serum method, which has been previously described for TCDD[25] and validated for the other PCDDs and PCDFs.[26] A mixture of carbon 13-labeled PCDDs and PCDFs was added to the serum, and the mixture was allowed to equilibrate at room temperature for 30 min. The serum was next extracted with a mixture of saturated ammonium sulfate, ethanol, and hexane, and the hexane layer was separated.

The aqueous phase was further extracted with fresh hexane, and the combined hexane, and the combined hexane layers were then treated with concentrated sulfuric acid. The hexane extract was then washed with water, dried, and applied to column 1 of a five-column cleanup procedure developed by Smith et al.[27] and modified by the Centers for Disease Control[28,29] for human samples. The total lipid content was calculated by summing the analytically determined concentrations of the individual lipids (total cholesterol, free cholesterol, phospholipids, and triglycerides).[30]

The same mixture of carbon 13-labeled standards was added to each sample, which was then analyzed with two blank samples, two unknown samples, and one quality control sample. The samples were analyzed by high-resolution gas chromatography (model 5890; Hewlett-Packard Co., Palo Alto, CA) and by high-resolution mass spectrometry (model 70S; Fisons Instruments, Manchester, England) by the same operator. Concentrations were calculated, using the standard curves developed for each congener by the isotope-dilution mass spectrometry technique.[25]

Results

In 1988, the average age of the nine applicators was 53 years (range, 45–62 years), which was similar to that of the control subjects (average age, 53 years; range, 44–64 years). Results of the analyses for PCDDs and PCDFs in serum for the applicators and the matched control subjects are summarized in Table 1. Average concentrations are expressed as parts per trillion (ppt), on a lipid-adjusted basis, and standard error of each congener and isomer measured is presented for each group. "Lipid-adjusted basis" means that the level of each congener or isomer is expressed as grams of congener or isomer per gram of total lipids in serum.

The ratio of the averages for applicators to those for matched control subjects for each congener and isomer measured is also presented in Table 1. The average level of TCDD measured among the nine applicators was greater than that measured among the matched control subjects by a factor of 9.5. The average levels of the other congeners and isomers measured among the two groups did not differ by more than a factor of 1.4.

Serum levels of TCDD for applicators and matched control subjects are presented in Table 2. For purposes of identification, each applicator was assigned a letter (A through I), according to decreasing concentration of TCDD in the serum. As indicated in Table 2, the levels of TCDD in the serum of eight of the nine applicators were higher than those of matched control subjects. The mean difference for all nine pairs was 47.7 ppt ($p < 0.01$).

In Table 3, TCDD levels for the applicators are presented with the duration of work exposure to 2,4,5-T. Initial applicators selection for the study was based on exposure to pesticides in general, since information for 2,4,5-T use was unavailable. As indicated in Table 3, some applicators did not start spraying 2,4,5-T until after 1960, and they reported a wide variation in total months of spraying.

Figure 1 depicts the relationship between TCDD level in the serum of the applicators and their total duration of work exposure to 2,4,5-T. In general, serum levels of TCDD in applicators increased linearly with total duration of exposure to 2,4,5-T ($r = 0.72$, $p = 0.03$, and slope =

Table 1. Levels on Lipid-Adjusted Basis of PCDDs and PCDFs in Serum of Nine 2,4,5-T Applicators and Nine Matched Control Subjects

Congener[a]	Average level, ppt ± SE[b]		Ratio[c]
	Applicator	Matched control	
Dibenzodioxins			
TCDD	53.3 ± 16.1	5.6 ± 1.1	9.5
1,2,3,7,8-PnPCDD	12.4 ± 1.1	8.8 ± 0.7	1.4
1,2,3,4,7,8-HxCDD	6.8 ± 0.5	5.7 ± 0.4	1.2
1,2,3,6,7,8-HxCDD	28.6 ± 5.1	23.3 ± 4.9	1.2
1,2,3,7,8,9-HxCDD	9.9 ± 0.9	8.2 ± 0.6	1.2
1,2,3,4,6,7,8-HpCDD	121.9 ± 28.5	119.4 ± 18.4	1.0
OCDD	788.6 ± 82.3	758.7 ± 92.8	1.0
Dibenzofurans			
2,3,7,8-TCDF	1.6 ± 0.3	1.7 ± 0.3	0.9
1,2,3,7,8-PnCDF	<2.1 ± 0.2[d]	<2.0 ± 0.2[d]	1.1
2,3,4,7,8-PnCDF	8.0 ± 0.9	74 ± 0.8	1.1
1,2,3,4,7,8-HxCDF	5.4 ± 0.3	5.1 ± 0.5	1.1
1,2,3,6,7,8-HxCDF	5.5 ± 0.4	5.6 ± 0.6	1.0
1,2,3,7,8,9-HxCDF	<0.8 ± 0.1[d]	<0.8 ± 0.1[d]	1.0
2,3,4,6,7,8-HxCDF[e]	<1.1 ± 0.4[d]	<1.7 ± 0.2[d]	1.1
1,2,3,4,6,7,8-HpCDF	14.2 ± 0.7	16.0 ± 2.3	0.9
1,2,3,4,7,8,9-HpCDF[e]	<1.6 ± 0.1[d]	<1.9 ± 0.3[d]	0.8

[a]1,2,3,7,8-PnCDD, 1,2,3,7,8-pentachlorodibenzodioxin; 1,2,3,4,7,8-HxCDD, 1,2,3,4,7,8-hexachlorodibenzodioxin; 1,2,3,6,7,8-HxCDD, 1,2,3,6,7,8-hexachlorodibenzodioxion; 1,2,3,7,8,9-HxCDD, 1,2,3,7,8,9-hexachlorodibenzodioxin; 1,2,3,4,6,7,8-HpCDD, 1,2,3,4,6,7,8-heptachlorodibenzodioxin; OCDD, octachlorodibenzodioxin; 2,3,7,8-TCDF, 2,3,7,8-tetrachlorodibenzofuran; 1,2,3,7,8-PnCDF, 1,2,3,7,8-pentachlorodibenzofuran; 2,3,4,7,8-PnCDF, 2,3,4,7,8-pentachlorodibenzofuran;1,2,3,4,7,8-HxCDF, 1,2,3,4,7,8-hexachlorodibenzofuran;1,2,3,6,7,8-HxCDF, 1,2,3,6,7,8-hexachlorodibenzofuran; 1,2,3,7,8,9-HxCDF, 1,2,3,7,8,9-hexachlorodibenzofuran; 2,3,4,6,7,8-HxCDF, 2,3,4,6,7,8-hexachlorodibenzofuran; 1,2,3,4,6,7,8-HpCDF, 1,2,3,4,6,7,8-hepatochlorodibenzofuran; 1,2,3,4,7,8,9-HpCDF, 1,2,3,4,7,8,9-heptachlorodibenzofuran.
[b]Values are adjusted for total lipids in serum.
[c]Ratio, average for applicators/average for matched control subjects.
[d]Not detected. Values are detection limits.
[e]A number of positive signals were below limit of quantification.

Table 2. Levels of TCDD in Serum of Nine 2,4,5-T Applicators and Nine Matched Control Subjects

Applicator designation	TCDD level in serum (ppt)[a]		Absolute difference
	Applicator	Matched control	
A	131.0	8.8	122.2
B	113.0	2.9	110.1
C	94.8	3.6	91.2
D	55.5	5.2	50.3
E	37.6	9.3	28.3
F	21.8	3.1	18.7
G	14.1	3.6	10.5
H	8.5	2.4	6.1
I	3.0	11.3	−8.3
Mean	53.3	5.6	47.7

[a]Values are adjusted for total lipids in serum.

0.39). We do not know why the level for applicator B is elevated more than expected—113 ppt after 137 months of exposure.

Discussion

The levels of TCDD measured in the applicators in our study were clearly elevated, although they were lower than those reported for other subjects with known exposure. In a U.S. study, 2,4,5-T production workers with heavy exposure to TCDD had an average lipid-adjusted serum

Table 3. Levels of TCDD in Serum of Nine Professional Applicators with History of Work Exposure to 2,4,5-T[a]

Applicator designation	TCDD level in serum (ppt)	Years sprayed 2,4,5-T		Total months sprayed 2,4,5-T
		Started	Stopped[b]	
A	131.0	1953	1988	372
B	113.0	1954	1988	137
C	94.8	1960	1988	278
D	55.5	1960	1985	232
E	37.6	1951	1987	165
F	21.8	1958	1987	165
G	14.1	1959	1988	180
H	8.5	1959	1988	121
I	3.0	1961	1988	83

[a]Values are adjusted for total lipids in serum.
[b]1988 indicates applicator was still spraying at the time of study.

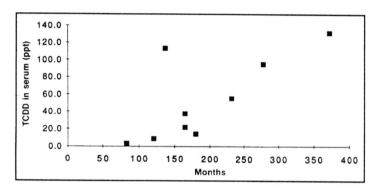

Figure 1. Concentration of TCDD in serum of applicators in relation to total months spent spraying 2,4,5-T. Level of TCDD is adjusted for total level of lipids in serum.

level of about 250 ppt, measured 15–37 years after exposure had ceased.[31–33] Based on a 7-year half-life, the mean level at last occupational exposure for the production workers was estimated at about 2240 ppt.[33] Vietnam war veterans involved with Operation Ranch Hand had a median level of about 12 ppt (range, 0–618 ppt) at least 15 years after the last exposure of each.[34] Although most of the applicators in our study were still spraying 2,4,5-T at the time of the blood donations, the levels measured are predominantly from exposures occurring prior to 1970 when TCDD contamination was more than 200 times the present level. Based on a 7-year half-life, the serum TCDD levels of these applicators may have averaged about 300 ppt 18 years ago.

The levels of TCDD in our control group were similar to those found in most population studies.[31] The means and ranges of the levels of other PCDDs and PCDFs for the applicators and matched control subjects were somewhat lower, however, than those reported for the general population in other industrial countries.[31]

The variation in levels of TCDD in the serum of professional applicators was related to the duration of work exposure of each to 2,4,5-T. Fingerhut et al.[32] and Sweeney et al.[33] also correlated the duration of workplace exposure with the serum levels of TCDD in U.S. workers who produced 2,4,5-T and 2,4,5-trichlorophenol.

Our findings have important implications concerning the hypotheses that exposure to TCDD though use of phenoxyherbicides results in various cancers such as soft-tissue sarcoma and non-Hodgkin's lymphoma. Our results indicate that the use of 2,4,5-T over many years in New Zealand resulted in substantial exposure to TCDD.

In this study, applicators had average serum TCDD levels 47.7 ppt higher than control subjects (Table 2), resulting from an average of 193 months of spraying 2,4,5-T (Table 3). Thus, on average, each 4 months of spraying resulted in a 1-ppt increase of TCDD in the blood of the applicators, up to 35 years later. On the basis of these figures, we can infer that after 1 year of spraying, the applicators' TCDD levels would increase by just 3 ppt, still below the 5.6-ppt average found in the control subjects. The increase would be greater if the extrapolation incorporated a half-life estimate for TCDD in humans, back to the time when the applicators had their heaviest exposures. Even so, since the half-life for TCDD is 5–10 years,[35] these results suggest that the increase in serum TCDD would be modest among workers who sprayed for only 1 year. Thus, if exposure to 2,4,5-T containing TCDD causes cancer in humans, the epidemiologic evidence would likely come only from individuals with exposure occurring over many years.

Evidence that exposure to TCDD causes human cancer, however, is largely based on studies of subjects with astonishingly short exposure. Swedish case-control studies have involved very short duration of exposure. (1) In one case-control study,[1] only two of 13 persons with soft-tissue sarcoma were exposed to phenoxyherbicides for more than 1 year, producing a relative risk estimate of 5.3. (2) In another study of soft-tissue sarcoma,[2] a relative risk estimate of 5.7 was reported for persons with less than 30 days of exposure. (3) In the most recent case-control study,[4] the relative risk estimate was 2.1 for persons with the longest duration of exposure—more than 8 weeks. (4) The Swedish case-control study of malignant lymphoma[6] yielded a relative risk estimate of 4.3 for persons with fewer than 90 days of exposure. On the basis of our present findings, it seems unlikely that exposures of such short duration would substantially increase serum levels of TCDD beyond background levels.

Our conclusions are supported by at least two other research efforts. First, Norstrom et al.[36] found no evidence that the 2,4,5-T used in Sweden had higher levels of TCDD contamination than that used in New Zealand. The levels of TCDD contamination in Sweden were measured at 1.1 ppm in 1952 and 0.4 ppm in 1971 in New Zealand.[10] Second, a little-cited Swedish study[37] of dioxin and furan levels in human adipose tissue reported no difference between the mean level of TCDD in patients with cancers attributed to phenoxyherbicide exposure (2 ppt) and that in other cancer patients (3 ppt).

Case-control studies conducted in New Zealand did not identify any professional pesticide applicators with soft-tissue sarcoma or malignant lymphoma, although brief exposure to 2,4,5-T was reported in some case subjects and control subjects. The overall relative risk estimate for soft-tissue sarcoma with any exposure to phenoxyherbicides was 1.1 (90% confidence limits, 0.7–1.8),[10] and no evidence of an increased risk associated with duration of exposure was found. The corresponding relative risk estimate for non-Hodgkin's lymphoma was 1.3 (90% confidence limits, 0.7–2.3).[13] The results of this study indicate that the subjects in these earlier case-control studies probably had not experienced substantial exposure to TCDD.

If TCDD causes human cancer, epidemiologic associations should be sought in persons with substantial occupational exposures. Although some Operation Ranch Hand personnel have experienced substantial exposures, only the cohort of chemical plant workers in the National Institute of Occupational Safety and Health (NIOSH) study[38] is known to have had substantial exposure to TCDD as well as increased cancer risks. When measured several years after exposure had ceased, average levels of TCDD were high—about 250 ppt.

Fingerhut et al.[39] previously reported a cluster of soft-tissue sarcoma cases in the NIOSH cohort. Their recent report[38] confirmed an increased risk of soft-tissue sarcoma with an estimated relative risk of 9.22 (95% confidence limits, 1.90–27.0) for more than 1 year of exposure. The average duration of work in processes involving exposure to TCDD contamination was 6.8 years. The results of the Swedish case-control studies[1,2,4] are inconsistent with these findings with respect to the hypothesis that TCDD causes cancer in humans. The Swedish studies reported relative risk estimates between 2 and 6 for exposures to TCDD that were trivial compared with those of the chemical production workers.

The NIOSH cohort study[38] did not reveal an increased risk of other cancers that have been postulated to be related to TCDD exposure,[3,15] including cancers of the lymphatic and hematopoietic tissues. The possibility, however, that persons with TCDD exposure have an increased risk of cancers at a variety of sites, e.g., the lung, raises an important hypothesis. Results that could support this hypothesis were reported in a study of German workers exposed to TCDD after a 2,4,5-trichlorophenol reactor accident in 1953.[40] For a subset of workers who had developed chloracne and had at least a 20-year latent period, a relative risk of 2.01 (90% confidence limits, 1.22–3.15) was reported for all cancers combined.

CONCLUSION

In this study of professional applicators of pesticide in New Zealand, we have documented substantial exposure to TCDD from use of 2,4,5-T over many years. Our finding that serum levels of TCDD increase only after several years of 2,4,5-T use suggests that exposure to TCDD was probably negligible in the Swedish studies reporting increased risks of soft-tissue sarcoma and malignant lymphomas with short duration of phenoxyherbicide exposures. To date, only one study[38] involving substantial exposures to TCDD has revealed an increased risk of soft-tissue sarcoma.

This lack of evidence does not mean that TCDD does not cause cancer in humans. It may mean, rather, that, except for the NIOSH cohort study, the numbers of persons with substantial occupational exposures have been too small to reveal any effects. We cannot determine whether TCDD exposure from prolonged use of 2,4,5-T poses significant health risks, but we can conclude that previous reports in other countries of increased cancer risks from brief exposure to phenoxyherbicides are probably not attributable to the TCDD that contaminates 2,4,5-T.

REFERENCES

1. **Hardell, L., and Sandstrom, A.,** Case-control study; Soft tissue sarcomas and exposure to phenoxyacetic acids or chlorophenols, *Br. J. Cancer,* 39, 711–717, 1979.
2. **Eriksson, M., Hardell, L., Berg, N. O., et al.,** Soft-tissue sarcomas and exposure to chemical substances; A case-referent study, *Br. J. Ind. Med.,* 38, 27–33, 1981.
3. **Hardell, L.,** Relation of soft tissue sarcoma, malignant lymphoma and colon cancer to phenoxy acids, chlorophenols and other agents, *Scand. J. Work Environ. Health,* 7, 119–130, 1981.
4. **Hardell, L., and Eriksson, M.,** The association between soft tissue sarcomas and exposure to phenoxyacetic acids: A new case-referent study, *Cancer,* 62, 652–656, 1988.
5. **Eriksson, M., Hardell, L., and Adami, H. L.,** Exposure to dioxins as a risk factor for soft tissue sarcoma: A population-based case-control study, *J. Natl. Cancer Inst.,* 82, 486–490, 1990.
6. **Hardell, L., Eriksson, M., Lenner, P., et al.,** Malignant lymphoma and exposure to chemicals, especially organic solvents, chlorophenols and phenoxy acids: A case-control study, *Br. J. Cancer,* 43, 169–176, 1981.
7. **Wiklund, K., and Holm, L. E.,** Soft tissue sarcoma risk in Swedish agricultural and forestry workers, *J. Natl.Cancer Inst.,* 76, 229–234, 1986.
8. **Wiklund, K., Dich, J., and Holm, L. E.,** Soft tissue sarcoma risk in Swedish licensed pesticide applicators, *J. Occup. Med.,* 30, 801–804, 1988.
9. **Wiklund, K., Lindefords, B. M., and Holm, L. E.,** Risk of malignant lymphoma in Swedish agricultural and forestry workers, *Br. J. Ind. Med.,* 45, 19–24, 1988.
10. **Smith, A. H., and Pearce, N. E.,** Update on soft tissue sarcoma and phenoxyherbicides in New Zealand, *Chemosphere,* 15, 1795–1798, 1986.
11. **Smith, A. H., Fisher, D. O., Giles, H. J., et al.,** The New Zealand soft tissue sarcoma case-control study: Interview findings concerning phenoxyacetic acid exposure, *Chemosphere,* 12, 565–571, 1983.
12. **Smith, A. H., Pearce, N. E., Fisher, D. O., et al.,** Soft tissue sarcoma and exposure to phenoxyherbicides and chlorophenols in New Zealand, *J. Natl. Cancer Inst.,* 73, 1111–1117, 1984.
13. **Pearce, N. E., Smith, A. H., Howard, J. K., et al.,** Non-Hodgkin's lymphoma and exposure to phenoxyherbicides, chlorophenols, fencing work, and meat works employment: A case-control study, *Br. J. Ind. Med.,* 43, 75–83, 1986.

14. **Pearce, N. E., Sheppard, R. A., Smith, A. H., et al.,** Non-Hodgkin's lymphoma and farming: An expanded case-control study, *Int. J. Cancer,* 39, 151–161, 1987.
15. **Zahm, S. K., Weisenburger, D. D., Babbitt, P. A., et al.,** A case-control study of non-Hodgkin's lymphoma and the herbicide 2,4-dichlorophenoxyacetic acid (2,4-D) in eastern Nebraska, *Epidemiology,* 1, 349–356, 1990.
16. **Woods, J. S., Polissar, L., Severson, R. K., et al.,** Soft tissue sarcoma and non-Hodgkin's lymphoma in relation to phenoxyherbicide and chlorinated phenol exposure in western Washington, *J. Natl. Cancer Inst.,* 78, 899–910, 1987.
17. **Vineis, P., Terracini, B., Ciccone, G., et al.,** Phenoxy herbicides and soft-tissue sarcomas in female rice weeders. A population-based case-referent study, *Scand. J. Work Environ. Health,* 13, 9–17, 1987.
18. **Lynge, E.,** A follow-up study of cancer incidence among workers in the manufacture of phenoxy herbicides in Denmark, *Br. J. Cancer,* 52, 259–270, 1985.
19. **Kogan, M. D., and Clapp, R. W.,** Soft tissue sarcoma mortality among Vietnam veterans in Massachussets. 1972 to 1983, *Int. J. Epidemiol.,* 17, 39–43, 1988.
20. **Greenwald, P., Kovasznay, B., Collins, D. N., et al.,** Sarcomas of soft tissues after Vietnam service, *J. Natl. Cancer Inst.,* 73, 1107–1109, 1984.
21. **Hoar, S. K., Blair, A., Holmes, F. F., et al.,** Agricultural herbicide use and risk of lymphoma and soft-tissue sarcoma, *JAMA,* 256, 1141–1147, 1986.
22. **Smith, A. H., and Bates, M. N.,** Epidemiology studies of cancer and pesticide exposure, in *Carcinogenicity and Pesticides. ACS Symposium Series 314,* Ragsdale, N. N., and Menzer, R. E., Eds., American Chemical Society, Washington, D.C., 1989, 207–222.
23. **Smith, A. H., Matheson, D. P., Fisher, D. O., et al.,** Preliminary report of reproductive outcomes among pesticides applicators using 2,4,5-7, *NZ Med. J.,* 93, 177–179, 1981.
24. **Smith, A. H., Fisher, D. O., Pearce, N. E., et al.,** Congenital defects and miscarriages among New Zealand 2,4,5-T sprayers, *Arch. Environ. Health,* 37, 197–200, 1982.
25. **Patterson, D. G., Jr., Hampton, L., Lapeza, C. R., Jr., et al.,** High-resolution gas chromatographic/high-resolution mass spectrometric analysis of human serum on a whole-weight and lipid basis for 2,3,7,8-tetrachlorodibenzo-*p*-dioxin, *Anal. Chem.,* 59, 2000–2005, 1987.
26. **Patterson, D. G., Jr., Furst, P., Henderson, L. O., et al.,** Partitioning of in vivo bound PCDDs/PCDFs among various compartments in whole blood, *Chemosphere,* 19, 135–142, 1989.
27. **Smith, L. M., Stalling, D. L., and Johnson, J. L.,** Determination of part-per-trillion: Levels of polychlorinated dibenzofurans and dioxins in environmental samples, *Anal. Chem.,* 56, 1830–1842, 1984.
28. **Patterson, D. G., Jr., Holler, J. S., Lapeza, C. R., Jr., et al.,** High-resolution gas chromatographic/high-resolution mass spectroscopic analysis of human adipose tissue for 2,3,7,8-tetrachlorodibenzo-*p*-dioxin in human adipose tissue, *Anal. Chem.,* 58, 713–716, 1986.
29. **Lapeza, C. R., Jr., Patterson, D. G., Jr., and Liddle, J. A.,** Automated apparatus for the extraction and enrichment of 2,3,7,8-tetrachlorodibenzo-*p*-dioxin in human adipose tissue, *Anal. Chem.,* 58, 713–716, 1986.
30. **Patterson, D. G., Jr., Furst, P., and Alexander, L. R.,** Analysis of human serum for PCDDs/PCDFs: A comparison of three extraction procedure, *Chemosphere,* 19, 89–96, 1989,
31. **Patterson, D. G., Jr., Fingerhut, M. A., Roberts, D. W., et al.,** Levels of polychlorinated dibenzo-*p*-dioxins and dibenzofurans in workers exposed to 2,3,7,8-tetrachlorodibenzo-*p*-dioxin, *Am. J. Ind. Med.,* 16, 135–146, 1989.
32. **Fingerhut, M. A., Sweeney, M. H., Patterson, D. G., Jr., et al.,** Levels of 2,3,7,8-tetrachlorodibenzo-*p*-dioxin in the serum of U.S. chemical workers exposed to dioxin contaminated products: interim results, *Chemosphere,* 19, 835–840, 1989.
33. **Sweeney, M. H., Fingerhut, M. A., Patterson, D. G., Jr., et al.,** Comparison of serum levels of 2,3,7,8-TCDD in TCP production workers and in an unexposed comparison group, *Chemosphere,* 20, 993–100, 1990.
34. **Wolfe, W. H., Michalek, J. E., Miner, J. C., et al.,** Health status of Air Force veterans occupationally exposed to herbicides in Vietnam. I. Physical health, *JAMA,* 264, 1824–1831, 1990.
35. **Pirkle, J. L., Wolfe, W. H., Patterson, D. G., Jr., et al.,** Estimates of the half-life of 2,3,7,8-tetrachlorodibenzo-*p*-dioxin in Vietnam veterans of Operation Ranch Hand, *J. Toxicol. Environ. Health,* 27, 165–171, 1989.

36. **Norstrom, A., Rappe, C., Lindahl, R., et al.,** Analysis of some older Scandinavian formulations of 2,4-dichlorophenoxy acetic acid and 2,4,5-trichlorophenoxy acetic acid for contents of chlorinated dibenzo-*p*-dioxins and dibenzofurans, *Scand. J. Work Environ. Health,* 5, 375–378, 1979.
37. **Nygren, M., Rappe, C., Lindstrom, G., et al.,** Identification of 2,3,7,8-substituted polychlorinated dioxins and furans in environmental and human samples, in *Chlorinated Dioxins and Dibenzofurans in Perspective,* Rappe, C., Choudhary, G., and Keith, L. H., Eds., Lewis Publishers, Chelsea, MI, 1986, 17–34.
38. **Fingerhut, M. A., Halperin, W. E., Marlow, D. A., et al.,** Cancer mortality in workers exposed to 2,3,7,8-tetrachlorodibenzo-*p*-dioxin, *N. Engl. J. Med.,* 324, 212–218, 1991.
39. **Fingerhut, M. A., Halperin, W. E., Honchar, P. A., et al.,** An evaluation of reports of dioxin exposure and soft tissue sarcoma pathology among chemical workers in the United States, *Scand. J. Work Environ. Health,* 10, 299–303, 1984.
40. **Zober, A., Messerer, P., and Huber, P.,** Thirty-four-year mortality follow-up of BASF employees exposed to 2,3,7,8-TCDD after the 1953 accident, *Int. Arch. Occup. Environ. Health,* 62, 139–157, 1990.

CHAPTER **18**

Recombinant Human Papilloma Type 16 DNA Induces Progressive Changes in Mouse 3T3 Cells and Human Epithelial Cells

Joseph A. DiPaolo

The ultimate goal of experimental cancer research is to provide fundamental information on the nature of the primary insult responsible for the transformation of normal to malignant cells. Such enlightenment would result in clarifying the subsequent steps that lead to cancer so that interventive and preventive measures could be taken against this disease. Dose-dependent transformation frequencies coupled with the formation-transformed, tumor-producing lines have provided evidence that chemically and virally induced carcinogenesis can be studied *in vitro* with animal cells. Unfortunately, a reproducible model in which normal human cells are converted to permanent malignant cell lines in a predictable fashion does not exist.

An approach for overcoming the refractiveness of human cells to carcinogens involves cocarcinogenesis experiments in which either human papillomavirus (HPV) or HPV DNA and diverse carcinogens are used sequentially. HPV-16 and HPV-18 DNAs, which have been isolated and molecularly cloned from cervical carcinoma cells and which have been classified as HPVs on the basis of DNA sequence homology, are strongly associated with cervical carcinomas. Attempts to correlate cervical dysplasia with HPV infection dates from the mid-1970s.[1] Molecular biological studies have demonstrated the presence of HPV-16 and -18, as well as other related viral DNAs in cervical intraepithelial neoplasia lesions, invasive cervical cancer, and cervical carcinoma cell lines.[2] Furthermore, recombinant HPVs associated with invasive cervical cancer can reproducibly immortalize keratinocytes derived from foreskin and from the transition zone of the cervix.[3-7] Consistent with the basal cell origin of squamous metaplastic lesions, epithelium cells are probably the primary target for anogenital HPV. Immortalization of cervical epithelial cells was independent of the age of the donor. The HPV-immortalized cells have integrated HPV DNA, express RNA and specific proteins, and are aneuploid with structural chromosomal alterations but nontumorigenic as indicated by failure to form tumors after subcutaneous injection into immunosuppressed athymic mice.

Some HPVs transfected into established rodent cell lines resulted in tumor-producing cells two to three weeks after injection into nude mice.[8-10] Later, we developed a two-stage model using the recombinant plasmid (pMHPV16d) carrying a head-to-tail dimer of HPV-16 DNA and a neomycin resistant gene.[11] After antibiotic selection of HPV-containing cells, a number of resistant colonies with a flat morphology similar to that obtained by mock transfection were isolated. Only some of these colonies gave rise to tumorigenic cell lines. The latter indicate that HPV-16 DNA possesses oncogenic potential, but its presence alone is insufficient to produce tumors. The fully transformed state requires a unique interaction between HPV and host cells.

The role of virulent HPVs in human cancer induction today is unclear. There is ample evidence indicating that cervical cancer is a sexually transmitted disease. Previously it was accepted that a small percentage of women (about 10%) had detectable levels of HPV DNA without cytological alterations.[12] Now it appears possible that the incidence of HPV in normal women has been underestimated. Amplification of HPV DNA sequences (one virus per 10^6 cells sensitivity) by polymerase chain reaction indicates the rate of infection in scrapes of normal cervices is currently reported to be 70–80%.[13,14] Whether this percentage would be the same worldwide is unknown. Moreover, it should be kept in mind that whereas premalignant tumors are characterized as having primarily episomal viral DNA, anogenital cancer commonly has the viral DNA integrated into the host genome. Epidemiological studies concerning the association of HPVs and cervical cancer are reminiscent of studies previously reported involving herpes simplex virus type 2 and cervical cancer. A Latin American study found HPV-16 or -18 in controls and cancers, 43 and 67%, respectively.[15] These studies suggest that HPVs are only one factor in cancer. In another study it was found that there is a 5.7-fold higher incidence of cervical cancer in Greenland compared to Denmark that is independent of the rate of infection.[16] Thus, there is a need for clarification of the oncogenic role of HPVs in malignancies.

Currently, 11 human keratinocyte (HKc) lines have been established in serum-free medium after transfection of 13 normal HKc strains (each derived from neonatal foreskin from a different individual) with a recombinant HPV-16 dimer with a neoresistant gene.[16] Vector-transfected HKc senesced. Chromosome analysis of transfected HKc indicated drastic alterations: breaks, pulverizations, endoreduplication, dicentric, and double minutes. Early passages of all lines exhibited gene amplification as indicated by double minutes or homogeneous-staining regions. DNA analysis of transfected HKc after *Bam*HI digestion indicated the presence of HPV-16 DNA sequences that had become progressively amplified and rearranged (integrated); these HKc expressed several HPV-16 RNA species similar to RNA species detected in NIH 3T3 cells transfected with the same HPV-16. The integration of HPV-18 DNA sequences in HeLa and C755 carcinoma cell chromosomes primarily at fragile sites and/or sites of protooncogenes rather than at random sites may indicate that integration at a specific locus is critical for conversion to malignancy. The two-step transition to tumorigenicity associated with integration of HPV-16 DNA into the host genome of transfected NIH 3T3 cells supports the importance of a specific host/viral interactions. Thus far, the HPV-16-transfected human cells mimic premalignant NIH 3T3 cells because no tumor has been found in nude mice. Transfected human cells with an indefinite life span provide a model for studying the molecular biology of HPV-16 in a human genetic background as well as its role in human cancer.

REFERENCES

1. **Rotkin, I. D.,** A comparison review of key epidemiological studies in cervical cancer related to current searches for transmissible agents, *Cancer Res.,* 33, 1353, 1973.
2. **zur Hausen, H.,** Papillomaviruses in anogenital cancer as a model to understand the role of viruses in human cancers, *Cancer Res.,* 49, 4677, 1989.
3. **Pirisi, L., Yasumoto, S., Feller, M., Donigar, J., and DiPaolo, J. A.,** Transformation of human fibroblasts and keratinocytes with human papillomavirus type 16 DNA, *J. Virol.,* 61, 1061, 1987.
4. **Durst, M., Dzarlieva-Petrusevska, R. T., Boukamp, P., Fusenig, N. E., and Gissmann, L.,** Molecular and cytogenetic analysis of immortalized human keratinocytes obtained after transfection with human papillomavirus type 16 DNA, *Oncogene,* 1, 251, 1987.

5. **Kaur, P., and McDougall, J. K.,** Characterization of primary human keratinocytes transformed by human papillomavirus type 18, *J. Virol.,* 62, 1917, 1988.
6. **Woodworth, C. D., Bowden, P. E., Doniger, J., Pirisi, L., Barnes, W., Lancaster, W. D., and DiPaolo, J. A.,** Characterization of normal human exocervical epithelial cells immortalized *in vitro* by papillomavirus types 16 and 18 DNA, *Cancer Res.,* 48, 4620, 1988.
7. **Pecorado, G., Morgan, D., and Defendi, V.,** Differential effects of human papillomavirus types 6, 16, and 18 DNAs on immortalization and transformation of human cervical epithelial cells, *Proc. Natl. Acad. Sci. U.S.A.,* 86, 563, 1989.
8. **Yasumoto, S., Burkhardt, A. L., Doniger, J., and DiPaolo, J. A.,** Human papillomavirus type 16 DNA-induced malignant transformation of NIH 3T3 cells, *J. Virol.,* 57, 572, 1986.
9. **Tsunokawa, Y., Takebe, N., Kasamatsu, T., Terada, M., and Sugimura, T.,** Transforming activity of human papillomavirus type 16 DNA sequences in a cervical cancer, *Proc. Natl. Acad. Sci. U.S.A.,* 83, 2200, 1986.
10. **Bedell, M. A., Jones, K. H., and Laimins, L. A.,** The E6-E7 region of human papillomavirus type 18 is sufficient for transformation of NIH 3T3 and rat-1 cell, *J. Virol.,* 61, 3635, 1987.
11. **Yasumoto, S., Doniger, J., and DiPaolo, J. A.,** Differential early viral gene expression in two stages of human papillomavirus type 16 DNA-induced malignant transformation, *Mol. Cell Biol.,* 7, 2165, 1987.
12. **Lorincz, A. T., Lancaster, W. D., Kurman, R. J., Jenson, A. B., and Temple, G. F.,** *Banbury Report 21: The Viral Etiology of Cervical Cancer,* Cold Spring Harbor Laboratory, Cold Spring Harbor, 1986, 225.
13. **Young, L. S., Bevan, I. S., Johnson, M. A., Bloomfield, P. I., Bromidge, T., Mailtand, N. J., and Woodman, C. B. J.,** The polymerase chain reaction: A new epidemiological tool for investigating cervical human papillomavirus infection, *Br. Med. J.,* 298, 14, 1989.
14. **Tidy, J. A., Parry G. C. N., Ward, P., Coleman, D. V., et al.,** High rate of human papillomavirus type 16 infection in cytologically normal cervices, *Lancet,* i, 816, 1989.
15. **Reeves, W. C., Caussy, D., Brinton, L. A., Brenes, M. M., Montalvan, P., et al.,** Case-control study of human papillomaviruses and cervical cancer in Latin America, *Int. J. Cancer,* 40, 450, 1987.
16. **Pirisi, L., Creek, K. E., Doniger, J., and DiPaolo, J. A.,** Continuous cell lines with altered growth and differentiation properties originate after transfection of human keratinocytes with human papillomavirus type 16 DNA, *Carcinogenesis,* 9, 1573, 1988.

CHAPTER **19**

Spatial Distribution of Dose in the Human Lung after Inhalation of Poorly Soluble Radionuclides and Its Sequelae

Hans Cottier, Arne Burkhardt, Rainer Kraft, Fritz Meister, and Arthur Zimmermann

This is a short overview of certain effects of inhaled radionuclides, in particular the poorly soluble ^{239}PuO$_2$, on the mammalian lung. Since human data on this subject are virtually lacking, most of our knowledge in the field is based on animal experimentation. In the present context, emphasis is placed on the spatial distribution of poorly soluble radionuclides in the pulmonary tissue, which may be expected to be highly heterogeneous. The latter assumption appears to be justified in view of the accumulation, in older age, of the insoluble anthracotic pigment along the lymph vessels of the human lung. Although the almost inert carbon particles cannot readily be compared with the highly toxic ^{239}PuO$_2$, both materials have in common a poor solubility and a propensity for being translocated, over the years, via pulmonary lymphatics to regional lymph nodes. Following inhalation of ^{239}PuO$_2$, so-called "hot spots" of α-emitting material are thus apt to develop. Possible consequences of this highly unequal distribution of α-radiation dose in the pulmonary tissue comprise both stochastic, e.g., cancerogenic, and nonstochastic effects, such as lymphocytopenia and progressive lung fibrosis.

INTRODUCTION

The extended use of nuclear energy for peaceful purposes, in particular the production of electricity, necessitates the participation of medical research in evaluating possible hazards involved. Human data on the effects of inhalation of radionuclides, except for miners who were exposed to high concentrations of radon and radon decay products, are very scarce. Therefore, much of our present knowledge in this field relates to animal experimentation. A comprehensive report on this subject has been compiled by a task group [of which one of the authors (H. Cottier) was a member] of Committee 1 of the International Commission on Radiological Protection (ICRP) in 1980.[1] As emphasized in this publication, different short-term and late effects of inhaled radionuclides must be considered together with a multitude of other variables. Among the most prominent factors that influence the outcome of such a study, the following should be mentioned: species, sex, age, specific activity of the respective radioactive compound, the type of radiation emitted, radioactive decay, concentration of radionuclides in particles, size and degree of aggregation of the latter, solubility of the inhaled material, particle translocation within the body, presence or absence of additional toxic substances, particle–tissue interactions, and the nature and vulnerability of the tissue components concerned.

DAMAGING EFFECTS ON ANIMALS OF INHALED POORLY SOLUBLE RADIONUCLIDES

As one might suspect, the greatest hazards per unit dose might arise from inhalation of radionuclides that are both long-lived, potent sources of radiation and poorly soluble. The latter property will tend to accumulate large doses of radiation in the lung tissue, while more soluble radionuclides are more easily subject to distribution throughout the body and eventual excretion. Results of animal experimentation seem to confirm this notion. In fact, observations on dogs exposed to various doses of inhaled plutonium oxides have established the high toxicity of these long-lived α-emitting particles. Of the numerous elements produced in the fission process, only a limited number of isotopes are of major medical concern (Table 1, see Ferlic[2]). Among these, plutonium compounds, especially the poorly soluble plutonium oxides, are among the most hazardous, if not the most dangerous of all. After exposure to either $^{239}PuO_2$ or $^{238}PuO_2$, beagle dogs with initial lung burdens (ILB) of 2960 Bq (80 nCi!) developed a long-standing lymphocytopenia as the most subtle change. Animals with higher ILBs exhibited a dose-dependent life-shortening due essentially to the development of lung tumors and radiation pneumonitis, although the pattern of lesions was different for $^{239}PuO_2$ and $^{238}PuO_2$, respectively. Dogs that had inhaled the more soluble $^{238}PuO_2$ showed less lung lesions but had more bone tumors and liver damage than those exposed to a comparable ILB of $^{239}PuO_2$.[3,4] With very high ILBs, the prominent feature of exposed beagles consisted in pulmonary edema and hemorrhage, as an expression of severe radiation-induced diffuse alveolitis ("radiation pneumonitis"), which—depending on dose—led to death within weeks to months after exposure.[1]

DOSE DISTRIBUTION IN THE LUNGS AFTER INHALATION OF POORLY SOLUBLE α-EMITTING RADIONUCLIDES: PROBLEMS INVOLVED

In order to better understand pathogenetic mechanisms underlying the lesions mentioned above, it is mandatory to know more details about the spatial distribution of dose following inhalation of poorly soluble radionuclides. Kinetic models concerning deposition, retention, and translocation of inhaled particles as guidelines for dosimetry of the human respiratory tract were proposed years ago.[5] However, these are mathematical simulations based on the assumption that the respiratory tract and the lung constitute a limited number of "compartments." Possible sequelae of unequal distribution of dose within any one of these compartments and corresponding estimates of tissue components predominantly at risk are not contained in these models. The most important factor, yet the one with the greatest uncertainty, is thus the spatial distribution of dose as a function of time. Although lung cancer mortality after exposure to inhaled radionuclides "may be adequately accounted for by the conventional method of averaging the absorbed dose over the entire lung,"[1] many problems, including the formation of "hot spots" and their possible consequences, cannot be resolved by this type of dosimetry alone. What one would wish to do, is an analysis of events following inhalation of radionuclides that resembles more the modern toxicological rather than the usual radiological approach. In other words, we may distinguish between an exposition phase (uptake of the material, i.e., by inhalation), a toxokinetic phase (so-called invasion, i.e., "resorption" and distribution, including the elimination of the toxon), and a toxodynamic phase, which in the present context relates to radiation effects sensu strictiori. One would also have to consider

Table 1. Selected List of Radionuclides Relevant in a Fission Accident

Isotope	Major type of emission	Physical half-life	Retention half-life in man	Remarks
^{131}I	Beta	8 days	8–140 days (30% enter thyroid:120 days)	Thyroid
^{90}Sr	Beta	28 yr	49 yr	Mostly in bone
^{137}Cs	Beta, gamma	30 yr	10%: 2 days 90%: 110 days	Similar to K
^{239}Pu	Alpha	2.4×10^4 yr	Bone: 100 yr Liver: 40 yr	Soluble: bone Insoluble inhaled in lung and lymph
^{235}U	Alpha	1.5×10^5 yr	Bone 90%: 20 days 10%: 14 yr	Possible damage to kidney and colon
^{14}C	Beta	5.7×10^3 yr	18%: 5 min 81%: 60 min 1%: 83 days	Mostly inhaled as ^{14}CO or ^{14}CO$_2$
^{3}H	Beta	12 yr	10 days for bulk of ^{3}H	Mostly ingested ^{3}H$_2$O

Note: Of the 36 elements produced by the fission process, only a few are of major concern.[2] In addition to those listed above, ^{89}Sr, ^{140}Ba, ^{144}Ce, ^{91}Y, and activated materials of the device, such as isotopes of Zn, Cu, Mg, and Fe, may be mentioned. Ground-induced radionuclides comprise, among other, ^{22}Na, ^{38}Cl, ^{56}Mn, ^{28}Al, and ^{31}Si. These have a physical half-life of less than 3 h. Insoluble compounds, after inhalation, pose a particular problem (see text).

terms or phenomena such as ''biological availability'' and ''biological half-life'' of the toxon as well as cumulation and sequestration phenomena. Ideally, the ''CT-product'' (''concentration-time-product'') at any given site within the human lung should be known as a function of time, and this separately for each isotope in all its chemical forms and physical states (see Cottier and Zimmermann[6]). It is quite clear that we are still very far from this ultimate goal. For the time being one is forced to include a number of assumption into our estimates on the spatial distribution of inhaled radionuclides and its changes as a function of time. Evidently, the often discussed questions of tissue components and cells most at risk cannot be answered without penetrating more deeply into this problem. Since almost nothing is known about the effects of inhaled poorly soluble radionuclides on the human lung, predictions must to a large extent remain speculative. Extrapolation of animal data on man should be considered with caution: profound interspecies differences are known to exist, not only as regards the frequency and type of radiation-induced neoplasia but also with respect to biological responsiveness and tissue structure.

ANTHRACOSIS AS A MODEL FOR STUDYING THE RETENTION PATTERN OF POORLY SOLUBLE PARTICLES IN THE HUMAN LUNG

Although translocation kinetics of the essentially inert carbon particles probably differ considerably from those of the highly toxic, poorly soluble α-emitting particles mentioned above, both materials have in common that large fractions of the particulates that are not removed by the mucociliary transport system of bronch(iol)i are ultimately subject to drainage via the pulmonary lymphatics. We have studied in more detail and on large sections of the human lung the retention pattern of anthracotic pigment. In doing so, we did not take into account carbon particles contained in free alveolar macrophages because these could still have been subject to clearance via the bronchial system. Tracheobronchial lymph nodes were also excluded from the counts since their progressive load with inhaled insoluble particles is well established. With these restrictions in mind, it was concluded that the retention pattern of dense anthracotic particle aggregates in the human lung is highly nonuniform. Impressive accumulations of this material were found along the pulmonary lymphatics, i.e., the deep (peribronchial), septal (perivenous), and superficial (pleural) networks, while the lung parenchyme sensu strictiori (i.e., alveolar walls) contained only a few percent of the total load (Table 2).[7] In view of this pronounced inhomogeneity of carbon

Table 2. Example of Frequency Distribution of Dense Carbon Particle Aggregates of Different Sizes in the Human Lung[7]

Localization (left upper lobe, lower portion)[a]	Fraction (%) of all aggregates per lung section (size classes according to longest diameter)[b]			
	<15 µm	16–100 µm	101–250 µm	251–500 µm
In alveolar walls (A)	2.3	0	0	0
Along pleural lymphatics (P)	2.6	12.7	7.9	0.8
Along septal lymphatics (S)	2.3	26.9	4.8	0
Along deep lymphatics (D)	1.4	22.1	13.6	2.6

[a]Lung was fixed *in situ*, i.e., without collapse, and a large section (12.8 cm^2) was prepared perpendicular to the pleural surface.
[b]Frequency distribution of the carbon aggregates of various size classes assessed with the help of the point counting technique (37,500 test points). Volume densities (fractions of measured lung profile) of heavily anthracotic foci were as follows: A, 0.2%; P, 0.23%; S, 0.32% and D, 0.37%, i.e., a total of 0.94%.

particle aggregates in the human lung, it was also of interest to measure the distances between the border of the latter and tissue components in their vicinity. Considering the range of α-particles emitted by plutonium (about 45 µm in solid soft tissue and 4 cm in air) and assuming ^{239}PuO$_2$ rather than inert particle aggregates, it could be estimated that practically all cell types of the human lung, including those circulating in the blood, would have been reached by α-radiation. However, the probability of being hit would have differed enormously from site to site, depending on the presence or absence of a larger deposit in the immediate neighborhood.

EVENTS TO BE EXPECTED IN THE PULMONARY PARENCHYMA FOLLOWING INHALATION OF ^{239}PuO$_2$

The rather rapid clearance of the respiratory system from the major fraction of inhaled particles with the help of macrophages and the mucociliary transport mechanism has long been well documented.[5] The ILB, i.e., the radioactive material that is deposited on alveolar walls, may constitute less than 1% of the total inhaled amount. Most of the particles are then endocytosed by alveolar macrophages. To better understand the complex interactions between the α-emitting material and lung tissue, one must be aware of the very high doses of radiation delivered on short distance. It can be calculated, for instance, that a small particle of ^{239}PuO$_2$ with an activity of 2.6 mBq (0.07 pCi) in a stationary situation will expose a sphere with about 45 µm radius of the surrounding solid soft tissue to a dose rate of 168 Gy per year. If a single α-track traverses the nucleus of a cell, the latter will be killed with a 50–100% probability. It is thus easy to visualize what is apt to happen in pulmonary alveoli containing such material. Stimulated and/or killed macrophages can initiate an inflammatory response and thus contribute to lung damage (for review, see Brain[8]), in addition to direct lesions in the alveolar wall due to radiation. As a consequence, a dose-dependent alveolitis will develop, with all its possible sequelae. It should be recalled in this context that some degree of alveolar denudation may be expected to occur and that this process is usually associated with some fibrin deposition. In case such alveoli collapse, atelectatic induration with apposition and coalescence of neighboring interalveolar walls is a possible hazard contributing to progressive lung fibrosis.[9] In the course of such events, radioactive material in small amounts may become entrapped in interalveolar walls and thus constitute small "hot spots."

THE IMPORTANCE OF PULMONARY LYMPHATICS IN THE CAUSATION OF AN INHOMOGENEOUS DISTRIBUTION OF INHALED ^{239}PuO$_2$

The studies on anthracosis of the human lung mentioned above demonstrated that most of the insoluble material retained over the years in the lungs is located along pulmonary lymphatics and in regional lymph nodes. In the case of ^{239}PuO$_2$ inhalation, the inflammatory reaction due to the killing effect of α-particles most probably damages also the lymphatics of the organ. This can be expected since the long-term, progressive translocation of poorly soluble α-emitting material from the lungs to tracheobronchial lymph nodes is well documented.[1] Small lymphoid nodules along the lymphatics, which are normally found in the human lung,[10] may be preferential sites of particle arrest. We are thus confronted with a situation that is apt to cause "hot spots," "hot streaks," and "hot plates" associated with pulmonary lymphatics in a distribution similar to that found for anthracotic particles. It may be added that radiation-induced damage to the delicate walls of the lymphatics can be expected to result in focal obliteration, consecutive lymphangiectasis and lymphedema, and fibrosis. The latter, in this complex scheme of events, can thus ensue from atelectatic induration, reparative scarring, and/or lymphsclerema.

THE "HOT SPOT" PROBLEM IN RELATION TO EFFECTS OF INHALED POORLY SOLUBLE α-EMITTING PARTICLES SUCH AS ^{239}PuO$_2$

Radiobiological problems concerning "hot spots" of radioactive material are known to be highly complex. It may be emphasized that the questions to be discussed in the present context should not be confounded with the so-called "hot particle" issue created by the introduction, from the exterior, of particles each containing ≳ about 0.6 pCi (for overview, see Bair[11]). Rather, what we should like to focus on are particle aggregates forming after inhalation exposure. These aggregates are apt to be rather small in alveolar macrophages and interalveolar septa, but large along pulmonary lymphatics and even greater in regional lymph nodes. In view of the high dose rate delivered by such "hot spots" of α-emitting ^{239}PuO$_2$ (see above), most if not all, cells in the immediate vicinity would, in a stationary situation, be killed. This sterilizing effect is apt to greatly reduce the number of surviving target cells that could be subject to neoplastic transformation. If the latter were solely due to radiation-induced somatic mutation, one might expect the tumor yield to be smaller, if a given lung burden of poorly soluble α-emitting material is preferentially concentrated in "hot spots" rather than diffusely. However, with an observation period of 11 years, ^{239}PuO$_2$—per unit estimated cumulative lung dose—was more effective than the more soluble ^{238}PuO$_2$ in causing lung tumors in dogs.[3] This apparent puzzle of radiation carcinogenesis raises a number of unresolved questions. Various possibilities should be considered to explain the high carcinogenicity of multiple α-emitting "hot spots" as compared to a diffuse distribution of the radioactive material in the lung:

1. "Hot spots" may not remain stationary and could thus continuously expose new target cells to radiation. One mechanism of migration of "hot spots" possibly lies in the formation of small cavities around the radioactive particle aggregates, in which the latter could be displaced. Constant lymph drainage of loosened particles should also be mentioned in this context.
2. Diffusion of oxygen-derived radicals and their secondary products into the zone of tissue surrounding "hot spots," in addition to direct hits by paricles, may induce neoplastic transformation in target cells.
3. Transfection of target cells by nucleic acid fragments released from disintegrated cells constitutes a possibility that has not yet been tested following inhalation of radionuclides.
4. The same is true for other carcinogenic substances that may be freed from such sites.
5. Little is known on the possible role of viruses in these situations.
6. Enhanced proliferative activity of cells adjacent to the necrotic zone could favor neoplastic transformation.

All the mechanisms mentioned above may, for instance, activate oncogenes. Obviously, further studies are needed to explore the carcinogenic effect of α-emitting "hot spots."

NONSTOCHASTIC EFFECTS OF INHALED RADIONUCLIDES

Traditionally, risk estimates for inhaled radionuclides have almost exclusively been based on their leukemogenic or carcinogenic action. However, a considerable amount of tissue damage results from nonneoplastic changes caused by the radioactive material. As mentioned in the ICRP report 31,[1] very high doses of inhaled plutonium oxides have a destructive effect on the lung tissue, which expresses itself in a severe diffuse, exudative alveolitis, lung edema, and hemorrhage. But even rather low ILBs of, e.g., ^{239}PuO$_2$ or ^{238}PuO$_2$ can initiate progressive, nonstochastic damage, in particular pulmonary fibrosis and a certain propensity for developing lung edema. In fact, in the ICRP report 31, these changes are listed — after lymphocytopenia — as the second most sensitive biological effect of inhaled plutonium. Since impairment of pulmonary functions contributes to a reduction of the quality of life, such tests should be carried out more frequently and in more detail.

Future studies will show if the value for the quality factor of α-radiation as it is accepted by the ICRP,[12] i.e., 20, will be maintained or changed.

REFERENCES

1. **International Commission on Radiological Protection,** *Biological Effects of Inhaled Radionuclides,* Publication 31, Pergamon Press, Oxford, 1980.
2. **Ferlic, K. P.,** *Fallout: Its Characteristics and Management,* Armed Forces Radiobiology Research Institute, Bethesda, MD, 1983, AFRRI TR 83–85.
3. **Park, J. F., Apley, G. A., Buschbom, R. L., Dagle, G. E., Fisher, D. R., Gideon, K. M., Gilbert, E. S., Kashmitter, J. D., Powers, G. J., Ragan, H. A., Weller, R. E., and Wieman, E. L.,** Inhaled plutonium oxide in dogs, in *Pacific Northwest Laboratory Annual Report for 1985 to the DOE Office of Energy Research, Part 1, Biomedical Sciences,* National Technical Information Service, Springfield, VA, 1986, PNL 5750: 3–17.
4. **Sanders, C. L., and McDonald, K. E.,** Low-level ^{239}PuO$_2$ lifespan studies, in *Pacific Northwest Laboratory Annual Report for 1985 to the DOE Office of Energy Research, Part 1, Biomedical Sciences,* National Technical Information Service, Springfield, VA, 1986, PNL 5750: 31–35.
5. **International Commission on Radiological Protection Task Group on Lung Dynamics,** Deposition and retention models for internal dosimetry of the human respiratory tract, *Health. Phys.,* 12, 173, 1966.
6. **Cottier, H., and Zimmermann, A.,** Cellular and tissual aspects of radiation damage with special reference to radionuclides, in *IVth Boettsteiner Colloquium on "Radionuclides for Therapy,"* Schubiger, P. A., and Hasler, P. H., Eds., Roche, Basel, 1986, 22.
7. **Cottier, H., Meister, F., Zimmermann, A., Kraft, R., Burkhardt, A., Gehr, P., and Poretti, G.,** Accumulation of anthracotic particles along lymphatics of the human lung: relevance to "hot spot" formation after inhalation of poorly soluble radionuclides, *Radiat. Environ. Biophys.,* 26, 275, 1987.
8. **Brain, J. D.,** Macrophages damage in relation to the pathogenesis of lung disease, *Environ. Health Perspect.,* 35, 21, 1980.
9. **Burkhardt, A.,** Pathogenesis of pulmonary fibrosis, *Hum. Pathol.,* 17, 971, 1986.
10. **Spencer, H.,** *Pathology of the Lung,* 4th ed., Pergamon Press, Oxford, 1985.
11. **Bair, W. J.,** Current status of the hot particle issue (a review of relevant experimental and theoretical approaches), in *Recueil des Communications,* Vol. 3, IRPA Proceedings, Paris, 1977, 703.
12. **International Commission on Radiological Protection,** *Recommendations of the International Commission on Radiological Protection,* Publication 26, Pergamon Press, Oxford, 1977.

CHAPTER 20

The Fate of Inhaled Toxic Substances

Hans Cottier and Rainer Kraft

It is common usage to differentiate, in the context of an intoxication by a foreign (xenobiotoc) hazardous chemical substance (toxon), three phases or groups of events:

1. The exposition phase, which includes the way of access of the toxon to the body and processes preceding resorption.
2. The toxokinetic phase, which encompasses events such as resorption, invasion, distribution, biotransformations (activation or inactivation), and elimination (evasion) of the toxic material.
3. The toxodynamic phase, which concerns the interaction of the toxon with its target structure(s), i.e., the toxic effect sensu strictiori.

Quite obviously, this classification is to some extent artificial since overlaps are well known, however, it helps to better understand the sequence of the immensely complex processes that are associated with intoxications.[2,3] The magnitude of a toxic effect is primarily determined by the CT (concentration of the toxon × time) product at the target structures. We may also distinguish different types of interactions between the toxon and the target, in particular reversible effects (e.g., those of many medicinal drugs in overdose) and irreversible "chemical lesions," which often result from strong covalent bonds between the xenobiotic substance and target molecules.

As regards inhalation of toxic substances, the physicochemical properties of the latter play a dominant role, in particular the phase (gaseous, fluid, or solid), the solubility in water or in oil ("lipophilia"), with vapors the size distribution of droplets and with dusts that of particles. Inhaled toxic materials in high concentrations may exert a corrosive (necrotizing) action at the site of first contact and thus cause a severe exudative and even hemorrhagic alveolitis, as has been the case in chemical warfare and in accidental severe intoxications by inhalation. If the primary damage of the pulmonary parenchyma does not lead to a peracute respiratory insufficiency, other systemic or organic signs of intoxication may dominate the clinical picture. Gaseous and water-soluble (droplets!) toxons rapidly gain, primarily via diffusion, access to the circulating blood and can thus be transported to other organs. Oil droplets and solid particulates will to a great extent be endocytosed by alveolar macrophages, which they may damage and thus elicit a milder form of alveolitis. Depending on the solubility of this material, systemic effects can also ensue. Lipophilic substances have a good chance to be transported in the blood bound to proteins, i.e. albumin for moderately lipophilic materials and lipoproteins for lipids. Most insoluble particles that have been deposited in the lung parenchyma will be phagocytosed and transported by macrophages to the mucociliary excalator of bronchioles and the tracheobronchial tree, from where they are eliminated by coughing. Only a minute fraction of inhaled insoluble particulate material gains access to pulmonary lymphatics and to regional lymph nodes.

Once the toxon is in the blood stream, its further fate depends on a number of factors.[4] Hydrophilic toxons circulate predominantly in a free form and have a tendency to bind to glycosaminoglycans in the intercellular tissue spaces. Protein-bound lipophilic toxons, however, are known to accumulate in fat tissue, where they may be retained for long periods of time. In such cases, acute starvation with lipolysis can result in the emergence of clinical signs of intoxication. Biotransformation of toxons primarily occurs in the liver: so-called phase I reactions, which encompass bioactivations or inactivations, are often associated with a transformation of lipophilic into polar substances; polar xenobiotics are particularly prone to so-called phase II reactions, i.e., conjugations with compounds such as glucuronic acid, which results in the production of even more hydrophilic substances.

Substances conjugated in the liver can be excreted with the bile, however, hydrolysis and reabsorption of the toxic substances may take place in the gut. As a rule, hydrophilic toxons or conjugates are easily eliminated via the kidneys. Volatile substances quite rapidly leave the body with the expired air. Conversely, repetitive exposure to certain toxons may lead to their accumulation, especially by "physical sequestration" of lipophilic substances in the fat tissue or, less frequently, by "chemical sequestration" through covalent binding to host molecules. Thus, the biological half-life time of toxons varies greatly. It should be added at this point that the host is able to modify, in particular accelerate, the metabolism of toxons, especially via enzyme induction in the liver.

Toxodynamic processes, as a rule, set in with the binding of the toxon to a "specific" or to a "silent" receptor on the target structure, with the effect of a particular, more or less specific, response by the cell or some ill-defined change of the latter, respectively. Depending on the toxicity of the xenobiotic substance and the local CT product, the result of and interaction between the toxon and the target can range from some sort of "adaptation" over "irritation" to "lesions" of cells. The latter can best be recognized as "single cell necrosis." Mechanisms of cell damage include reversible or irreversible enzyme inhibition (e.g., irreversible inhibition of acetylcholinesterase by organophosphates, which are used as pesticides), reaction with SH-groups of proteins (e.g., heavy metals), decoupling of oxidation and phosphorylation (e.g., by dinitrophenol), so-called "lethal synthesis" (e.g., transformation of fluoroacetate into fluorocitrate), removal of metal ions by chelating agents, inhibition of electron transport in the respiratory chain (e.g., by binding of HCN, CO, or H_2S to cytochrome iron), formation of carboxyhemoglobin (by CO) or methemoglobin (by a number of xenobiotica), and other. High grade lipophilic xenobiotics often exert a more or less pronounced narcotic effect, probably via accumulation in the lipid phase of the neuronal cell membrane. Toxons with mutagenic and/or carcinogenic effects usually act via alterations of nucleic acids, especially DNA. The same is true for a number of fetotoxic and/or teratogenic xenobiotics, although necrotizing effects are also important in this group of disorders.

In view of the complex processes alluded to above, it comes as no surprise that the clinical symptoms and signs of intoxications can vary greatly. This is also true for the great number of pesticides presently in use. Acute intoxications by such compounds are not frequent (accidental or suicidal uptake per os, rarely by inhalation), however, long-term and/or repeated exposure to such toxons is quite common and should cause concern. If we consider the clinical pictures observed after acute intoxication by pesticides [e.g., halobenzene derivatives and analogues, benzene hexachloride (lindane), chlorinated camphenes, polycyclic chlorinated insecticides, organic phosphates, carbamates, and others], we are impressed with the large variety of symptoms and signs. If the acute intoxication was due to inhalation of dusts (powders) or sprays, pulmonary disease may be a prominent finding, in addition to more systemic, especially central nervous, neuromuscular, hepatic, gastrointestinal, renal, and/or other alterations. There is only scarce information on long-term effects of these xenobiotics: this is a vast field of research that should not be neglected.

REFERENCES

1. **Ariens, E. J., Mutschler, E., and Simonis, A. M.,** *Allgemeine Toxicologie,* Eine Einführung, Thieme, Stuttgart, 1978.
2. **Bretherick, L.,** *Handbook of Reactive Chemical Hazards,* (an indexed guide to published data), Butterworth, London, 1978.
3. **Mehlman, M. A., Shapiro, R. E., and Blumenthal, H.,** Advances in modern toxicology, in *New Concepts In Safety Evaluation,* Vol. 1, Wiley, New York, 1976.
4. **Stolman, A.,** The absorption, distribution, and excretion of drugs and poisons and their metabolites, *Prog. Chem. Toxicol.,* 5, 1, 1974.

CHAPTER 21

The Pathogenesis of Pulmonary Alveolitis

Hans Cottier and Rainer Kraft

The term "alveolitis" denotes diffuse inflammatory changes of the lungs that do not result from local bacterial growth. Histological types of such disorders range from "luminal phagocytic" or "mural lymphoplasmacellular" over "exudative" to "fibrosing" alveolitis.[3] The pulmonary parenchyma, being the most delicate inner body surface constantly exposed to the potentially hazardous environment, is protected by efficacious defense mechanisms. Most of the inhaled particulate matter becomes impacted in, and removed by, the mucociliary escalator of the tracheobronchial system. The small fraction of inhaled particulates that reaches the pulmonary parenchyma is to a great extent endocytosed by alveolar macrophages, which are mobile scavenger cells and can gain access to the mucociliary transport system in great numbers. Normally, only a minute portion of inhaled insoluble material becomes translocated, via the lymphatics, to regional lymph nodes.[8] Nonspecific and specific, cellular and humoral host defenses can rapidly be mobilized and focused on the lung. Most often, these reactions are successful inasmuch as they succeed in degrading or removing potentially hazardous materials. However, discrete remnants of such reparative processes can accumulate throughout life and may in part be responsible for the well-known fact that pulmonary functions in man deteriorate with age at a faster pace than those of other major organs.

There is a vast array of agents that are able to initiate alveolitis. Exogenous noxious materials include toxic gases (e.g., hyperbaric oxygen, oxides of nitrogen, and ozone), vapors (e.g., droplets containing H_2SO_4 and other), fumes (containing diverse anorganic and/or organic substances), dusts[4] (for instance, those with quartz or asbestos particles), poisons with particular action on the lungs (e.g., paraquat, certain chemotherapeutic agents), ionizing radiation,[1] and viruses. As regards inhalation of organic particles, much depends on the question of whether these have antigenic properties. If they do, they may induce a "hypersensitivity alveolitis," characterized by the appearance of lymphocytes and/or plasma cells.

Particularly severe, acute forms of alveolitis may develop as a consequence of incidents/disorders such as trauma and shock, especially in association with abdominal disease ("septic shock syndrome"), severe burns, or aspiration of gastric contents: in this group of patients, lethality may exceed 50%. The release of endotoxin from Gram-negative bacteria in such conditions appears to be an important pathogenetic factor, however, the question is still open if these lipopolysaccharides damage the lungs more by direct toxic action or by the stimulation of phagocytes to release damaging mediator substances, such as a tumor necrosis factor and interleukin-1.[11]

It appears from comparing the various types of alveolitis and their pathogenesis that the reaction pattern depends more on the severity of damage than on the nature of causative agents. There is good evidence to suggest that the further course of the process is to a great extent determined by the degree to which the alveolar epithelium has been impaired. Death of pneumocytes, in particular the type I cells, results in a more or less pronounced alveolar denudation, which may be likened to an erosive lesion, concomitant with partial disintegration of the underlying basement membrane, acute inflammatory response, and fibrin desposition ("hyaline membranes"). If the blood flow through capillaries becomes reduced, e.g., due to endothelial swelling and/or formation of microthrombi, hypoxic damage may also play a role. At this stage, and with limited initial damage, reconstitution of alveolar structures, by resolution of the exudate and regeneration of pneumocytes via proliferation of type II cells and their transformation into type I cells, still seems possible.[6,7] With even lesser initial damage, the defense reaction may restrict itself to an accumulation essentially of alveolar macrophages, which by themselves may cause some additional tissue injury. If, however, for one reason or another (e.g., secondary bacterial infection) the fibrin-rich exudate within the alveoli is not resolved in due time, organization by granulation tissue ("onion-scale type of fibrosing alveolitis") and focal scar formation can follow. Such events on a small scale are probably quite frequent, but need not result in an appreciable impairment of lung functions. If, however, the pulmonary parenchyma has undergone severe and/or repetitive damage, diffuse fibrosing alveolitis (diffuse pulmonary fibrosis) may develop.[5,9,10] Observations made on humans and in animal experiments indicate that in most—if not all—such cases "cuboid metaplasia" of the alveolar epithelium is a prominent feature indicative of intense pneumocyte regeneration. This in turn can be taken as evidence for a preceding, partial or complete, alveolar denudation with fibrin deposition. Studies on cultured mouse lung have also led to the conclusion that severe alveolar epithelial damage is needed for fibrosing alveolitis to develop. Less important lesions, involving the capillary endothelium only, seem not to be associated with marked fibrosis.

The pathogenesis of fibrosing alveolitis is quite complex.[3] A number of lymphokines and monokines are instrumental in fibroblast chemotaxis, proliferation and synthesis of matrix, as well as of collagen. If this were the only mechanism involved, one would expect a truly diffuse fibrosis to ensue. However, the histological picture of "diffuse" fibrosing alveolitis is characterized by a marked focal pattern, with coexistence of virtually unchanged alveoli and thick fibrotic septa. This patchy type of fibrosis seems to be due to three different processes:[2]

1. Appearance of "onion-scale" type of granulations tissue (see above) with consecutive scarring.
2. Incorporation of murally attached fibrinous masses into the denuded alveolar wall with fibroplasia as a result of such an event.
3. Collapse (atelectatic) induration, a process that has largely been neglected in the international literature and that consists in the coalescence, by fibrin, of neighboring walls of collapsed alveoli.

In this way, "thick" septa can form within a short time, a phenomenon that has in the past often been mistaken for "fibrosis" and that may in fact be reversible.

Since fibrosing alveolitis is more patchy than truly diffuse, compliance of lung tissue cannot be expected to be uniformly reduced. For the same reason, the imbalance between ventilation and perfusion may vary markedly from one small area to another. Furthermore, alveoli not affected by fibrosis are apt to expand, producing the well-known pattern of alternating small foci of emphysematous change and fibrosis. In fact, even a fibrosing lung is subject to constant remodeling.

From a clinical point of view, the recognition of collapse induration as an essential mechanism in the pathogenesis of pulmonary fibrosis should further encourage forced inspiratory breathing exercises to be included in the therapeutic program in such conditions, as some physicians have always done.

REFERENCES

1. **Adamson, I. Y. R., and Bowden, D. H.,** Endothelial injury and repair in radiation-induced pulmonary fibrosis, *Am. J. Pathol.,* 112, 224, 1983.
2. **Burkhardt, A.,** Pathogenesis of pulmonary fibrosis, *Hum. Pathol.,* 17, 971, 1986.
3. **Burkhardt, A., and Cottier, H.,** Cellular events in alveolitis and the evolution of pulmonary fibrosis, *Virchows Arch. B,* 58, 1, 1989.
4. **Cottier, H., Meister, F., Zimmermann, A., Kraft, R., Burkhardt, A., Gehr, P., and Poretti, G.,** Accumulation of anthracotic particles along lymphatics of the human lung: Relevance to "hot spot" formation after inhalation of poorly soluble radionuclides, *Radiat. Environ. Biophys.,* 26, 275, 1987.
5. **Hammar, S. P., Winterbauer, R. H., Bockus, D., Remington, F., and Friedman, S.,** Idiopathic fibrosing alveolitis: A review with emphasis on ultrastructural and immunohistochemical features, *Ultrastruct. Pathol.,* 9, 345, 1985.
6. **Katzenstein, A. L. A., Myers, J. L., and Mazur, M. T.,** Acute interstitial pneumonia. A clinicopathologic, ultrastructural, and cell kinetic study, *Am. J. Surg. Pathol.,* 10, 256, 1986.
7. **Kuroki, Y., Mason, R. J., and Voelker, D. R.,** Alveolar type II cells express a high-affinity receptor for pulmonary surfacant protein A, *Proc. Natl. Acad. Sci. U.S.A.,* 85, 5566, 1988.
8. **Morgan, W. K. C.,** On dust, disability, and death, *Am. Rev. Respir. Dis.,* 134, 639, 1986.
9. **Snyder, G. L.,** Interstitial pulmonary fibrosis, *Chest,* 89 (Suppl.), 115, 1986.
10. **Spencer, H.,** Chronic interstitial pneumonia, in *The Lung,* Liebow, A. A., and Smith, D. E., Eds., Williams & Wilkins, Baltimore, 1968, 134.
11. **Ward, P. A., and Johnson, K. J.,** Adult respiratory distress syndrome and neutropenia, *N. Engl. J. Med.,* 316, 413, 1987.

CHAPTER 22

Environmental Epidemiology

Pietro Comba

FOREWORD

The objects of epidemiology are the study of the occurrence of disease in populations, and the search for associations between the occurrence of disease and its casual factors. Environmental epidemiology is concerned with the patterns of distribution of disease in relation to exposure to environmental factors. The acquisition of knowledge on these matters has, in the first place, a noticeable etiological interest, providing criteria for the implementation of ad hoc preventive action. Further epidemiological studies may contribute to the evaluation of the effectiveness of prevention, and to the planning of optimal resource allocation. It should be stressed, at this point, that many well-established environmental hazards can be directly dealt with, developing appropriate remedial action, without waiting for the completion of further epidemiologic studies. Epidemiological investigations are part of public health strategy, with regard to environmental risk, and in the following presentation it will be assumed that epidemiologists cooperate with other experts, within the framework of a multidisciplinary approach, to study the relationship between environment and health. For an overview of environmental epidemiology, see the contributions by Goldsmith[1] and Gardner.[2]

Two questions are currently considered by epidemiologists in the context of environmental health. The first question is, "Is there an increase in the occurrence of disease X in the area Y, characterized by environmental pollution due to the presence of agents A, B, C?" and the second is, "In case of a positive answer to the first question, can the increased occurrence of the disease be associated with the industrial setting of the area at study? Is there an increased risk for subjects with occupational exposures only, or also for subjects resident in the area?" In order to respond to these questions, a biologic model of the phenomenon is needed, and the feasibility of an epidemiologic investigation has to be evaluated.

STUDY DESIGN

Biological modeling requires knowledge about sources and modes of diffusion of the agents of interest and the patterns of exposure of the population. An estimate of the length of time

Table 1. Steps in Planning an Epidemiologic Study

- Definition of the study population
- Identification of cases
- Ascertainment of exposure
- Evaluation of confounding variables (especially occupational exposures)
- Identification of a suitable reference entity
- Evaluation of study precision

between the beginning of exposure and the commencement of observations is required, and biological plausibility that disease X may be induced by agents A, B, and C needs to be critically considered. There should be particular caution about those studies that were started as a consequence of the detection of a cluster of cases of the disease in a given area. It should be considered, in fact, that the occurrence of rare events may show time-spatial clustering due to random variation alone. The observation of an apparent excess of cancer or congenital malformations in an area does not necessarily mean that an epidemic of environmentally induced disease is taking place; for a review of this topic, see Schulte et al.[3]

When an epidemiological investigation is regarded as being desirable, a feasibility analysis is generally undertaken, in order to asses whether valid information can be collected at a reasonable cost. Table 1 shows the main issues that should be taken into account when planning a study.

Health events that are the object of an epidemiological investigation have to be defined in a standardized way and it is required that their occurrence be exhaustively recorded. Mortality data and incidence data concerning well identified and notified disease fit in well with this definition. Subclinical lesions, whose study would be of the greatest interest, may not be suitable for epidemiologic investigation, in light of the aforementioned points. In the feasibility analysis it is necessary to evaluate not only whether the association between exposure and disease is real, but also if it is detectable, that is, if the existing methods enable us to demonstrate its presence.

Assessment of exposure is a crucial point in study design. A categorization of the study population in terms of exposure requires knowledge about absorption and distribution of the agent in the organism, biotransformation, excretion, and dose to target organs. Furthermore, information on the duration and intensity of exposure is required. Exposure to environmental pollutants affects some sectors of the population more than others. Epidemiological investigations of high risk groups have a twofold meaning: to provide information on the particular health experience of these groups and to provide the initial data for extrapolations aimed at estimating risk associated with low doses. High risk groups may be defined both in terms of acute exposures, mainly accidental releases of chemicals into the environment, and chronic exposures, mainly in the occupational setting.

The study of the health effects of chemical accidents has contributed to the development of toxicological knowledge, especially when large amounts of chemicals have been released, resulting in massive exposures (for a review, see Silano and Comba[4]). Pollution episodes associated with less severe exposures require the development of particularly sensitive methods, in order to detect causal links.[5]

Notwithstanding the favorable evolution of environmental conditions in most occupational settings, workers still represent a section of population that is exposed to toxic agents at significantly higher doses than the general population. Exposure limits have decreased in recent decades in most industrialized countries, but the enforcement of standards for the workplace in Europe and North America has contributed to the selective export of hazardous technologies and materials to developing countries. The study of the occurrence of neoplasms among workers exposed to defined chemical agents has provided much of the currently available evidence of the carcinogenicity of chemicals in humans.[6]

A favorable situation for epidemiologic studies is the presence of a strong association between a rare disease and a specific exposure. On the other hand, the epidemiologic approach is not

Table 2. Methods Commonly Adopted in Epidemiologic Investigation

- Geographic study
- Case-control study
- Cross-sectional study
- Cohort study

particularly suitable for the study of weak associations, especially if the disease of interest is relatively common, and the association can be affected by confounding variables. Low relative risks (orientatively, less than 1.5) are often encountered in environmental epidemiology, especially when random misclassification of exposure occurs, resulting in underestimation of the relative risk itself. The presence of multiple factors of potential etiological meaning, and the operation of confounding variables both contribute to the loss of sensitivity of the epidemiological approach.

Most epidemiological investigations are comparative, dealing with the incidence of disease in various exposure categories. Thus, the choice of an appropriate reference entity is crucial, since it provides the nuisance parameter of interest, that is the incidence of disease among the unexposed or the prevalence of exposure among healthy subjects.

While aspects of validity so far discussed have to do with the control of systematic error (bias), the issue of precision relates to control of the statistical error associated with random variation. This is essentially a matter of adequacy of sample size.

In light of these points, it can be agreed that a comprehensive feasibility study is needed before any investigation is started.[7]

METHODOLOGIC APPROACHES

The choice of the optimal procedure depends on the specific problem and of the availability of information. Table 2 shows the procedures most commonly adopted in epidemiology. A description of the various types of study goes beyond the objectives of this paper, and readers are referred to textbooks of epidemiology. What should be stressed in this context is that while geographic studies utilize routinely collected data, the other approaches require ad hoc collection of information.

Geographic, or "ecological," studies are based on the contrast of mortality data from areas that contrast with another in regard to certain environmental factors; in those countries served by cancer registries, geographic studies may utilize incidence data, as well. Mapping the occurrence of disease can be helpful in order to generate etiological hypotheses that will subsequently be tested within the framework of analytical, or "individual," studies. Geographic studies are in fact subject to the "ecological fallacy": even if disease rates are higher in areas with higher exposures, it is not certain that cases will occur at a higher rate among exposed, rather than nonexposed, individuals. Analytical studies may provide more convincing evidence, since the establishment of both exposure status and health outcome takes place at individual level (for a comparison of the results of ecological and individual studies, see Richardson et al.[8]). It should, at the same time, be stressed that analytical studies tend to be costly and time-consuming, and they necessarily address a limited number of questions. It is thus desirable to develop both geographic and analytical studies, in order to optimize the allocation of resources.

In order to clarify this discussion, four examples of studies are now briefly presented.

Geographic approach:[9] Mortality from malignant neoplasms of the pleura closely reflects the incidence of pleural mesothelioma, a disease that is specifically induced by exposure to asbestos. Annual mortality rates in Italy are approximately 1.5×10^{-5} in males and 0.8×10^{-5} in females. The geographic distribution of pleural neoplasms has shown that the higher rates occur in provinces characterized by the presence of the naval industry and the asbestos-cement manufacture.

Cross-sectional approach:[10] A health survey has been conducted in the Latium Region in order to investigate the occurrence of respiratory problems in children, in relation to environmental pollution. School children from different areas performed respiratory tests, and a questionnaire was administered to their parents. The contrast between a town where a major thermoelectric power plant is located and a relatively unpolluted neighboring area demonstrated a significantly increased occurrence of respiratory disease among children living in the vicinity of the power plant.

Cohort approach:[11] A mortality study has been performed in the Seveso area. The cohort included all subjects who were resident in the affected area in 1976 and in the subsequent years, to 1986. The population living in the surrounding municipalities was taken as reference entity. An increased mortality from cardiovascular causes, especially in the first years after the accident, has been interpreted as being stress-related. Increases in some neoplasms, namely lymphohematopoietic malignancies and soft tissue sarcomas, have been detected and associated with the action of TCDD.

Case-control approach:[12] An increased risk of lung cancer, in relation to the general population, was investigated among men living in the surroundings of an arsenic-emitting copper smelter in Northern Sweden. Information on residency, occupation, and smoking habits was collected for subjects who died from lung cancer (cases) and from other causes (controls). A doubled risk of lung cancer was detected for the subjects living within 20 km of the smelter, contrasted to those staying in other parts of the county. A three- and fourfold increase in risk was observed, respectively, for smelter workers and miners.

CONCLUDING REMARKS

Environmental epidemiology is a discipline at the boundary between ecology and the health sciences, and its range of applications is rapidly growing, as a consequence of the broadening of our awareness of environmental problems.

Environmental epidemiological studies require a multidisciplinary approach, especially when dealing with the relatively unknown health effects of complex exposures, as those resulting from hazardous sites.[13,14]

The study of high-risk groups, such as those experiencing occupational exposures, is of the greatest interest in environmental epidemiology, since it provides the initial data in terms of evaluation of risk for the general population.

Various epidemiological methods may be adopted when studying the role of environmental risk factors; assessment of exposure is the crucial point with regard to validity. Various issues associated with environmental epidemiology, such as confidentiality of information, dissemination of results, and subsequent decision-making processes in terms of public health, raise ethical problems for health authorities, researchers, and the general public. Such aspects deserve, in the future, a much greater concern than has been given them so far.

REFERENCES

1. **Goldsmith, J. R.,** Improving the prospects for environmental epidemiology, *Arch. Environ. Health,* 43, 69–74, 1988.
2. **Gardner, M. J.,** Epidemiological studies of environmental exposure and specific diseases, *Arch. Environ. Health,* 43, 102–108, 1988.
3. **Schulte, P. A., Ehrenberg, R. L., and Singal M.,** Investigation of occupational cancer clusters: Theory and practice, *Am. J. Publ. Health,* 77, 52–56, 1987.
4. **Silano, V., and Comba, P.,** Chemical accidents: Long-term health issues, in *Methods for Assessing and Reducing Injury from Chemical Accidents,* Bordeau, P. and Green, G., Eds., John Wiley, New York, 1989, 211–222.

5. **Health, C. W.,** Use of epidemiologic information in pollution episode management, *Arch. Environ. Health,* 43, 75–80, 1988.
6. **IARC 1987,** Monographs on the evaluation of carcinogenic risk to humans, (Suppl. 7), 1–440, 1987.
7. **Bender, A. P., Williams, A. N., Sprafka, J. M., Mandel, J. S., and Straub, C. P.,** Usefulness of comprehensive feasibility studies in environmental epidemiology investigations: A case study in Minnesota, *Am. J. Publ. Health,* 78, 287–290, 1988.
8. **Richardson, S., Stucker, I., and Hemon, D.,** Comparison of relative risk obtained in ecological and individual studies: Some methodological considerations, *Int. J. Epidemiol.,* 16, 111–120, 1987.
9. **Bruno, C., Comba, P., De Santis, M., and Malchiodi, F.,** Mortalita per tumore maligno della pleura in Italia: 1980–83, *Rapporti ISTISAN,* 24, 1988.
10. **Regione Lazio-O. E. R.,** Rapporto preliminare dell'indagine "La salute dei bambini e l'ambiente," Osservatorio Epidemiologico Regionale, Rome, 1988.
11. **Bertazzi, P. A., Zocchetti, C., Pesatori, A. C., Guercilena, S., Sanarico, M., and Radice, L.,** Ten-year mortality study of the population involved in the Seveso incident in 1976, *Am. J. Epidemiol.,* 129, 1187–1200, 1989.
12. **Pershagen, G.,** Lung cancer mortality among men living near an arsenic-emitting smelter, *Am. J. Epidemiol.,* 122, 684–694, 1985.
13. **Grisham, J. W., Ed.,** *Health Aspects of the Disposal of Waste Chemicals,* Pergamon Press, New York, 1986.
14. **Marsh, G. M., and Caplan, R. J.,** Evaluating health effects of exposures at hazardous waste sites, in *Health Effects from Hazardous Waste Sites,* Andelman and Underhill, Eds., Lewis Publishers, Chelsea, MI, 1987, 3–80.

CHAPTER 23

Exposure to Phenoxy Herbicides and Chlorinated Dioxins and Cancer Risks: An Inconsistent Pattern of Facts and Frauds?

Olav Axelson

CANCER RISKS FROM PHENOXY HERBICIDES?

In the late 1970s and early 1980s indications were obtained from clinical observations and epidemiologic studies in Sweden that exposure to phenoxy herbicides and chlorophenols could play a role in the development of soft-tissue sarcomas and lymphomas.[1-3] Risk ratios in the range of 3 to 6 were obtained for these disorders and a combined exposure to pesticides and solvents resulted in a more or less multiplicative effect.[4] A somewhat increased risk was also seen for nasopharyngeal cancer and phenoxy acid exposure, although chlorophenol exposure in sawmill work appeared as the main risk factor, also after adjustment for wood dust exposure.[5] Both the phenoxy herbicides and the chlorophenols might have contained chlorinated dibenzodioxins as contaminants.

The economic importance of phenoxy acids in agriculture and their questionable role as defoliants in the Vietnam War may explain why there was considerable, but rather imprecise, and not clearly justified critiques of these various studies. In particular, the critiques addressed the possibility of recall bias.[6,7] Measures had been undertaken to avoid and reveal such bias, however, including the application of a special technique in this respect.[8]

Further Studies

In particular, the indicated relation between phenoxy acids and soft-tissue sarcoma and lymphoma prompted a number of studies in many countries, some of which were adequately designed to further evaluate the possibility of a carcinogenic effect of the phenoxy acids, whereas others had rather poor data on exposure. The results of these studies, including Vietnam veterans, have been summarized in several reviews, which also include some aspects on exposure.[9-18] These reviews also reflect the continuing debate caused by some puzzling inconsistencies between the results obtained in the different studies.

Most of the studies of phenoxy acid exposure and soft-tissue sarcoma or lymphoma have been of the case-control design. With regard to non-Hodgkin's lymphoma and phenoxy herbicide exposure, the risk ratios have been ranging from slightly more than one to almost five, when particular subpopulations were considered, whereas no clear excess has been found for Hodgkin's disease.[19,20] In particular, some early studies from New Zealand appeared contradictory to the Swedish findings, whereas more recent observations on forestry workers from this country seem to be mainly supportive of an effect both for soft-tissue sarcoma and non-Hodgkin's lymphoma as well as for nasopharyngeal cancer.[21] A cohort study on phenoxy herbicide producers from Denmark may be mentioned also, since five cases of soft-tissue sarcoma were observed versus 1.8 expected.[22]

Recent Case-Control Studies

In the late 1980s and early 1990s, several more studies on the effects of phenoxy acids have been reported.[23–25] Hence, Hardell and co-workers again found a rather strong effect with regard to phenoxy acid exposure and soft-tissue sarcomas with risk ratios of up to about three.[24,25] A less consistent effect was seen for chlorophenol exposure with an increased risk present in one study but absent in the other.

A further study on soft-tissue sarcoma in the southeast of Sweden[26] and still another study with a somewhat complex design, involving cases from Finland as well as Sweden,[27] have shown increased odds ratios for railroad and construction work and also for gardening, farming, or forestry work. Also the relationship between lymphomas and phenoxy herbicide exposure has appeared in another Swedish study.[28]

Recent Cohort Studies

An important report concerns a cohort study of mortality among 5172 workers at 12 plants in the United States that produced chemicals contaminated with 2,3,7,8-tetrachlorodibenzo-p-dioxin (TCDD).[29] In a subcohort of 1520 workers with exposure for 1 year or more, and with 20 years or more of latency, there were three soft-tissue sarcoma deaths and the SMR was 922 with lower 95% confidence limit at 190. Mortality from respiratory cancer, as well as from all cancers combined, was also significantly elevated, but no clear excess appeared for lymphomas.

Similarly a multinational European study including 18,910 production workers and sprayers of phenoxy herbicides showed an excess of soft-tissue sarcomas, in total four cases.[30] Considering a latency time of 10 to 19 years since first exposure, the risk was sixfold and significant; for the subgroup of sprayers the risk was ninefold. There also appeared to be some excess risk for cancers of the testicle, thyroid, other endocrine glands, nose, and nasal cavity, but the numbers were small. Lymphomas did not appear in any excess, however.

Comments on Exposure Assessment

It may also be noted that proper exposure data are necessary to permit the observation of an effect. This is well illustrated by the fact that using occupational titles only, the high risk ratios seen in the original Swedish case-control studies of soft-tissue sarcoma and lymphoma were reduced to 1.4 or less.[18] Poor information on exposure, as characteristic of many studies, may well explain some of the inconsistent findings and also why a record-linkage study from Sweden using census data on occupation could not confirm any cancer risk.[23] Another aspect is that the relative rareness of especially the soft-tissue sarcomas would necessarily lead to small expected numbers of cases in any population with exposure so that lack of (an excess of) these rather rare tumours in a study would merely mean a nonpositive and inconclusive finding rather than an indication of no effect.[31]

MECHANISMS

There is only weak support for a carcinogenic effect of the phenoxy acids from experimental data whereas TCDD exerts a promoting effect in animal studies.[32,33] However, the peroxisome proliferation induced by chlorinated phenoxy compounds may be recalled, and such effects have also been seen for the lipid-lowering drug clofibrate, a chlorinated phenoxypropionic acid derivative.[34] This effect on the peroxisomes is associated with increased free radical formation, as also connected with cancer promotion.

For some compounds a relationship has been found between the induction of free radical formation, cancer-promoting activity, and hindering of intercellular communication; the latter effect has also been observed for phenoxy herbicides[35] as well as for some other pesticides.[36] It is certainly also of great interest in this context that clofibrate in a large-scale randomized trial was found to increase the occurrence of cancer among those treated with the drug.[37]

Possible Interactions and Other Exposures

Interaction phenomena of some sort may be part of the explanation of why there are differences in risk between the various studies. In particular, the combination of exposure to solvents or other pesticides and phenoxy acids may carry a risk. The rather strong effect of combined exposure to meat work and fencing in the data on non-Hodgkin's lymphoma from New Zealand is certainly of great interest in this context as well, and may include a viral factor, although the main chemical exposure might have been to chlorophenols rather than to phenoxy acids.[38]

The appearance of soft-tissue sarcomas in excess in relation to insecticide use in Kansas as well as among gardeners in Sweden and Denmark is another puzzling observation that, in view of the results from other studies, may indicate either unknown exposure to dioxins or to some other important factor closely associated with the use of pesticides.[27,39,40] Gardeners are certainly likely to be exposed to phenoxy acids to some extent, but especially to various insecticides, among which is nicotine, as discussed as an agent of interest in connection with the findings from Kansas.[39]

The possibility of a generally increased cancer risk from exposure to phenoxy acids might be considered in view of the observations from the clofibrate trial as well as with regard to some, but not all, cohort studies with mixed exposure as reported in the past, e.g., in railroad workers and other groups engaged in spraying operations.[20,37,41,42] Also the elevated risk ratios reported for various cancer types among forestry workers in New Zealand might fit into this pattern.[21] It is unclear though, how these observations may relate to the generally increased cancer risk seen in connection with TCDD exposure in two of the aforementioned studies.[29,43]

TCDD

A general carcinogenic effect in humans from TCDD exposure would certainly be consistent with the experimental data suggesting a promoting effect of this compound.[32,33] 2,3,7,8-Tetrachlorodibenzo-p-dioxin (TCDD) and two congeners of hexachlorinated dibenzodioxins have been shown to be carcinogenic in experimental animals.[44,45] From the epidemiologic data available at this time, it seems as if phenoxy herbicides could exert an effect as well. On the other hand, both TCDD and other dioxins could be responsible, as these impurities have occurred from time to time in phenoxy herbicide preparations.[46] However, some manipulation of data and misinterpretation of results have gone into the literature, which might have confused the question of a carcinogenic effect from phenoxy herbicides and their dioxin contaminations; therefore, this issue may deserve a comment.

THE DEBATE ABOUT CARCINOGENICITY OF PHENOXY ACIDS

A rather intense discussion about the potential cancer hazard from phenoxy herbicides started with the first reports in this respect and has continued for more than a decade. By 1991, there was certainly increasing evidence of a carcinogenic effect both with regard to soft-tissue sarcoma and lymphoma, but it is puzzling why studies in different populations so often have been only partially consistent as clearly confirming an effect only for one or the other type of tumors but so rarely for both.

Another aspect contributing to the inconsistent pattern of results from studies in this field seems to be that earlier so-called negative studies on dioxin-exposed populations have turned out to be questionable as indicated elsewhere.[47] Others have interpreted their data as negative in spite of an excess of one of the two types of malignancy of interest, i.e., lymphomas or of all malignancies, as in the BASF cohort.[13,43,48] A further aspect relates to the use of phenoxy acids in the Vietnam War, which might have caused illnesses in the Vietnamese population and has also resulted in compensation claims from veterans suffering from cancer and other disorders.[11]

Secondary Manipulation of Data and Distorted Information

This latter aspect has become even more intricate and has contributed to the confusion as a result of manipulations and falsifications of figures and facts from some of the published studies by a Royal Commission on the Use and Effects of Chemical Agents on Australian Personnel in Vietnam.[49] Instead of making an independent evaluation, this Australian commission simply copied several hundred pages of a submission, the conclusions included, that had been compiled by counsellors of Monsanto Australia.[50,51] Large parts of this questionable material have now also been published as a book chapter by O'Keefe, one of the Monsanto counsellors in the hearings before the Australian Royal Commission.[52] This chapter is not only misleading, but may cause concerns about how some chemical companies may deal with scientific matters.[47,52]

Surprisingly, the view taken by the Commission was even supported by professor Richard Doll in a polite letter from 1985 to The Honorable Mr. Justice Phillip Evatt, DSC, LLB, the Commissioner. The letter, which might have been written on request, has by now appeared in the literature.[53] A remarkable part of it may be quoted, namely that "His [Dr. Hardell's] conclusions cannot be sustained and in my opinion, his work should no longer be cited as scientific evidence. It is clear, too, from your review of the published evidence relating to 2,4-D and 2,4,5-T (the phenoxy herbicides in question) that there is no reason to suppose that they are carcinogenic in laboratory animals and that even TCDD (dioxin), which has been postulated to be a dangerous contaminant of the herbicides, is at the most, only weakly and inconsistently carcinogenic in animal experiments. I am sorry only that your review has had to be published in book form and not in a scientific journal, as books are so much less readily available to the majority of scientists. I am sure, however, that it will be widely quoted and that it will come to be regarded as the definitive work on the subject."

In the retrospect, it seems as if even the most experienced epidemiologist can easily go wrong when using second-hand information on scientific work. However, Professor Doll, at the time probably not aware of the manipulation of facts by Monsanto and the Commission, has not followed his own advice, since more recently he cites some of the Hardell papers in discussing the increasing mortality from non-Hodgkin's lymphoma and soft-tissue sarcoma.[54]

Nevertheless, the opinion expressed by Professor Doll in his letter, that the conclusions by the Commission should be made "readily available" and "widely quoted," has been adopted by some authors in spite of the fact that the Department of Veteran's Affairs in Australia has admitted a great number of serious errors in the Final Report by the commission.[55,56]

Epidemiologic Malpractice or Fraud?

However, not only the secondary manipulation or, at best, misinterpretation of published material may have contributed to the seemingly inconsistent findings about the carcinogenicity of phenoxy acids and chlorinated dioxins but an even more remarkable scenario seems now underway in this respect. Hence, published reports involving Monsanto employees have now been seriously questioned in a private litigation against the Monsanto Corporation and the head physician of the company had to concede that the results in two studies were biased due to misclassification of exposure.[57-59] Especially noteworthy is that cases of soft-tissue sarcoma and malignant lymphoma were admitted to have been wrongly excluded from the exposed group. At the time of this study, these malignant tumor types had already been associated with exposure to chlorinated phenols and phenoxy herbicides in the Swedish studies.

EPILOGUE

Based on several of the newer epidemiologic studies, as confirming earlier observations, a conclusion for practical purposes could be that some commercially available phenoxy herbicides have been carcinogenic, but probably only under certain circumstances. It may well be that the risk of soft-tissue sarcomas is mainly related to chlorinated dibenzodioxins, as especially indicated by one study, whereas the lymphomas might depend more specifically on phenoxy acids, particularly in combinations with other agents, such as other pesticides, solvents, and probably also some other chemicals.

The inconsistencies that have appeared between some of the studies from various parts of the world might perhaps be thought of as reflecting the effect of the differently combined, more or less complex exposures rather than stemming from bias. Also some ethnic differences might be involved as indicated in a study from Canada on non-Hodgkin's lymphoma and use of phenoxy herbicides.[60]

In summary, and for the time being, it may be concluded that there now remains less contradiction in the available observations on cancer and exposure to phenoxy acids and chlorinated dioxins, since the data provided to refute a carcinogenic risk to humans from chlorinated dioxins have been revealed as highly questionable if not fraudulent.

REFERENCES

1. **Hardell, L., and Sandström, A.,** Case-control study: Soft-tissue sarcomas and exposure to phenoxyacetic acids or chlorophenols, *Br. J. Cancer,* 39, 711, 1979.
2. **Eriksson, M., Hardell, L., Berg, N. O., Möller, T., and Axelson, O.,** Soft tissue sarcomas and exposure to chemical substances: A case-referent study, *Br. J. Industr. Med.,* 38, 27, 1981.
3. **Hardell, L., Eriksson, M., Lenner, P., and Lundgren, E.,** Malignant lymphoma and exposure to chemicals, especially organic solvents, chlorophenols and phenoxy acids: A case-control study, *Br. J. Cancer,* 43, 169, 1981.
4. **Axelson, O.,** Occupational cancer: Interaction of factors, in *Primary Prevention of Cancer,* Eylenbosch, W. J., Van Larebeke N., and Depoorter, A. M., Eds., Raven Press, New York, 1988.
5. **Hardell, L., Johansson, B., and Axelson, O.,** Epidemiological study of nasal and nasopharyngeal cancer and their relation to phenoxy acid or chlorophenol exposure, *Am. J. Ind. Med.,* 3, 247, 1982.
6. **Coggon, D., and Acheson, E. D.,** Do phenoxy herbicides cause cancer in man?, *Lancet,* 1, 1057, 1982.
7. **Colton, T.,** Herbicide exposure and cancer, *J. Am. Med. Assoc.,* 256, 1176, 1986.
8. **Axelson, O.,** A note on observational bias in case-referent studies in occupational health epidemiology, *Scand. J. Work Environ. Health,* 6, 80, 1980.
9. **Axelson, O.,** Pesticides and cancer risks in agriculture, *Med. Oncol. Tumour Pharmacother.,* 4, 207, 1987.

10. **Blair, A., Malker, H., Cantor, K. P., Burmeister, L., and Wiklund, K.,** Cancer among farmers. A review, *Scand. J. Work Environ. Health,* 11, 397, 1985.
11. **Sterling, T. D., and Arundel, A.,** Review of recent Vietnamese studies on the carcinogenic and teratogenic effects of phenoxy herbicide exposure, *Int. J. Health Serv.,* 16, 265, 1986.
12. **Sterling, T. D., and Arundel, A.,** Health effects of phenoxy herbicides. A review, *Scand. J. Work Environ. Health,* 12, 161, 1986.
13. **Bond, G. B., Bodner, K. M., and Cook, R. K.,** Phenoxy herbicides and cancer: Insufficient epidemiologic evidence for a causal relationship, *Fund. Appl. Toxicol.,* 12, 172, 1989.
14. **Sharp, D. S., Eskenazi, B., Harrison, R., Callas, P., and Smith, A. H.,** Delayed health hazards of pesticide exposure, *Annu. Rev. Public Health,* 7, 441, 1986.
15. **Lilienfeld, D. E., and Gallo, M. A.,** 2,4-D, 2,4,5-T, and 2,3,7,8-TCDD: An overview, *Epidemiol. Rev.,* 11, 28, 1989.
16. **Johnson, C. C., Feingold, M., and Tilley, B.,** A meta-analysis of exposure to phenoxy acid herbicides and chlorophenols in relation to risk of soft tissue sarcoma, *Int. Arch. Occup. Environ. Health,* 62, 513, 1990.
17. **Boyle, C. A., Decoufle, P., and O'Brian, T. R.,** Long term health consequences of military service in Vietnam, *Epidemiol. Rev.,* 11, 1, 1989.
18. **Hardell, L., and Axelson, O.,** Phenoxyherbicides and other pesticides in the etiology of cancer: Some comments on Swedish experiences, in *Cancer Prevention. Strategies in the Workplace,* Becker, C. E., and Coye, M. J., Eds., Hemisphere Publishing Corporation, Washington, D.C., 1986.
19. **Woods, J. S., Polissar, L., Severson, R. K., Heuser, L. S., and Kulander, B. G.,** Soft tissue sarcoma and non-Hodgkin's lymphoma in relation to phenoxy herbicide and chlorinated phenol exposure in western Washington, *J. Natl. Cancer Inst.,* 78, 899, 1987.
20. **Hoar, S. K., Blair, A., Holmes, F. F., Boysen, C. D., Robel, R. J., Hoover, R., and Fraumeni, J. F., Jr.,** Agricultural herbicide use and risk of lymphoma and soft-tissue sarcoma, *J. Am. Med. Assoc.,* 256, 1141, 1986.
21. **Reif, J., Pearce, N., Kawachi, I., and Fraser, J.,** Soft-tissue sarcoma, non-Hodgkin's lymphoma and other cancers in New Zealand forestry workers, *Int. J. Cancer,* 43, 49, 1989.
22. **Lynge, E.,** A follow-up study of cancer incidence among workers in manufacture of phenoxy herbicides in Denmark, *Br. J. Cancer,* 52, 259, 1985.
23. **Wiklund, K., and Holm, L-E.,** Soft tissue sarcoma risk in Swedish agricultural and forestry workers, *J. Natl. Cancer Inst.,* 76, 229, 1986.
24. **Eriksson, M., Hardell, L., and Adami, H-O.,** Exposure to dioxins as a risk factor for soft tissue sarcoma: A population-based case-control study, *J. Natl. Cancer Inst.,* 82, 486, 1990.
25. **Wingren, G., Fredriksson, M., Noorlind Brage, H., Nordenskjöld, and Axelson, O.,** Soft tissue sarcoma and occupational exposures, *Cancer,* 66, 806, 1990.
26. **Hardell, L., and Eriksson, M.,** The association between soft tissue sarcomas and exposure to phenoxyacetic acids. A new case-referent study, *Cancer,* 62, 652, 1988.
27. **Olsson, H., Alvegard, T. A., Harkonen, H., Brandt, L., and Möller, T.,** Epidemiological studies of high-grade soft tissue sarcoma within the framework of a randomised trial in Scandinavia, in *Management and Prognosis of Patients with High-Grade Soft Tissue Sarcomas. An Evaluation of a Scandinavian Joint Care Program,* Alvegard, T. A., Ed., University of Lund, Lund, 1989.
28. **Persson, B., Dahlander, A-M., Fredriksson, M., Noorlind Brage, H., Ohlson, C-G., and Axelson, O.,** Malignant lymphomas and occupational exposures. *Br. J. Ind. Med.,* 46, 516–520, 1989.
29. **Fingerhut, M. A., Halperin, W. E., Marlow, D. A., Piacitelli, L. A., Honchar, P. A., Sweeney, M. H., Greife, A. L., Dill, P. A., Steenland, K., and Suruda, A. J.,** Cancer mortality in workers exposed to 2,3,7,8-tetrachlorodibenzo-p-dioxin, *N. Engl. J. Med.,* 324, 212, 1991.
30. **Saracci, R., Kogevinas, M., L'Abbe, K., Bertazzi, P. A., Bueno de Mesquita, H. B., Coggon, D., Green, L. M., Kauppinen, T., L'Abbé, K. A., Littorin, M., Lynge, E., Mathews, J. D., Neuberger, M., Osman, J., Pearce, N., and Winkelman, R.,** Cancer mortality in workers exposed to chlorophenoxy herbicides and chlorophenols, *Lancet,* 338, 1027, 1991.
31. **Ahlbom, A., Axelson, O., Hansen, E. S., Hogstedt, C., Jensen, U. J., and Olsen, J.,** Interpretation of "negative" studies in occupational epidemiology, *Scand. J. Work Environ. Health,* 16, 153, 1990.
32. **Poland, A., Falen, D., and Glover, E.,** Tumour promotion by TCDD in skin of HRS/J hairless mice, *Nature (London),* 300, 271, 1982.

33. **Pitot, H. C., Goldsworthy, T., Campbell, H. A., and Poland, A.,** Quantitative evaluation of the promotion of 2,3,7,8-tetrachlorodibenzo-*p*-dioxin of hepatocarcinogenesis from diethylnitrosamine, *Cancer Res.,* 40, 3616, 1980.
34. **Vainio, H., Nickels, J., and Linnainmaa, K.,** Phenoxy herbicides cause peroxisome proliferation in Chinese hamsters, *Scand. J. Work Environ. Health,* 8, 70, 1982.
35. **Rubinstein, C., Jone, C., Trosko, J. E., and Chang, C. C.,** Inhibition of intercellular communication in cultures of Chinese hamster V79 cells by 2,4-dichlorophenoxy acetic acid and 2,4,5-trichloroacetic acid, *Fund. Appl. Toxiciol.,* 4, 731, 1984.
36. **Werngard, L.,** *Studies on the Chemically Induced Inhibition of Intercellular Communication,* Karolinska Institutet, Stockholm, 1988.
37. **Committee of principal investigators,** WHO cooperative trial on primary prevention of ischaemic heart disease using clofibrate to lower serum cholesterol: Mortality follow up, *Lancet,* ii, 379, 1980.
38. **Pearce, N. E., Smith, A. H., Howard, J. K., Sheppard, R. A., Giles, H. J., and Teague, C. A.,** Non-Hodgkin's lymphoma and exposure to phenoxy herbicides, chlorophenols, fencing work and meat works employment. A case-control study, *Br. J. Ind. Med.,* 43, 75, 1986.
39. **Hoar Zahm, S., Blair, A., Holmes, F. F., Boysen, C. D., and Robel, R. J.,** A case-referent study of soft tissue sarcoma and Hodgkin's disease. Farming and insecticide use, *Scand. J. Work Environ. Health,* 14, 224, 1988.
40. **Lander, F., Hansen, E. S., and Hasle, H.,** A cohort study on cancer incidence among gardeners, 23rd Congress on Occupational Health, Montreal, Canada, 22–28 September, (Abstracts), 1990, 601.
41. **Axelson, O., Sundell, L., Andersson, K., Edling, C., Hogstedt, C., and Kling, H.,** Herbicide exposure and tumour mortality. An updated epidemiologic investigation on Swedish railroad workers, *Scand. J. Work Environ. Health,* 6, 73, 1980.
42. **Riihimäki, V., Asp, S., Pukkala, E., and Hernberg, S.,** Mortality of 2,4-dichlorophenoxyacetic acid and 2,4,5-trichlorophenoxyacetic acid herbicide applicators in Finland, *Chemosphere,* 12, 779, 1983.
43. **Zober, A., Messerer, P., and Huber, P.,** Thirty-four-year mortality follow-up of BASF employees exposed to 2,3,7,8-TCDD after the 1953 accident, *Int. Arch. Occup. Environ. Health,* 62, 139, 1990.
44. **Lilienfeld, D. E., and Gallo, M. A.,** 2,4-D, 2,4,5-T, and 2,3,7,8-TCDD: An overview, *Epidemiol. Rev.,* 11, 28, 1989.
45. **National Toxicology Program, National Cancer Institute,** NIH bioassay of a mixture of 1,2,3,6,7,8- and 1,2,3,7,8,9-hexachlorodibenzo-*p*-dioxins for carcinogenicity (gavage study). Natl. Toxicol. Program Tech. Rep. Ser. No. 198, DHHS Publ. No. (NIH)80–198, Natl. Toxicol. Program, Research Triangle Park, NC, 1980.
46. **Hagenmaier, H.,** Determination of 2,3,7,8-tetrachlorodibenzo-*p*-dioxin in commercial chlorophenols and related products, *Fresenius Z. Anal. Chem.,* 325, 603, 1986.
47. **Hardell, L., and Eriksson, M.,** The association between cancer mortality and dioxin exposure: a comment on the hazard of repetition of epidemiological misinterpretation, *Am. J. Ind. Med.,* 19, 547, 1991.
48. **Rohleder, F.,** Dioxins and cancer mortality—reanalysis of the BASF-cohort, in *Occupational Epidemiology,* Sakurai, H. et al., Eds., Elsevier Science Publishers, Amsterdam, 1990.
49. **Royal Commission on the Use and Effects of Chemical Agents on Australian Personnel in Vietnam,** Final Report, Vol. 4, Cancer, Australian Government Publishing Service, Canberra, 1985.
50. **Axelson, O., and Hardell, L.,** Storm in a cup of 2,4,5-T, *Med. J. Aust.,* 144, 612, 1986.
51. **Hardell, L.,** Letter to the Editor: Serious errors in new volume of on Agent Orange and dioxin, *Am. J. Ind. Med.,* 17, 261, 1990.
52. **O'Keefe, B.,** Soft tissue sarcoma: Law, science and logic, an Australian perspective, in *Agent Orange and Its Associated Dioxin: Assessment of a Controversy,* Young, A. L., and Reggiani, G. M., Eds., Elsevier Science Publishers, Amsterdam, 1988.
53. **Young, A. L., and Reggiani, G. M., Eds.,** *Agent Orange and its Associated Dioxin: Assessment of a Controversy,* Elsevier Science Publishers, Amsterdam, 1988.
54. **Doll, R.,** Are we winning the fight against cancer? An epidemiological assessment, *Eur. J. Cancer,* 26, 500, 1990.
55. **Comments by Department of Veterans' Affairs Attachment C, Woden,** Australia, May 19, 1988 (unpublished).

56. **Hardell, L.,** Letter to the Editor: Agent Orange controversy: A response, *Am. J. Ind. Med.,* 19, 403, 1991.
57. **Kemner, et al.,** versus Monsanto Company, Civil No. 80-L-970, Circuit Ct., St Clair County, Illinois. Report of Proceedings. Testimony of Dr. Roush, 1985, July 8, pp. 1–147; July 9, pp. 1–137; July 22, pp. 1–76 (unpublished).
58. **Zack, J. A., and Gaffey, W. R.,** A mortality study of workers employed at the Monsanto company plant in Nitro, West Virginia, *Environ. Sci. Res.,* 26, 575, 1983.
59. **Suskind, R. R., and Hertzberg, V. S.,** Human health effects of 2,4,5-T and its contaminants, *J. Am. Med. Assoc.,* 251, 2372, 1984.
60. **Wigle, D. T., Semenciw, R. M., Wilkins, K., Riedel, D., Ritter, L., Morrison, H. I., and Mao, Y.,** Mortality study of Canadian male farm operators: Non-Hodgkin's lymphoma mortality and agricultural practices in Saskatchewan, *J. Natl. Cancer Inst.,* 82, 575, 1990.

CHAPTER 24

Converging Epidemiologic Findings on Radon in Mines and Homes as a Risk of Lung Cancer

Olav Axelson

RADON IN MINES AND HOMES

It has long been known that radon, a naturally occurring radioactive gas, can reach high concentrations in uranium and other mines and cause lung cancer in miners. It has now also been found in many countries that high levels of radon occur in many dwellings and possibly impose a health hazard to the general population. Changes in the construction and ventilation of houses in some parts of the world may even have caused the radon levels to increase over time, which to some extent may have contributed to the increasing lung cancer rates in this century. Furthermore, epidemiological observations in the past also suggest an influence on lung cancer from widespread factors other than smoking. For example, the urban–rural difference in lung cancer rates has been difficult to fully explain by differences in smoking.[1] The influence of immigration on lung cancer morbidity is another matter that has not yet been quite well understood; nor is the reason clear for the varying lung cancer rates throughout the world, even after allowing for smoking.[2-4]

A reduction of indoor radon might decrease the incidence of lung cancer, as a supplement to decreasing smoking. A pertinent question is whether the risk estimates of lung cancer in miners would be applicable also to the general population with exposure to indoor radon. A brief review will be provided about the existing knowledge on exposure to radon and lung cancer as well as a comparison of some risk estimates for exposure in mines and in houses.

Some Characteristics of Radon

Radon (more precisely, radon-222) is a noble gas that is created by the decay of uranium through radium. Radon itself decays into a series of radioactive isotopes of polonium, bismuth, and lead. The first four of these isotopes are referred to as short-lived radon progeny (or radon daughters) with half-lives from less than a millisecond up to almost 27 min. There is also radon-220, or thoron, originating from the thorium decay chain, but usually this series is of less hygienic concern. Like radon-222 itself, the decay products polonium-218 and polonium-214 also emit α-particles. Since these decay products get electrically charged when created, they tend to attach to surfaces and dust particles in the air, but some also remain unattached. When the air is dusty, the unattached fraction tends to decrease, which may be important when comparing the effects of exposure in mines and in homes. The unattached progeny is usually considered to be responsible for the major part of the α-irradiation to the bronchial epithelium, at least in work situations with mouth-breathing, when these isotopes are inhaled and deposited, in the bronchia. Radon is not deposited, so that the contribution of α-irradiation from the gas itself is relatively marginal.

The α-particles travel less than 100 μm into the tissue, but their high energy causes an intense local ionization, damaging the tissue with a subsequent risk for cancer development. β- and γ-Radiation is also present from some of the decay products, but the effect in this respect is considered to be marginal due to the much lower energy content compared to the α-radiation.

Measuring Exposure to Radon

Very high levels of radon may occur in uranium mines as well as sometimes in metal mines, as originating from the decay of trace amounts of uranium present in many types of mineral. Since the 1950s, exposure to radon progeny in mines has been measured in working levels (WL).[5] One working level is any combination of short-lived radon progeny in 1 L of air that will ultimately release 1.3×10^5 MeV of α energy by decay through polonium-214. This amount of radon progeny may also be taken as equivalent to 3700 Bq/m^3 EER (equilibrium equivalent radon) or 2.08×10^{-5} J/m^3.[6] The accumulated exposure to radiation is expressed in terms of working level months (WLM), the month in this context corresponding to 170 h of exposure. The corresponding SI unit is the joule-hour per cubic meter, and 1 WLM is equal to 3.6×10^{-3} Jh/m^3 and may also be taken as 72 Bq-years/m^3.

Lung Cancer in Miners

In the 16th century, both Paracelsus and Agricola described a high mortality from pulmonary disorders among miners, presumably also including lung cancer. The first clear-cut scientific report on lung cancer in miners appeared in 1879 from Schneeberg in eastern Germany, and some decades later an excess of lung cancer in miners was also discovered in Joachimsthal in Czechoslovakia.[7,8] The etiological role of radon and its decay products was suggested in the 1920s but not fully understood and agreed on until the 1950s and 60s.

Many mining populations with exposure to radon and its decay products have been investigated since the early 1960s, both by cohort and case-control studies. Some of the main results of these studies are summarized in Tables 1 and 2. It may be noted that there is a remarkable consistency between the results obtained from these various mining populations, even if the overall risk ratios range from about 1.5 to 15. It may be noted in this context also that malignant disorders other than lung cancer have not yet been clearly demonstrated to depend on radon progeny exposure in mines. A few studies have shown a tendency toward an excess of stomach cancer, however.[30]

Many agents other than radon and its decay products are present in the mine atmosphere and might be responsible to some extent for the lung cancer risk among miners, e.g., carcinogenic trace metals in the dust. Low levels of arsenic might have been present in some mines and asbestiform fibers have occurred in some Swedish mines at least, but are considered less likely to have played any substantial role in lung cancer in this context.[31]

It has been suggested, however, that miners in Lorraine have had rather low exposure to radon progeny (about 0.03 to 0.07 WL) but they have nevertheless a clearly increased proportional mortality of lung cancer, especially among those with pneumoconiosis.[32] Therefore, a series of reports in recent years on silica dust exposure as a probable cause of lung cancer, especially among silicotics,[33] is of some interest in this context. Nevertheless, agents other than radon, including silica, do not seem to explain the lung cancer risk of miners, at least not to any greater extent.[34]

In support of a causal connection between exposure to radon and its decay products, it may also be noted that miners with a very low exposure have had little or no excess of lung cancer,

Table 1. Main Results of Some Cohort Studies of Lung Cancer in Miners Exposed to Radon and Radon Progeny[89]

Type of mining; country	Exposure or concentration (means)	Person-years	Lung cancer deaths			Reference
			Observed	Expected	SMR[a]	
Metal, U.S.A	0.05–0.40 WL	23,862	47	16.1	2.92	9
Uranium, U.S.A	821 WLM	62,556	185	38.4	4.82	10,11
Uranium, Czechoslovakia	289 WLM	56,955	211	42.7	4.96	12,13
Tin, U.K.	1.2–3.4 WL	27,631	28	13.27	2.11	14
Iron, Sweden	0.5 WL	10,230	28	6.79	4.12	15
Iron, Sweden	81.4 WLM	24,083	50	12.8	3.90	16
Fluorspar, Canada	Up to 2040 WLM	37,730	104	24.38	4.27	17
Uranium, Canada	40–90 WLM	202,795	82	56.9	1.44	18
Uranium, Canada	17 WLM	118,341	65	34.24	1.90	19
Iron, U.K.	0.02–3.2 WL	17,156	39	25.50	1.53	20
Uranium non-smokers, U.S.A	720 WLM	7,861	14	1.1	12.70	21
Pyrite, Italy	0.12–0.36 WL	29,577	47	35.6	1.31	22

[a]SMR, standardized mortality ratio.

Table 2. Main Results of Some Case-Control Studies of Lung Cancer in Miners Exposed to Radon and Radon Progeny

Type of mining; country	Exposure or concentration (means)	Number of cases to controls	Number of exposed cases	Rate ratio (max.)	Reference
Zinc-lead, Sweden	1 WL	29/174	21	16.4	23
Iron, Sweden	0.1–2.0 WL	604/(467 × 2 + 137)	20	7.3	24
Iron, Sweden	0.3–1.0 WL	38/403	33	11.5	25
Uranium, U.S.A.	30–2698 WLM	32/64	23	Infinite	26
Uranium, U.S.A.	472 WLM in cases	65/230 (nested)	All	1.5% per WLM	27
Tin, China	515 WLM in cases	107/107	7	(20.0)	28
Tin, China	373 WLM	74/74	5	(13.2) 1.7% per WLM	29 (subset of 28)

i.e., in coal and potash mining and also in iron mining.[35–37] This also supports the view that agents other than radon would be of little etiologic importance as likely to occur also in these low-risk mines.

INDOOR RADON

Both building material and ground conditions as well as the ventilation of a house tend to determine the indoor concentrations of radon and its decay products. The leakage of radon from the ground is a somewhat irregular phenomenon and very high indoor concentrations can occur in one house but not in another even if located nearby. In general, the leakage of radon from the ground is usually more important than its emanation from stony building materials of a house.[38] Air pressure, temperature, and wind conditions, as well as various behavioral factors influence ventilation and the concentrations that may build up in a room. Efforts to improve insulation and preserve energy may have impaired the situation.[39–41]

Hultqvist seems to have been the first to measure indoor radon,[42] but his observations in Swedish dwellings attracted little interest from the hygienic point of view. The levels found were in the range of 20–69 Bq/m^3, whereas recent measurements of indoor radon in Swedish homes have revealed higher levels, i.e., 122 Bq/m^3 as an average in detached houses and 85 Bq/m^3 in apartments, but with such great variations, as from 11 to 3300 Bq/m^3.[43] The differences found between the earlier and the more recently measured concentrations may suggest a general increase in the levels over time.

Indoor radon concentrations in the range of 40–100 Bq/m^3 have been reported as an average from many countries, e.g., the United States,[44] Norway,[45] Finland,[46] and the Federal Republic of Germany.[47] Considerably higher levels, such as 2000 or 3000 Bq/m^3, may occur in many houses, and this is about double the level tolerated in mines in most countries (about 1100 Bq/m^3 or 0.3 WL).

Problems in Exposure Assessment

Epidemiologic studies of indoor radon and lung cancer involve great problems with regard to assessment of the exposure. There is no perception of exposure to radon and its decay products and almost everybody has lived in several houses with varying exposure levels. Further difficulties derive from the fact that some people spend most of their time at home, whereas others prefer to more often be out in the open air or have indoor activities in various other houses with different radon concentrations. Only the exposure relating to the home environment could be estimated reasonably accurately in retrospect, whereas exposure obtained in other houses can hardly be accounted for.

Measurements of indoor radon may help in obtaining information on exposure, but even extensive, current measurements in a number of subsequently used homes of an individual could not provide any particularly good estimate of his or her accumulated exposure over many decades. Nevertheless, a combination of measurements and judgments with regard to pertinent characteristics of a house might be assumed to give a usable estimate of radon progeny exposure.

Cellulose nitrate film has been used for measuring radon decay products (or indirectly radon, which now is thought to be preferable to direct measurements of radon progeny). Such measurements have also been found to agree relatively well with the exposure estimates based on various characteristics of the houses and geological features as likely to have determined indoor radon concentrations.

Studies of Indoor Radon and Lung Cancer

A number of epidemiologic studies have been undertaken since 1979 to evaluate the possible role of indoor radon in terms of lung cancer risk of the general population. Almost exclusively, the studies have been of the case-control type and the results are summarized in Table 3. In the aggregate, these studies seem to suggest an effect of indoor radon with regard to lung cancer, even if the results in some of them provide rather weak evidence in this respect. In particular, a study from China on women shows no clear overall effect, but the odds ratio amounted to 1.7 for the small cell lung cancers,[58] although it may be noted that this study was conducted in an area with an unusually high risk of lung cancer in women. This high background may have masked an effect of radon and it is of some interest also that in some studies, a less clear effect has been obtained for smokers as well as in an urban population.[55,57] Two studies of cohort character have also been reported, both being fairly inconclusive, but at least one of them showing tendencies consistent with an effect of indoor radon.[59,60]

Table 3. Main Results of Case-Control Studies of Lung Cancer and Exposure to Indoor Radon and Progeny

Reference and year	Number of cases per controls	Rate ratio	Remarks
48; 1979	37/178	1.8	Significant trend crude exposure assessment
49; 1982	50/50	2.1	Published abstract only
50; 1984	23/202	Up to 4.3	Special account for geology and Rn emanation
51; 1984	Two sets of 30/30		Significantly high exposure for smoking cases
52; 1987	604/(467 × 2 + 137)	Up to 2.0	For more than 20 years in wooden houses
53; 1987	292/584	2.2	Women with oat cell cancer
54; 1987	27/49	Up to 11.9	Risk of 11.9 for 10 WLM
55; 1988	177/673	Up to 2.0	Clear effect for rural residents only
56; 1989	210/209 + (191 hospital controls)	Up to 1.8 (in middle exposure category)	Females; rate ratio 3.1 for small cell cancers
57; 1989	433/402	Up to 7.2	Females; little effect except in highest exposure category
58; 1990	308/356	Up to 1.7 (small cell cancer)	Females in high risk area (China); slight effect for small cell cancer, only

Note: In addition two cohorts, one with questionable effect and one with a risk ratio of 1.7 (CI 95% 0.8–3.2).[59,60]

With few exceptions, the histologic types of the lung cancer cases have not been considered in the studies referred to. However, one of them specifically considered oat-cell and other anaplastic lung cancers in women only. This study showed an association with indoor radon,[53] and another demonstrated an interesting predominance of squamous and small cell carcinomas among those who had lived in nonwooden houses, which likely have higher radon levels than wooden houses.[52] This seems to be in agreement with the fact that a relative excess of small cell undifferentiated lung cancers has appeared, especially in the early studies on uranium miners, but the spectrum of histologic types has changed toward the more normal in course of time.[61–63]

In some correlation studies relating to indoor radon, a more or less clear association has been obtained between lung cancer and indirect measures of potential radon emission from the ground. For example, phosphate deposits, that had been worked,[64] or volcanic versus sedimentary structures,[65] have been utilized to create a contrast in exposure. The occurrence of granite with increased radioactivity has provided another opportunity for contrasting the lung cancer rates of the populations in and out of areas with high radon emanations.[66] Estimated averages of background γ-radiation levels per county have also been used as a surrogate for potential indoor radon since there tends to be a rather strong correlation between γ-radiation and emanation of radon from the ground.[67] Other somewhat crude but more or less positive studies have been based on measurements of radium-226 in water or investigated levels of radon in water and indoor air.[68,69]

Radon and Cancers Other Than in the Lung

In some correlation studies, a relationship has been obtained for cancers other than lung cancer, e.g., for pancreatic cancer and male leukaemia,[67] bladder and breast cancer,[68] and for reproductive cancer in males as well as for all cancers taken together.[69] A high mortality rate from stomach cancer has been reported from New Mexico in an area with uranium deposits.[70]

A rather recent study that has attracted considerable interest reports a correlation of the incidence of myeloid leukemia, cancer of the kidney, melanoma, and certain childhood cancers with average radon exposure in the homes in a number of countries.[71] It should be noted though that lung cancer did not show any significant correlation, as would have been expected rather than the other positive correlations. Inhaled radon, as finally reaching the fat cells in contrast to the radon progeny as deposited in the lungs, was thought to be responsible for the induction of myeloid leukemia (through its further decay). Furthermore, the filtering of radon progeny through the kidney and the accumulation of these elements on the skin were suggested to explain the other correlations seen.

It is not yet clear, however, if the increased cancer risks seen for the various cancer sites other than lung can be attributed to radioactivity. Only the considerable interest in environmental epidemiology that can be anticipated for the next decade or so may finally clarify this question. As in the study by Henshaw et al., some other correlation studies have been negative with regard to lung cancer, e.g., from Canada, China, and France.[72–74]

SMOKING AND RADON

There has been a more or less multiplicative interaction or synergism in most studies of miners, where adequate information on smoking has been available.[24,30,63,75,76] In a few studies the data have suggested a merely additive relationship and sometimes even less than an additive effect.[16,23,67,77] Observations in the latter direction has also been observed regarding sputum cytology of uranium miners.[78]

Considering Human Data

An explanation of the rather inconsistent observations from the various studies could be that smoking might influence the dose received by the epithelium. Hence, especially in a dusty environment, smoking would increase mucous secretion and thereby the thickness of the mucous sheath, so that fewer α-particles would be able to penetrate to the basal cells of the epithelium from which cancer may develop.[79] The clearance of deposited particles carrying radon decay products may also be influenced by smoking, with consequences for the ultimate radiation dose delivered to the epithelium.

It may be recalled in this context that a considerably increased prevalence of bronchitis has been reported in miners who smoke,[80,81] but the criteria used for this diagnosis may not clearly distinguish a simple mucous hypersecretion from true bronchitis. An increase in thickness of the mucous layer of only about 10 μm would decrease the dose to the epithelium by the order of some 50%.[82,83]

Nevertheless, a synergism seems likely between chemicals in tobacco smoke and the actual dose of radiation to the epithelium, explaining the more or less multiplicative interaction between smoking and radon progeny exposure seen in most of the studies of miners, especially from the more modern and presumably less dusty mines. Similarly, for indoor radon progeny and smoking, a more or less multiplicative effect has been indicated, especially in rural areas. In some studies, however, the effect of this combined exposure has been less, and even much less, than expected,[53,55–57] i.e., the effect of radon progeny exposure has been weaker or absent among (especially heavy) smokers.

Some Experimental Observations

Experimental data seem to support the somewhat complex view given here on the interaction of smoking and radon progeny exposure in miners, since smoking dogs were less affected by respiratory cancer than nonsmoking dogs, when both groups were exposed to uranium ore dust and radon progeny.[84] On the other hand, experiments in rodents indicate that radon progeny exposure followed by exposure to cigarette smoke stimulated tumor development, whereas the reverse combination did not.[85] Smoking may therefore have a sort of double role, as both activating mucous secretion and subsequently influencing the dose to the epithelium along with exerting a carcinogenic effect itself. As a result, such a complex type of interaction would lead to almost any overall epidemiologic result from a multiplicative to an even less than additive effect depending on the particular circumstances under which the combined exposure occurs.

A further aspect on the interaction of smoking and radon progeny exposure may also be noted in this context, namely the tendency of radon progeny to attach to environmental tobacco smoke as to other particles in the air. The airborne radioactivity therefore tends to increase in the presence of tobacco smoke, as there is less plating out of radon progeny on walls, furniture, and other surfaces in the room.[86] The fraction of unattached progeny tends to be reduced, however, while the attached fraction increases proportionally.

The risk of lung cancer has usually been mainly tied to the unattached fraction, which is thought to contribute much more than the attached fraction to the radiation dose of the epithelium. However, the unattached radon progeny seems to be deposited in the nose to a great extent, i.e., some 50%, whereas the attached fraction is little affected by nose breathing.[87,88] Radon decay particles attached to smoke may therefore be deposited further down in the respiratory tract, or in those bronchial regions, where the epithelium is less thick than in the upper respiratory tract. In these regions, the α-particles may be able to penetrate to the basal cells, from which the cancer can develop.[83] The biological net effect of the increased radioactivity of smoke-polluted indoor air and the subsequent change of the proportion of unattached and attached radon progeny is not yet clear, and various explanations may remain unknown for the somewhat discrepant interactions observed with regard to smoking and exposure to radon decay products.

RADON AND LUNG CANCER RISK FOR MINERS AND THE GENERAL POPULATION

Risk estimates for exposure to radon progeny have been derived,[89] especially from uranium miners, and the increase in risk was calculated as 57% (i.e., a risk ratio of 1.57) over a 30-year working life and a cumulated exposure of 30 WLM (above background). This would correspond to an excess of 10 lung cancer deaths per 1000 miners. Similarly, the risk ratio was estimated as 1.31 for a cumulative exposure of 15 WLM, and the expected excess of cases per 1000 miners was then 4.9. In additive terms, the lifetime risk would be somewhat lower, and, for example, the BEIR IV committee arrived at 0.35 cases per WLM and 1000 individuals over a lifetime.

Preliminary Risk Estimates from Studies on Indoor Radon

Corresponding figures may be drawn from some of the studies of indoor radon exposure and will be in the range of 0.3–0.8 cases per WLM and 1000 individuals over a lifetime.[48,50,90] However, these estimates of the effect of indoor radon are quite uncertain due to the small numbers involved, even if there is good formal agreement with the figures from the miners.

Both NIOSH and the BEIR IV committee advocated a multiplicative risk model for lung cancer effect of radon progeny exposure, whereas earlier risk assessments usually have presumed additivity.[30,89] A multiplicative model implies synergistic effects with other causes of lung cancer as influencing the background rate of this disease; first and foremost is smoking, but in principle also including any preceding exposure to radon progeny, both occupational and background in character.

The multiplicative risk model may be applied to data from one of the case-control studies of indoor radon for risk comparisons with miners.[55] Some approximations would be necessary, however, because of missing information or differences with regard to stratification on age, etc.

The model proposed in the BEIR IV report is

$$R(a)/r(a) = 1 + 0.025M(a)(W + 0.5V)$$

where $R(a)$ is the lung cancer rate at age a among the exposed, $r(a)$ is the baseline rate (as in the 1980–1984 U.S. population); $M(a)$ is taken as 1.2 for ages less than 55 years, 1.0 for ages 55 to 64 years, and 0.4 for age 65 years or greater. W is the cumulated exposure in WLM from 5 to 15 years before age a and V is the cumulative exposure in WLM 15 years or more before age a. $R(a)/r(a)$ then represents the risk ratio at age a. The model implies a greater role, i.e., a late stage effect, of relatively recent exposure for lung cancer development, even if the animal experiments referred to also clearly suggest an early stage effect from exposure to radon decay products.[85]

The M-factor of the model may be taken as 0.8 for the house study referred to because of the relatively high ages of the lung cancer cases and the recent exposure.[55] W might be based on the measured average for elevated exposure, i.e., 151 Bq/m^3; then, at the age of 60 and with some 80% indoor occupancy time, 25 WLM is obtained for this recent exposure. For the earlier period of life, i.e., until age 45, V might be 50 WLM as accumulated at those probably somewhat lower exposure levels occurring in the past. Under these assumptions, the rate ratio, $R(a)/r(a)$, is obtained as 2.0, or just as found in the study for 151 Bq/m^3 as the average exposure. Similar comparisons based on other data would presumably give similar results, but it seems premature to go further into such uncertain calculations at this time. Instead, the results of ongoing larger studies might await a more definite, quantitated risk estimation.

Since the results available so far from epidemiologic studies of indoor radon seem to agree fairly well with the data from miners, the conclusion might be that indoor radon means a risk of lung cancer for the general population, but the quantitative aspect of this health hazard is not yet possible to assess more definitely. However, considering the lifetime risk of lung cancer in miners, one extra death per 1000 miners has been proposed as possibly acceptable, which would permit an exposure of only about 0.1 WLM per year.[89] The magnitude of the indoor radon problem might then be considered in relation to the fact that the average background exposure in the United States has been estimated as 0.2 WLM per year (and up to 0.4 WLM per year in the vicinity of radon emitting ore bodies). Furthermore, there are clear indications that the exposure levels may be even much higher for large population sectors in other countries.

Definite Risk Estimates Await On-going Studies

In view of the present interest in the potential lung cancer hazard from indoor radon, it is not surprising that studies in this respect have started in several countries or are being planned, e.g., in the United States, UK, and elsewhere in Europe. Many of these studies are quite large and rather expensive, but it seems likely that several studies on indoor radon should be reported by the mid- or late 1990s.

The first of these large scale studies has now been reported from Sweden.[91] This study involved 1360 lung cancer cases and 2847 controls. Measurements were conducted in 8992 dwellings, where the subjects had lived for two years or more since 1947. The geometrical mean radon concentration was 60.5 and the arithmetic mean was 106.5 Bq/m^3. Similar to earlier findings, this study also showed a moderate but significant effect of indoor radon on lung cancer with an odds ratio of 1.3 for a time weighted exposure to 140–400 Bq/m^3 and 1.8 at levels above 400 Bq/m^3 of radon gas. A multiplicative interaction with smoking was indicated. Interestingly, sleeping with an open window slit eliminated the risk. This finding certainly indicates the importance of ventilation in reducing especially the radon decay products. Overall about 9% of the lung cancers in the study were attributable to indoor radon.

REFERENCES

1. **WHO,** *Health Hazards of the Human Environment,* World Health Organization, Geneva, 1972.
2. **Dean, G.,** Lung cancer among white South Africans, *Br. Med. J.,* 2, 1599, 1961.
3. **Burch, P. R. J.,** Smoking and lung cancer. The problem of inferring cause, *J. R. Stat. Soc.,* 141, 4, 437, 1978.
4. **Dean, G.,** The effects of air pollution and smoking on health in England, South Africa and Ireland, *J. Ir. Med. Assoc.,* 72, 284, 1979.
5. **Holaday, D. A.,** Digest of the proceedings of the 7-state conference on health hazards in uranium mining, *Arch. Ind. Health,* 12, 465, 1955.
6. **ICRP, International Commission on Radiological Protection,** *Radiation Protection in Uranium and Other Mines,* ICRP publication No. 24, Pergamon Press, Oxford, 1976.
7. **Härting, F. H., and Hesse, W.,** Der Lungenkrebs, die Bergkrankheit in den Schneeberger Gruben, *Vierteljahrsschr. Gerichtl Med. Offentl. Gesundheitswesen,* 30, 296; 31, 102; 31, 313, 1879.
8. **Arnstein, A.,** Über den sogenannten "Schneeberger Lungenkrebs," *Verh. dt. Gesellsch. Pathol.,* 16, 332, 1913.
9. **Wagoner, J. K., Miller, R. W., Lundin, F. E., Jr., Fraumeni, J. F., and Haij, N. E.,** Unusual mortality among a group of underground metal miners, *N. Engl. J. Med.,* 269, 281, 1963.
10. **Lundin, F. E., Jr., Wagoner, J. K., and Archer, V. E.,** Radon daughter exposure and respiratory cancer. Quantitative and temporal aspects, NIOSH-NIEHS joint monograph No. 1, Public Health Service, Springfield, VA, 1971.
11. **Waxweiler, R. J., Roscoe, R. J., Archer, V. E., Thun, M. J., Wagoner, J. K., and Lundin, E.,** Mortality follow-up through 1977 of the white underground uranium miners cohort examined by the United States Public Health Service, in *Radiation Hazards in Mining: Control, Measurement and Medical Aspects,* Gomez, M., Ed., Society of Mining Engineers of the American Institute of Mining, Metallurgical, and Petroleum Engineers, Inc., New York, 1981, 823.
12. **Kunz, E., Sevc, J., and Placek, V.,** Lung cancer in uranium miners, *Health Phys.,* 35, 579, 1978.
13. **Placek, V., Smid, A., Sevc, J., Tomasek, L., and Vernerova, P.,** Late effects at high and very low exposure levels of the radon daughters, in *Radiation Research—Somatic and Genetic Effects. Proceedings of 70th International Congress of Radiation Research,* Broerse, J. J., Barendsen, G. W., Kal, H. B., Vanderkogel, A. J., Eds., Martinus Nijhoff, Amsterdam, 1983.
14. **Fox, A. J., Goldblatt, P., and Kinlen, L. J.,** A study of the mortality of Cornish tin miners, *Br. J. Ind. Med.,* 38, 378, 1981.
15. **Jörgensen, H. S.,** Lung cancer among underground workers in the iron ore mine of Kiruna based on thirty years of observation, *Ann. Acad. Med., (Singapore),* 13, 371, 1984.
16. **Radford, E. P., and Renard, K. G. St. C.,** Lung cancer in Swedish iron miners exposed to low doses of radon daughters, *N. Engl. J. Med.,* 310, 1485, 1984.
17. **Morrison, H. I., Semenciw, R. M., Mao, Y., and Wigle, D. T.,** Cancer mortality among a group of fluorspar miners exposed to radon progeny, *Am. J. Epidemiol.,* 128, 1266, 1988.
18. **Muller, J., Wheeler, W. C., Gentleman, J. F., Suranvi, G., and Kusiak, R.,** Study of mortality of Ontario miners, in *Occupational Radiation Safety in Mining. Proceedings of the International Conference,* Stocker, Ed., Canadian Nuclear Association, Toronto, 1985, 335.
19. **Howe, G. R., Nair, R. C., Newcombe, H. B., Miller, A. B., Frost, S. E., and Abbatt, J. D.,** Lung cancer mortality (1950–1980) in relation to radon daughter in a cohort of workers at the Eldorado Beaverlodge uranium mine, *J. Natl. Cancer,* 77, 357, 1986.
20. **Kinlen, L. J., and Willows, A. N.,** Decline in the lung cancer hazard: A prospective study of the mortality of iron ore miners in Cumbria, *Br. J. Ind. Med.,* 45, 219, 1988.
21. **Roscoe, R. J., Steenland, K., Halperin, W. E., Beaumont, J. J., and Waxweiler, R. J.,** Lung cancer mortality among nonsmoking uranium miners exposed to radon daughters, *J. Am. Med. Assoc.,* 262, 629, 1989.
22. **Battista, G., Belli, S., Garboncini, F., Comba, P., Giovanni, L., Sartorelli, P., Strambi, F., Valentini, F., and Axelson, O.,** Mortality among pyrite miners with low-level exposure to radon daughters, *Scand. J. Environ. Health,* 14, 280, 1988.
23. **Axelson, O., and Sundell, L.,** Mining, lung cancer and smoking, *Scand. J. Work Environ. Health,* 4, 46, 1978.

24. **Damber, L., and Larsson, L. G.,** Combined effects of mining and smoking in the causation of lung cancer, A case-control study, *Acta. Radiol. Oncol.,* 21, 305, 1982.
25. **Edling, C., and Axelson, O.,** Quantitative aspects of radon daughter exposure and lung cancer in underground miners, *Br. J. Ind. Med.,* 40, 182, 1983.
26. **Samet, J. M., Kutvirt, D. M., Waxweiler, R. J., and Key, C. R.,** Uranium mining and lung cancer in Navajo men, *N. Engl. J. Med.,* 310, 1481, 1984.
27. **Samet, J. M., Pathak, D. R., Morgan, M. V., Marbury, M. C., and Key, C. R.,** Valdivia, A. A., Radon progeny exposure and lung cancer risk in New Mexico U miners, *Health Phys.,* 56, 415, 1989.
28. **Qiao, Y., Taylor, P. R., Yao, S-X., Schatzkin, A., Mao, B-L., Lubin, J., Rao, J-Y., McAdams, M., Xuan, X-Z., and Li, J-Y.,** Relation of radon exposure and tobacco use to lung cancer among tin miners in Yunnan province, China, *Am. J. Ind. Med.,* 16, 511, 1989.
29. **Lubin, J. H., Qiao, Y., Taylor, P. R., Yao, S-X., Schatzkin, A., Mao, B-L., Rao, J-Y., Xuan, X-Z., and Li, J-Y.,** Quantitative evaluation of the radon and lung cancer association in a case control study of Chinese tin miners, *Cancer Res.,* 50, 174, 1990.
30. **BEIR IV,** Committee on the Biological Effects of Ionizing Radiations, U.S. National Research Council Health risk of radon and other internally deposited alpha-emitters, National Academy Press, Washington, D.C., 1988.
31. **Edling, C.,** Lung cancer and smoking in a group of iron ore miners, *Am. J. Ind. Med.,* 3, 191, 1982.
32. **Mur, J-M., Meyer-Bisch, C., Pham, Q. T., Massin, N., Moulin, J-J., Cavelier, C., and Sadoul, P.,** Risk of lung cancer among iron ore miners: A proportional mortality study of 1,075 deceased miners in Lorraine, France, *J. Occup. Med.,* 29, 762, 1987.
33. **IARC,** Monographs on the evaluation of carcinogenic risks to humans, Supplement 7, International Agency for Research on Cancer, Lyon, 1987.
34. **Archer, V. E., Roscoe, J., and Brown, D.,** Is silica or radon daughters the important factor in the excess lung cancer among underground miners?, in *Silica, Silicosis, and Cancer,* Goldsmith, D. F., Winn, D. M., and Shy, C. M., Eds., Praeger, New York, 1986, 375.
35. **IARC,** Monographs on the evaluation of carcinogenic risks to humans. Radon and Man-made mineral fibres, International Agency for Research on Cancer, Lyon, 1988.
36. **Waxweiler, R. J., Wagoner, J. K., and Archer, V. E.,** Mortality of potash workers, *J. Occup. Med.,* 15, 486, 1973.
37. **Lawler, A. B., Mandel, J. S., Schuman, L. M., and Lubin, J. H.,** A retrospective cohort mortality study of iron ore (hematite) miners in Minnesota, *J. Occup. Med.,* 27, 507, 1985.
38. **Åkerblom, G., and Wilson, C.,** *Radon—Geological Aspects of an Environmental Problem.* Rapporter och meddelanden nr 30, Sveriges geologiska undersökning, Uppsala, 1982.
39. **Dickson, D.,** Home insulation may increase radiation hazard, *Nature London,* 276, 431, 1978.
40. **Stranden, E., Berteig, L., and Ugletveit, F.,** A study on radon in dwellings, *Health Phys.,* 36, 413, 1979.
41. **McGregor, R. G., Vasudev, P., Létourneau, E. G., McCullough, R. S., Prantl, F. A., and Taniguchi, H.,** Background concentrations of radon and radon daughters in Canadian homes, *Health Phys.,* 39, 285, 1980.
42. **Hultqvist, B.,** Studies on naturally occurring ionizing radiations with special reference to radiation dose in Swedish houses of various types, *Kungl svenska vetenskapsakademiens handlingar,* 4:e serien, band 6, Nr 3, Almqvist och Wiksell Boktryckeri AB, Stockholm, 1956.
43. **Swedjemark, G. A., Buren, A., and Mjönes, L.,** Radon levels in Swedish homes: A comparison of the 1980s with the 1950s, in *Radon and Its Decay Products—Occurrence, Properties and Health Effects,* Hopke, P. K., Ed., American Chemical Society, Washington, D.C., 1987, 85.
44. **Nero, A. V., Schwehr, M. B., Nazaroff, W. W., and Revzan, K. L.,** Distribution of airborne radon-222 concentrations in U.S. homes, *Science,* 234, 992, 1986.
45. **Stranden, E.,** Radon-222 in Norwegian dwellings, in *Radon and Its Decay Products—Occurrence, Properties and Health Effects,* Hopke, P. K., Ed., American Chemical Society, Washington, D.C., 1987, 70.
46. **Castren, O., Mäkeläinen, I., Winqvist, K., and Voutilainen, A.,** Indoor radon measurements in Finland: A status report, in *Radon and Its Decay Products—Occurrence, Properties and Health Effects,* Hopke, P. K., Ed., American Chemical Society, Washington, D.C., 1987, 97.

47. **Schmier, H., and Wick, A.,** Results from a survey of indoor radon exposures in the Federal Republic of Germany, *Sci. Total Environ.*, 45, 307, 1985.
48. **Axelson, O., Edling, C., and Kling, H.,** Lung cancer and residency—A case referent study on the possible impact of exposure to radon and its daughters in dwellings, *Scand. J. Work Environ. Health,* 5, 10, 1979.
49. **Lanes, S. F., Talbott, E., and Radford, E.,** Lung cancer and environmental radon, *Am. J. Epidemiol.,* 116, 565, 1982.
50. **Edling, C., Kling, H., and Axelson, O.,** Radon in homes—A possible cause of lung cancer, *Scand. J. Work Environ. Health,* 10, 25, 1984.
51. **Pershagen, G., Damber, L., and Falk, R.,** Exposure to radon in dwellings and lung cancer: A pilot study, in *Indoor Air, Radon, Passive Smoking, Particulates and Housing Epidemiology,* Vol. 2, Berglund, B., Lindvall, T., and Sundell, J., Eds., Swedish Council for Building Research, Stockholm, 1984, 73.
52. **Damber, L. A., and Larsson, L. G.,** Lung cancer in males and type of dwelling. An epidemiological pilot study, *Acta Oncol.,* 26, 211, 1987.
53. **Svensson, C., Eklund, G., and Pershagen, G.,** Indoor exposure to radon from the ground and bronchial cancer in women, *Int. Arch. Occup. Environ. Health,* 59, 123, 1987.
54. **Lees, R. E. M., Steele, R., and Robert, J. H.,** A case-control study of lung cancer relative to domestic radon exposure, *Int. J. Epidemiol.,* 16, 7, 1987.
55. **Axelson, O., Andersson, K., Desai, G., Fagerlund, I., Jansson, B., Karlsson, C., and Wingren, G.,** A case-referent study on lung cancer, indoor radon and active and passive smoking, *Scand. J. Work Environ. Health,* 14, 286, 1988.
56. **Svensson, C., Pershagen, G., and Klominek, J.,** Lung cancer in women and type of dwelling in relation to radon exposure, *Cancer Res.,* 49, 1861, 1989.
57. **Schoenberg, J. B., Klotz, J. B., Wilcox, H. B., Gil-del-Real, M., Stemhagen A., and Nicholls, G. P.,** Lung cancer and exposure to radon in women—New Jersey, *M.M.W.R.,* 38, 715, 1989.
58. **Blot, W. J., Xu, Z-Y., Boice, J. D., Jr., Zhao, D.-Z., Stone, B. J., Sun, J., Jing, L-B., and Fraumeni, J. F., Jr.,** Indoor radon and lung cancer in China, *J. Natl. Cancer Inst.,* 82, 1025, 1990.
59. **Simpson, S. G., and Comstock, G. W.,** Lung cancer and housing characteristics, *Arch. Environ. Health,* 38, 248–51, 1983.
60. **Klotz, J. B., Petix, J. R., and Zagraniski, R. T.,** Mortality of a residential cohort exposed to radon from industrially contaminated soil, *Am. J. Epidemiol.,* 129, 1179, 1989.
61. **Archer, V. E., Sacomanno, G., and Jones, J. H.,** Frequency of different histologic types of bronchogenic carcinoma as related to radon exposure, *Cancer,* 34, 2056, 1974.
62. **Horacek, J., Placek, V., and Sevc, J.,** Histologic types of bronchogenic cancer in relation to different conditions of radiation exposure, *Cancer,* 34, 832, 1977.
63. **Saccomanno, G., Huth, G. C., Auerbach, O., and Kuschner, M.,** Relationship of radioactive radon daughters and cigarette smoking in the genesis of lung cancer in uranium miners, *Cancer,* 62, 1402, 1988.
64. **Fleischer, R. L.,** A possible association between lung cancer and phosphate mining and processing, *Health Phys.,* 41, 171, 1981.
65. **Forastiere, F., Valesini, S., Arca, M., Magliola, M. E., Michelozzi, P., and Tasco, C.,** Lung cancer and natural radiation in an Italian province, *Sci. Total Environ.,* 45, 519, 1985.
66. **Archer, V. E.,** Association of lung cancer mortality with precambrian granite, *Occup. Environ. Health,* 42, 87, 1987.
67. **Edling, C., Comba, P., Axelson, O., and Flodin, U.,** Effects of low-dose radiation—A correlation study, *Scand. J. Work Environ. Health,* 8(Suppl. 1), 59, 1982.
68. **Bean, J. A., Isacson, P., Hahne, R. M. A., and Kohler, J.,** Drinking water and cancer incidence in Iowa, *Am. J. Epidemiol.,* 116, 924, 1982.
69. **Hess, C. T., Weiffenbach, C. V., and Norton, S. A.,** Environmental radon and cancer correlation in Maine, *Health Phys.,* 45, 339, 1983.
70. **Wilkinson, G. S.,** Gastric cancer in New Mexico conties with significant deposits of uranium, *Arch. Environ. Health,* 40, 307, 1985.
71. **Henshaw, D. L., Eatough, J. P., and Richardson, R. B.,** Radon as a causative factor of myeloid leukaemia and other cancers, *Lancet,* 335, 1008, 1990.

72. **Letourneau, E. G., Mao, Y., McGregor, R. G., Semenciw, R., Smith, M. H., and Wigle, D. T.,** Lung cancer mortality and indoor radon concentrations in 18 Canadian cities, *Proceedings of the Sixteenth Midyear Topical Meeting of the Health Physics Society on Epidemiology Applied to Health Physics,* Albuquerque, NM, January 9–13. Health Physics Society, Ottawa, 1983, 470.
73. **Hofmann, W., Katz, R., and Zhang, C.,** Lung cancer incidence in a Chinese high background area —epidemiological results and theoretical interpretation, *Sci. Total Environ.,* 45, 527, 1985.
74. **Dousset, M., and Jammet, H.,** Comparison de la mortalitie par cancer dans le Limousin et le Poitou-Charentes, *Radioprotection GEDIM,* 20, 61, 1985.
75. **Archer, V. E., Wagoner, J. K., and Lundin, F. E.,** Uranium mining and cigarette smoking effects on man, *J. Occup. Med.,* 15, 204, 1973.
76. **Whittemore, A. S., and McMillan, A.,** Lung cancer mortality among U.S. uranium miners: A reappraisal, *J. Natl. Cancer Inst.,* 71, 489, 1983.
77. **Dahlgren, E.,** Lungcancer, hjärt-kärlsjukdom och rökning hos en grupp gruvarbetare (Lung cancer, cardiovascular disease and smoking in a group of miners), *Läkartidningen,* 76, 4811, 1979.
78. **Band, P., Feldstein, M., Saccomanno, G., Watson, L., and King, G.,** Potentiation of cigarette smoking and radiation. Evidence from a sputum cytology survey among uranium miners and controls, *Cancer,* 45, 1237, 1980.
79. **Axelson, O.,** Room for a role for radon in lung cancer causation?, *Med. Hypotheses,* 13, 51, 1984.
80. **Sluis-Cremer, G. K., Walthers, L. G., and Sichel, H. F.,** Chronic bronchitis in miners and non-miners: An epidemiological survey of a community in the goldmining area in Transvaal, *Br. J. Ind. Med.,* 24, 1, 1967.
81. **Jörgensen, H., and Swensson, A.,** Undersökning av arbetare i gruva med dieseldrift, särskilt med hänsyn till lungfunktion, luftvägssymtom och rökvanor (Investigation of workers in a mine with diesel drift, especially regarding lung function, respiratory symptoms and smoking habits), AI rapport No. 16. Arbetarskyddsverket, Stockholm (English summary), 1974.
82. **Altschuler, B., Nelson, N., and Kuschner, M.,** Estimation of lung tissue dose from the inhalation of radon and daughters, *Health Phys.,* 10, 1137, 1964.
83. **Walsh, P. J.,** Radiation dose to the respiratory tract of uranium miners—A review of the literature, *Environ. Res.,* 3, 14, 1970.
84. **Cross, F. T., Palmer, R. F., Filipy, R. E., Dagle, G. E., and Stuart, B. O.,** Carcinogenic effects of radon daughters, uranium ore dust and cigarette smoke in beagle dogs, *Health Phys.,* 42, 33, 1982.
85. **Chameaud, J., Masse, R., Morin. M., and Lafuma, J.,** Lung cancer induction by radon daughters in rats; present state of the data in low dose exposures, in *Proceedings of the International Conference on Occupational Radiation Safety in Mining,* Stocker, H., Ed., Canadian Nuclear Association, Toronto, 1985, 350.
86. **Bergman, H., Edling, C., and Axelson, O.,** Indoor radon daughter concentrations and passive smoking, *Environ. Int.,* 12, 17, 1986.
87. **George, A. C., and Breslin, A. J.,** Deposition of radon daughters in human exposed to uranium mine atmospheres, *Health Phys.,* 17, 115, 1969.
88. **James, A. C.,** A reconsideration of cells at risk and other key factors in radon daughter dosimetry, in *Radon and Its Decay Products—Occurrence, Properties and Health Effects,* Hopke, P. H., Ed., American Chemical Society, Washington, D.C., 1987, 400.
89. **NIOSH,** National Institute for Occupational Safety and Health, A recommended standard for occupational exposure to radon progeny in underground mines, U.S. Department of Health and Human Services, Washington, D.C., 1987.
90. **Edling, C., and Axelson, O.,** Radon daughter exposure, smoking and lung cancer, *Clin. Ecol.,* 5, 59, 1987.
91. **Pershagen, G., Åkerblom, G., Axelson, O., Clavensjö, B., Damber, L., Desai, G., Enflo, A., Lagarde, F., Mellander, H., Svartengren, M., and Swedjemark, G. A.,** Residential radon exposure and lung cancer in Sweden, *N. Engl. J. Med.,* 330, 159, 1994.

SECTION V

Environmental Management

CHAPTER 25

The Role of Microorganisms in the Environment Decontamination

Enrica Galli

Microorganisms play an important role in the natural cycling of elements. In fact, they are widely distributed in soil and water, where they are able to degrade and utilize for growth naturally occurring compounds synthetized by plants and animals, including the most complex compounds such as lignin, cellulose, terpenes, and flavonoids. The mineralization of organic carbon is essential to maintain life on earth, because it allows the release into the environment of carbon, nitrogen, sulfur, phosphorus, etc. at an inorganic state. This ability is due primarily to the vast array of microbial degradative activities.

In the last decades the release of large quantities of synthetic chemicals, such as solvents, plasticizers, insecticides, herbicides, and fungicides, into the environment, through industrial, agricultural, medical, and domestic activities, has created considerable ecotoxicological problems with heavy consequences for human health and generally for all the living organisms. Such man-made compounds are called xenobiotics, because microorganisms have never been exposed to their chemical structure in the course of evolution.

More than 1000 new compounds are marketed every year. The total annual world production of synthetic organic chemicals is over 300 million tons. Nobody knows how many or how much of these substances are dangerous to man and the environment. The U.S. Environmental Protection Agency's (EPA) list of priority pollutants includes pesticides, halogenated aliphatics and aromatics, nitroaromatics, polychlorinated biphenyls, and polycyclic aromatic hydrocarbons.[1]

In all these cases the fundamental problem is their recalcitrance or persistence, that is, the ability of a substance to remain in a particular environment in an unchanged form. On the basis of their persistence pesticides may be classified as follows:

Persistent
    Chlorinated insecticides
    Cationic herbicides
Moderately persistent
    Herbicides
        Triazines

Phenylureas
  Dinitroanilines
Not persistent
  Insecticides
      Organophosphates
    Carbamates
  Herbicides
      Phenoxyderivatives
    Phenylcarbamates
  Fungicides
      Carbamate derivatives

The resistance to microbial attack is mainly due to the chemical structure of these compounds, which does not resemble that of naturally occurring ones.[2] This implies that microorganisms have not yet evolved the genetic information for the synthesis of the appropriate catabolic enzymes. Another plausible explanation is that microorganisms may have the enzymatic capability to degrade the pollutant but the compound may not be present at sufficient levels to induce the necessary degradative enzymes in the microorganisms. As a consequence, a persistent compound, for instance, a chloroaromatic such as DDT or PCB, which are lipid soluble, accumulates in food chains causing toxicity; moreover, it has been demonstrated that many of these substances are mutagenic and carcinogenic.

On the contrary, organic compounds, which can be utilized by microorganisms as growth substrate, are called biodegradable. Therefore a man-made compound will be degradable only if it is susceptible to attack by the so-called "preexisting" enzymes whose genetic determinants were acquired by microorganisms during the course of evolution. This, in turn, depends on two factors; first, the ability of microbial enzymes to accept as substrate compounds having a chemical structure similar to, but not identical with those found in nature; and second, the ability of these novel substrates, when in the presence of microorganisms, to induce or derepress the synthesis of the necessary degradative enzymes.

Such enzymes show a relaxed substrate specifity and then may be used by microorganisms to degrade partially also those synthetic compounds structurally similar to the substrate supplied as carbon and energy source, but which cannot be utilized for growth. The process by which a substrate is modified but is not utilized for growth by an organism that is growing on or metabolizing another substrate is called cooxidation. Such a process allows a molecule, which could not be otherwise degraded, to be modified in its structure.[3]

We can conclude that a molecule is biodegradable if it allows the enrichment, i.e., the selection of those microorganisms, whose genetic inheritance allows adaptation to any environmental change, as the introduction of a new man-made substance.

Soil microorganisms that have been isolated on the basis of their ability to degrade natural or man-made organic substrates belong mainly to the genus *Pseudomonas, Alcaligenes, Nocardia, Flavobacterium, Arthrobacter,* and *Corynebacterium,* without forgetting the relevant role of fungi and other eukaryotic microorganisms in biodegradation and biotransformation processes.

In particular, *Pseudomonas* species are known to metabolize a broad range of organic compounds and therefore are an ideal choice as the bacteria to be used for degradative biotechnologies. They have, in fact, an extraordinary range of catabolic pathways; a single species such as *P. cepacia* utilizes more than 100 different substrates as the only C, N, or S source.

Microorganisms have taken milions and milions of years to evolve the ability to degrade natural organic compounds. Chemical industry has produced new molecules in the last century and the biosphere has received these new compounds in significant quantities in the past 30 years.

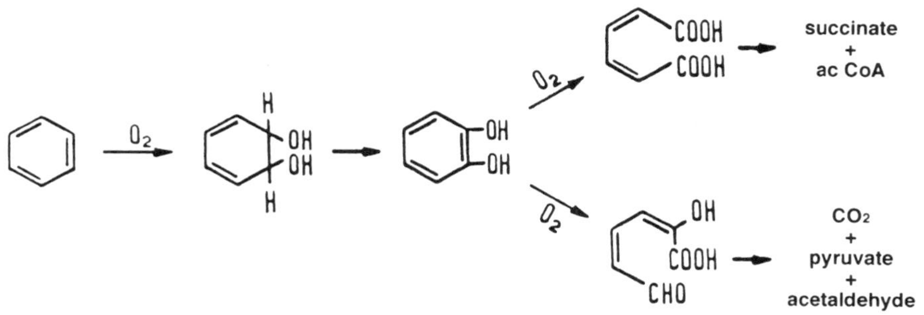

Figure 1. Degradative pathways of benzene.

BIODEGRADATION PATHWAYS

Knowledge on the degradative pathways of aromatic compounds has contributed toward an understanding of xenobiotic biodegradation. In fact, man-made compounds are mainly synthesized starting from aliphatic and aromatic hydrocarbons.

Aromatic hydrocarbons as well substituted derivatives must be modified into o-diphenols through different biochemical reactions before ring cleavage may occur.

The ability to perform these transformations is an important property of microorganisms, which utilize aromatic compounds as the sole carbon source, since the main energy yielding reactions take place only after ring fission. The microbial attack of aromatic hydrocarbons begins with the oxygenation of two adjacent carbon atoms leading to the formation of a dihydrodihydroxyderivative (diol), which is further dehydrogenated to an o-diphenol, prior to ring fission (Figure 1). The ring fission of o-diphenols is catalyzed by dioxygenases and may occur in different positions:

- between two hydroxylated carbon atoms with the formation of a *cis, cis*-muconic acid (*ortho* or intradiol cleavage)
- between two adjacent carbon atoms only one of which bears a hydroxyl group, with the formation of a muconic semialdehyde (*meta* or extradiol cleavage).

One of the most important features of the degradation of aromatic compounds is referred to the fate of substituent groups, which may be modified before or after the ring cleavage depending on the selected microorganism and the chemical structure of the substituent. In fact, the biodegradability of a substituted aromatic compound is affected by whether a substituent group is to be modified or removed, in order to allow hydroxylation and then cleavage of the aromatic nucleus. o-Diphenols are key intermediates for further metabolism of an aromatic compound leading to its complete oxidation (Figure 2).

The influence of chemical structure on the biodegradability of a synthetic compound will be discussed through some significant examples.

1. A very simple example is given by xylenes, which are produced in large amounts, and widely used as solvents and then must be considered as environmental pollutants. Many bacteria have been isolated capable of utilizing m- and p-xylene according to a main biodegradation scheme in which the corresponding alcohols, aldehydes, and acids are formed before the hydroxylation of the aromatic ring and its further cleavage.[4] None of these organisms has been shown to be able to oxidize the *ortho* isomer. A strain of *Pseudomonas* has been isolated, which is able to degrade o-xylene through a direct hydroxylation of the aromatic nucleus.[5] This example shows the influence of the relative

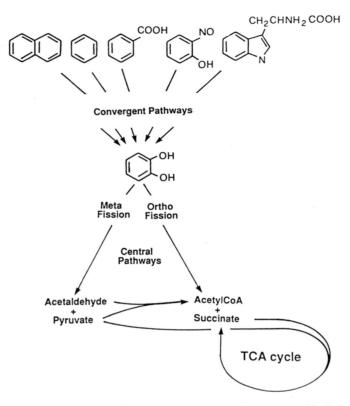

Figure 2. Convergent pathways of aromatic compounds catabolism (by courtesy of S. Harayama).

position of the methyl substituents on the selection of microorganisms capable of utilizing xylenes through different initial reactions (Figure 3).

2. Haloaromatic compounds represent a class of chemicals whose chemical structure strongly affects their biodegradability. Chloroaromatics are used in large quantities in a variety of industrial and agricultural situations such as solvents, insecticides, herbicides, and heat transfer media. Many of these compounds become widely dispersed into the biosphere either deliberately or accidentally and can be detected in water, soil, air, and even man and animals.

The crucial point in the biodegradability of chloroaromatic compounds is the removal of chlorosubstituents from the aromatic nucleus; this reaction may occur at an early stage of the degradative pathway with reductive, hydrolytic, or oxygenolytic elimination of the substituent. Alternatively, nonaromatic structures may be produced, which spontaneously lose chlorine by hydrolysis.[6]

Reductive dechlorination occurs in anaerobic conditions and allows the loss of the chloride without the alteration of the aromatic ring. When all the chlorine atoms are removed, ring fission leads to methane and carbon dioxide. This mechanism has been demonstrated for 3-chlorobenzoate, 2,4,5-thrichlorophenoxyacetic acid, chlorophenols, pentachlorophenol, and 1,2,4-thrichlorobenzene by a methanogenic consortium.

Replacement of chlorine from the aromatic ring through hydrolytic cleavage of the carbon–chloride bond has been shown for 4-chlorobenzoate, which is converted to 4-hydroxybenzoate and also for penthachlorophenol by a *Flavobacterium* sp.[7] (Figure 4).

Dehalogenation by dioxygenases is another mechanism to remove halogen substituents from haloaromatics, as has been shown for the initial dehalogenation of 4-chlorophenylacetate to homoprotocatechuate by a *Pseudomonas strain*.[8] As far as polychlorinated

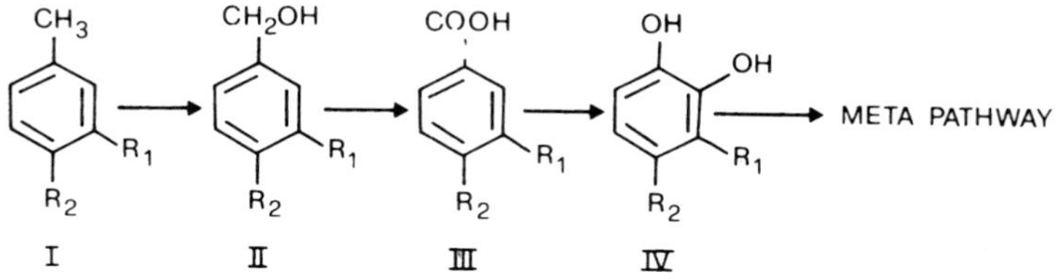

$R_1 = R_2 = CH_3$; 1,2,4-Trimethylbenzene

I ; 1,2,4-Trimethylbenzene

II ; 3,4-Dimethylbenzyl alcohol

III ; 3,4-Dimethylbenzoate

IV ; 3,4-Dimethylcatechol

$R_1 = H$ $R_2 = CH_3$; p-Xylene

I ; p-Xylene

II ; p-Methylbenzyl alcohol

III ; p-Toluate

IV ; 4-Methylcatechol

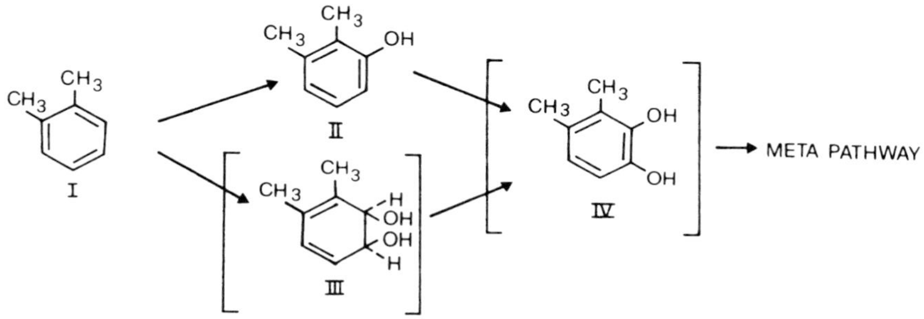

I ; o-Xylene

II ; 2,3-Dimethylphenol

III ; 3,4-Dimethyl dihydrodiol

IV ; 3,4-Dimethylcatechol

Figure 3. Degradative pathways of xylenes.

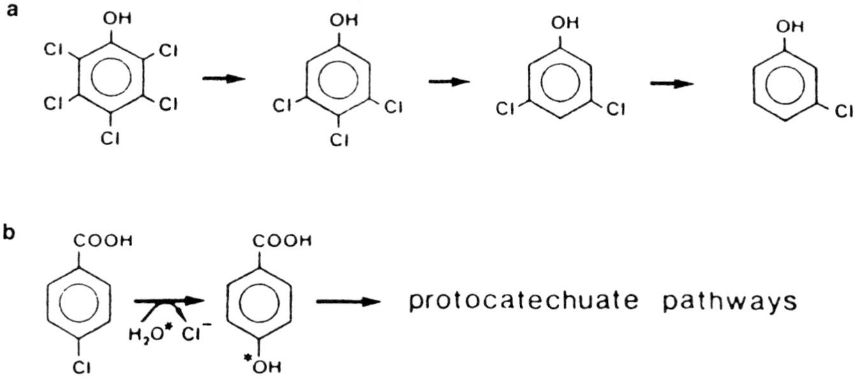

Figure 4. Chloroaromatic degradative pathways: (a) pentachlorophenol; (b) 4-chlorobenzoic acid.

biphenyls are concerned, their degradation rate decreases as the number of chlorosubstituents on both rings increases[9] (Figure 5).

3. Another interesting example is given by atrazine, 2-chloro-4-(ethylamino)-6-(isopropylamino)-*s*-triazine, which is a widely used selective herbicide for weed control in corn. Its recalcitrance toward microbial and physicochemical attack and its widespread use in

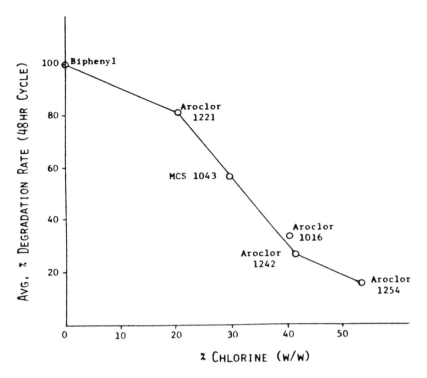

Figure 5. Biodegradation rate of polychlorinated biphenyls.

agriculture give rise to accumulation in the environment. It has been shown that the metabolism of atrazine in soil involves hydroxylation, dealkylation, and ring cleavage.[10] Phytotoxicity of the herbicide is destroyed by hydroxylation at the 2-position but not by dealkylation of either of the two alkylamino groups. Dechlorination of atrazine to a non-phytotoxic product, hydroxyatrazine, occurs in soil treated with atrazine, but this reaction has been attributed to chemical hydrolysis rather than to a microbial activity. Although biodegradation of atrazine in soil has been attributed mainly to fungi, bacteria belonging to genus *Nocardia* and *Pseudomonas* have been isolated, which caused N-dealkylation of atrazine. It was suggested that microbial dehalogenation could occur in soil after the removal of one of the alkyl group (Figure 6).

4. In recent years concern about the presence of polycyclic aromatic hydrocarbons (PAH) in air, soil, and water systems has increased, due to the fact that these chemicals represent a potential health risk to man[11] (Figure 7). Industrial elements from coal gassification and liquification processes, waste incinerination, coke, carbon black, and other petroleum-derived products contribute to the high input of PAH into the environment. Due to their hydrophobic properties and limited water solubility, PAH tend to adsorb to particulates and eventually migrate to the sediments in river, lake, estuarine and marine waters. The PAH levels ranged from 5 ppb for an undeveloped area in Alaska to 1.79×10^6 ppb for an oil refinery outfall in England.

PAH enter the biosphere through various routes such as accidental discharges of fossil fuels, direct aerial fallout, chronic leakage, industrial and sewage discharges, and surface water. Numerous studies have been done on the microbial degradation of PAH, from which the following statements may be derived:

- Prokaryotic microorganisms metabolize PAH by an initial dioxygenase attack to *cis*-dihydrodiols that are further oxidized to dihydroxy products.

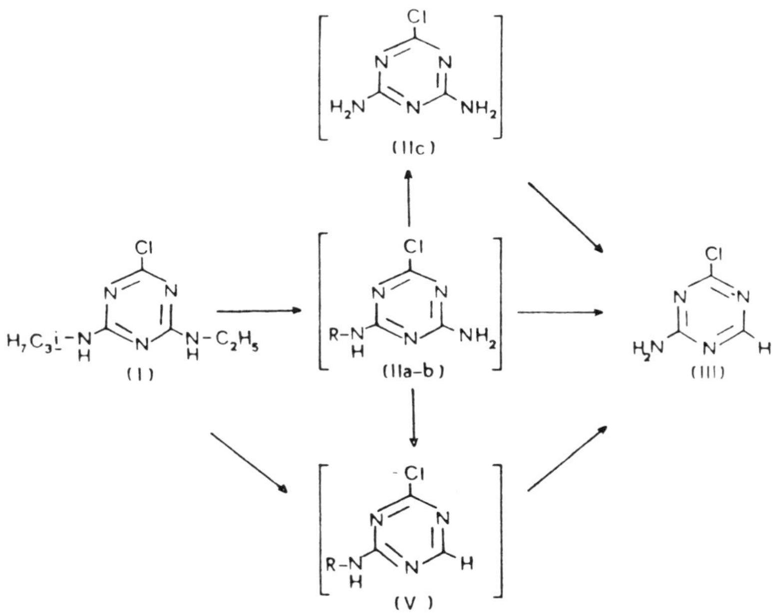

Figure 6. Microbial degradation pathways proposed for atrazine.

- Eukaryotic microorganisms use monooxygenases to initially attack PAH to form arene oxides followed by the enzymatic addition of water to yield *trans*-dihydrodiols.
- PAH with more than three condensed benzene rings do not serve as substrates for microbial growth, though they may be subject to cometabolic transformations.
- Lower weight PAH such as naphthalene are degraded rapidly, whereas higher weight PAH such as benz[*a*]anthracene or benzo[*a*]pyrene are quite resistant to microbial attack.
- Many of the genes coding for PAH degradation are plasmid associated.

The ability of microorganisms to completely degrade naphthalene, phenanthrene, and anthracene has been largely investigated. The degradative sequences are quite similar and lead to the formation of catechol as common intermediate. Little is known about the bacterial oxidation of PAH containing four or more fused benzene rings. Microorganisms can cometabolize these extremely insoluble PAH when grown on an alternative carbon source. Fungi have been shown to metabolize benzo[*a*]pyrene to a mixture of metabolites including epoxides, phenols, and diols, which are carcinogenic in higher organisms, but also detoxified derivatives, such as sulfate conjugates.

GENETICS OF BIODEGRADATION AND EVOLUTION OF NEW DEGRADATIVE PATHWAYS

Knowledge on the genetics of biodegradative pathways has been increasing in the last years, offering interesting and promising prospects of research and application.

It has been found that the large metabolic versatility of some bacteria, first of all *Pseudomonas*, is mainly due to the existence of plasmids, which code for the synthesis of enzymes involved in the degradation of a large number of aromatic compounds even if in many cases the same catabolic pathway may be encoded by chromosomal genes.

Table 1 reports the best known degradative plasmids, which encode all or part of hydrocarbon oxidation pathways that feed into central metabolism. Generally these pathways are inducible and

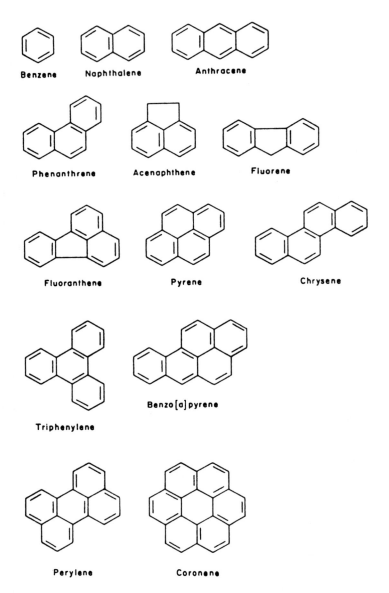

Figure 7. Polycyclic aromatic hydrocarbons.

comprise two sets of activities: *upper pathway* enzymes with specialized activities for conversion of the initial compound to a substrate for *lower pathway* enzymes with more generalized activities for conversion of this substrate to intermediates of central metabolism.

A well-characterized pathway is that for degradation of toluene and xylenes encoded by the TOL plasmid, pWWO, which represents the archetype of this class of plasmids[12] (Figure 8).

The widespread occurrence of transmissible plasmids carrying catabolic genes in members of this genus can contribute to the explanation of their enormous versatility, and also has implications for the evolution of new catabolic pathways.

When considering synthetic compounds that are recalcitrant to microbial attack, it is difficult and perhaps impossible to enrich cultures having any effect on those compounds. However, there is the theoretical possibility of constructing strains that can degrade recalcitrant molecules by

THE ROLE OF MICROORGANISMS IN THE ENVIRONMENT DECONTAMINATION

Table 1. Degradative Plasmids

	Plasmid	Degradative pathway
Pseudomonas	CAM	Camphor
	OCT	Octane
	SAL	SALSalicylate
	NAH	Naphthalene
	TOL	Xylene, toluene
	pEG	Styrene
	pCS1	Parathion hydrolysis
	pJP2	2,4-Dichlorophenoxyacetic acid (2,4-D)
	pWR1	3-Chlorobenzoic acid (3CB)
	pAC25	3-Chlorobenzoic acid (3CB)
	pAC29	3CB, 4CB, 3,5DCB benzoic acid
Alcaligenes	pJP1	2,4-D
	pJP3	2,4-D, 3CB
Klebsiella	pAC21	*p*-Chlorobiphenyl

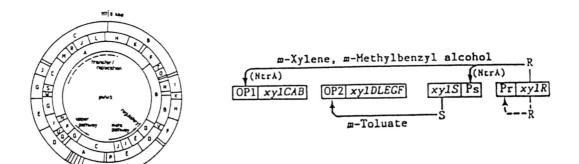

Figure 8. TOL plasmid pWWO.

using genetic engineering technology.[13] The hybrid metabolic pathways could be constructed in a number of ways, and two promising methods are (1) cloning multiple catabolic enzymes on appropriate vectors, and (2) the use of naturally cloned catabolic DNA located on degradative plasmids. The latter will be probably the more effective, as already suggested by Reineke and Knackmuss[6] and also by Chatterje and Chakrabarty[14] who have shown many significant examples of the role of plasmids in the degradation of some chlorosubstituted aromatic compounds as well their involvement in the "natural" construction of new catabolic functions through the continuous enrichment technique. The growth of mixed cultures and stable microbial communities establishes a situation in which organisms with different genetic backgrounds have the opportunity to exchange genetic information through different mechanisms, including plasmid-mediated transfer; the latter can be a highly significant process that contributes to the genetic flexibility of the soil and aquatic microbial communities.

244   CONTAMINANTS IN THE ENVIRONMENT

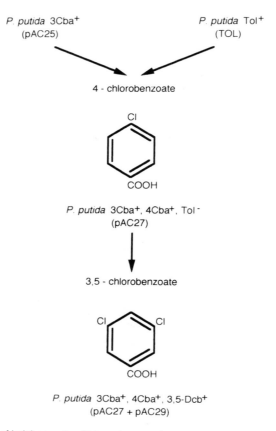

Figure 9. Construction of halobenzoate utilizing microorganisms.

The question is, how frequently do these events occur in the environment, in natural mixed populations? Degradative plasmids code for a variety of catabolic sequences, hence they are so widespread in nature; in fact, new catabolic plasmids continue to be discovered, including those coding for the degradation of chloroaromatics.

One possibility is that these catabolic functions have gradually been built up on a plasmid by sequestering genes from different organisms and leading to bacteria with new catabolic abilities. The "new" properties would have been selected by the new compounds introduced by man into the environment during the past 20 years: however, this is an extremely short period on an evolutionary time scale.

The production and release of large quantities of polychlorinated aromatics, as herbicides and insecticides, often characterized by persistence into the environment due to their low rates of biodegradation, have stimulated the development of strains that are capable of enhanced degradation of persistent toxic chemicals.

By exploiting appropriate combinations of degradative plasmids, strains have been constructed that affect total degradation of specific halogenated aromatics. Reineke and Knackmuss[6] demonstrated the transfer of TOL plasmid from the strain of *P. putida* mt-2 (WR101) to the strain of *Pseudomonas* B 13 (WR1), able to grow on 3-chlorobenzoate and 4-chlorophenol. Continuous enrichment culture allowed strains of *Pseudomonas* to be selected that had acquired the ability to grow on different chloroaromatics such as 4-chloro- and 3,5-dichlorobenzoates, which were not utilized by the parent strains.

Chatterje and Chakrabarty[14] demonstrated that plasmid pAC25, which permits total degradation of 3-chloro- but not of 4-chlorobenzoate, can cooperate with the TOL plasmid (specifying toluene and xylene degradation) for the complete metabolism of both these chlorinated compounds (Figure 9).

The TOL plasmid encodes a broad substrate-specific benzoate oxygenase, facilitating the conversion of 4-chlorobenzoate to 4-chlorocatechol, which can be degraded by pAC25-encoded enzymes. 4-Chlorobenzoate-degrading *Pseudomonas putida* strains have been isolated by growing cells harboring pAC25 with TOL-containing cells in a chemostat under selective conditions, with 4-chlorobenzoic acid as the major carbon and energy source. During enrichment, TOL sequences became transposed to the chromosome, and plasmid pAC27, carrying all the 3-chlorobenzoate degradative genes, was generated from pAC25 by deletion. In addition, 3,5-dichlorobenzoate-degrading strains could be obtained by introducing the TOL plasmid into cells capable of utilizing 4-chlorobenzoate and selecting with 3,5-dichlorobenzoate. These strains harbored both pAC27 and a second plasmid, pAC29, which comprised replication/incompatibility functions of TOL and duplicate copies of a part of pAC27, with modified chlorobenzoate degradative genes specifying utilization of 3,5-dichlorobenzoate.

The extended application of plasmid-assisted breeding under selective conditions in a chemostat should permit the isolation of organisms with an assortment of new degradative abilities for the utilization of particular toxic compounds. Gene cloning may be necessary for the construction of strains capable of degrading certain recalcitrant molecules that cannot be utilized by strains developed by plasmid-assisted breeding. Acquisition of novel degradative functions by a wide range of soil and aquatic microorganisms is likely to enhance the rate of degradation of pollutants and hasten their ultimate removal from the environment. Potential problems associated with this approach include competition by indigenous microflora and the presence of carbon sources in the soil that may be utilized in preference to the toxic chemicals. As an alternative to the use of engineered microorganisms for degradation of specific wastes, suitably manipulated microbial enzymes may be used as sprays or immobilized systems.

Knowledge of the fate and effects of manipulated microorganisms released into the enviroment is still scarce. It must be established which factors affect the persistence of the manipulated organisms into the environment, the rate of their growth, the rate of transfer of genes from manipulated organisms to indigenous ones, their spreading, and finally the impact on the ecosystems.

REFERENCES

1. **Fewson, C. A.,** Biodegradation of xenobiotic and other persistent compounds: The causes of recalcitrance, *Tibtec,* 6, 148, 1988.
2. **Gibson, D. T.,** Microbial transformation of aromatic pollutants, in *Degradation of Synthetic Organic Molecules in the Biosphere* Goldman, Ed., National Academy Science, Washington, D.C., 1982, 187.
3. **Perry, J. J.,** Microbial cooxidations involving hydrocarbons, *Microbiol. Rev.,* 43, 59, 1979.
4. **Galli, E., Barbieri, P., and Bestetti, G.,** Potential of Pseudomonads in the degradation of methylbenzenes, in *Pseudomonas, Molecular Biology and Biotechnology,* Galli, E., Silver, S. and Witholt, B., Eds., American Society for Microbiology, Washington, D.C., 1992, 268.
5. **Baggi, G., Barbieri, P., Galli, E., and Tollari, S.,** Isolation of a *Pseudomonas stutzeri* strain that degrades o-xylene, *Appl. Environ. Microbiol.,* 53, 2129, 1987.
6. **Reineke, W., and Knackmuss, H. J.,** Microbial degradation of haloaromatics, *Annu. Rev. Microbiol.,* 42, 263, 1988.
7. **Steiert, J. G., and Crawford, R. L.,** Catabolism of pentachlorophenol by a *Flavobacterium sp.,* *Biochem. Biophys. Res. Commun.,* 141, 825, 1986.
8. **Klages, U., Markus, A., and Lingens, F.,** Degradation of 4-chlorophenylacetic acid by a *Pseudomonas* species, *J. Bacteriol.,* 146, 64, 1981.
9. **Safe, S. H.,** Microbial degradation of polychlorinated biphenyls, in *Microbial Degradation of Organic Compounds,* Gibson, D.T., Ed., Marcel Dekker, New York, 1984, 361.

10. **Behki, R. M., and Shahamet, K. U.,** Degradation of atrazine by *Pseudomonas:* N-dealkylation and dehalogenation of atrazine and its metabolites, *J. Agric. Food Chem.,* 34, 746, 1986.
11. **Cerniglia, C. E., and Heitkamp, M. A.,** Microbial degradation of polycyclic aromatic hydrocarbons (PAH) in the aquatic environment, in *Metabolism of Polycyclic Aromatic Hydrocarbons in the Aquatic Environment,* Varanasi, U., Ed., CRC Press, Boca Raton, FL, 1987, 41.
12. **Worsey, M. J., Franklin, F. C. H., and Williams, P. A.,** Regulation of the degradative pathway enzymes coded for by the TOL plasmid (pWWO) from *Pseudomonas putida* mt-2, *J. Bacteriol.,* 134, 757, 1978.
13. **Mermod, N., Lehrback, P. R., Don, R. H., and Timmis, K. N.,** Gene cloning and manipulation in *Pseudomonas,* in *The Biology of Pseudomonas,* Sokatch, J.R., Ed., Academic Press, Inc., London, 1986, 325.
14. **Chatterjee, D. K., and Chakrabarty, A. M.,** Genetic rearrangements in plasmids specifying total degradation of chlorinated benzoic acids, *Mol. Gen. Genet.,* 188, 279, 1982.

CHAPTER 26

The Uses of Pesticides and Their Levels in Food in Eastern Europe: The Example of Poland

Jerzy Falandysz

The level of use of pesticides in Poland is low, e.g., from 0.6 to 1.1 kg of the active ingredient per hectare of soil yearly. Much higher application rates have been used by our neighbors and in many other European countries. The levels of pesticides found in tissues of slaughtered and game animals, and in vegetables and fruits in Poland are generally low. Among the organochlorine insecticides, DDT and analogues are the dominant residues. However, there is a marked decline in levels of the DDT complex in foods of animal origin and in soil. Apart from the organochlorines, the residues of organophosphorus, carbamate and pyrethroid insecticides, dithiocarbamate fungicides and benzimidazol (MBC) derivatives, as well as of herbicides were found in only a small percent of the vegetable and fruit samples or were nondetectable. From a hygienic point of view such a situation is favorable for consumers; however, it seems to reflect the relatively low application rate of pesticides.

INTRODUCTION

The use of pesticides in agriculture is one of the factors for crops husbandry. Beside the new, higher yielding varieties of crops irrigation, fertilizers, and rational tillage, the use of pesticides determines the intensity of agricultural production. In the opinion of experts in crop protection the needs of Polish agriculture for an increase in pesticide application are significant, because of a present limited protection program of cereals, potatoes, beets, and rape, as well as other species of cultivated plants. The advantages resulting from pesticide application are generally undisputed, but there is the question of the control of pesticide usage to minimize their unwanted side effects.

Systematic studies on pesticide residues in the tissues of animals, dairy produce, vegetables, fruits, and soil randomly taken throughout the country have been conducted since 1969–1971. The research program considers first of all organochlorines, organophosphorous, and other pesticides applied in large amounts on large areas and at late dates in relation to harvest. Information obtained is first of all used to review existing plant protection programs, giving consideration to possible changes in rates of pesticide application, waiting periods, and tolerance limits.

Table 1. Summary of the Data on Production and Consumption of Pesticides in Poland[1-5]

Year	Production	Consumption						Consumption (in the form of the active substance)	Production of DDT
		Total	Insecticides	Fungicides	Herbicides	Rodenticides	Others		
1938	ca. 470	—	—	—	—	—	—	ca. 20	—
1946	410	—	—	—	—	—	—	50	—
1950	3,800	—	—	—	—	—	—	1,790	147
1955	22,800	—	—	—	—	—	—	1,800	1,528
1960	51,600	45,013	41,883	1,969	1,031	60	151	4,370	3,142
1965	78,200	75,039	67,624	3,605	3,443	222	145	8,350	4,407
1970	59,900	53,728	42,841	4,039	6,522	123	203	7,691	3,487[a]
1975	70,100	59,971	39,812	3,818	15,933	193	216	11,201	—
1980	34,500	29,329	9,082	3,896	14,181	356	1,814	9,332	—
1985	37,400	36,526	11,459	4,525	18,942	241	1,357	12,398	—
1986	33,800	36,045	4,836	6,294	24,363	299	1,253	14,479	—

[a] = For 1969.
— = No data.

APPLICATION RATES

Years 1918–1939

The beginnings of the application of pesticides in Poland date back to the 1920s and 1930s. At first insecticides and fungicides were used. Compounds such as salts of arsenic and barium, Paris green, and fluorides were used as insecticides while sulfur, California mixture, copper sulfate, calomel, corrosive sublimate, and formaldehyde were used as fungicides. The first tests with herbicides were also undertaken. Sodium chlorate was used in nonagricultural areas, as well as kainite and lime nitrogen (calcium cynamide). Generally there is no quantitative data about the diffusion of pesticides in Poland in the 1920s and 1930s. Nevertheless, the volume of pesticides used seemed to be very small. Only 20 tons of pesticides (as active substances) were used in 1938 (Table 1).

Years 1946–1986

After the Second World War, the production and use of pesticides started from point zero. Soon after the war rodents—field mouse and field vole—were so plentiful that it was necessary to develop methods and agents against rodents. In the 1950s, the situation was very much the same. At the end of the 1940s an effort was also made to intensify the production and application of pesticides on a wide scale. The discovery of DDT and the great interest in that insecticide, the wish to intensify agricultural protection, and also the invasion of Colorado beetle (potato-beetle) in the 1950s stimulated capital expediture on the application of pesticides in Poland, as in other countries. Insecticides constituted the bulk of pesticides used in 1955–1975, and DDT was the major compound (Table 1). Up to 1956, arsenic insecticides were mainly used. After 1975 the application of organochlorine insecticides decreased, and the cyclodienes aldrin and dieldrin and DDT have all been banned, while melipax (toxaphene), Lindane, and methoxychlor are still used. The introduction of organophosphorous and carbamate insecticides into common agricultural practice allowed a decrease in the rate of application of the active ingredient per hectare of soil from about 2 kg (organochlorines) to 0.3–0.5 kg. Next, the pyrethoroids force out carbamates (carbaryl, propoxur, carbofuran, and pirimicarb) from the market and enable a further decrease of application rates.

The usage of fungicides was fairly constant between 1965 and 1980, but after that it progressively increased. Herbicides, when introduced into commom practice, became in the 1980s the main class of pesticides used in Poland, and their application rates increased rapidly, while the use of rodenticides and other groups of pesticides remained at the same level, with only minor

Table 2. Some Data on Consumption of Pesticides in Poland and Some other Countries in kg of the Active Substance/Hectare[6]

Year	Poland	Czecho-slovakia	Denmark	West Germany	Italy	USA
1962	0.36			2.76	2.11	0.57
1965				1.69		1.47
1970	0.59			2.43		1.79
1971	0.42	1.18				
1975	0.73	2.13	1.79	3.12		2.39
1978	0.65	3.13	2.21			
1979	0.62	3.24	2.28			
1980	0.62				4.12	2.68
1981	0.83					
1983	0.99					
1984	0.74					
1985						3.41

fluctuations. A large qualitative change has taken place during the past 40 years. During the 1950s, a dozen or so kinds of pesticides had been used but in the 1980s the number of active forms of pesticides used in Poland was more than 200. Nevertheless, the consumption of pesticides in Poland is low compared to many other countries and on the average is lower or much lower than 1.0 kg of the active ingredient/hectare/year (Table 2).

RESIDUES

Insecticides

Organichlorine Insecticides

Organochlorine insecticides constituted a large percent of the total quantity of pesticides used in Poland. DDT and analogues (ΣDDT) are the dominant residues found in tissues of domestic and game animals as well as in food products. α- and γ-hexachlorocyclohexane (HCH, BHC), and hexachlorobenzene (HCB) are also found, but in much lower levels and not in all the samples, while the residues of aldrin, dieldrin, heptachlor, heptachlor epoxide, methoxychlor, and toxahene are generally absent. The residues level of ΣDDT in the environment and in edible tissues of animals and food decreased during the past 19 years nearly 20-fold (Tables 3 and 4). This is also true for muscle samples from Baltic fish (Table 5). Nevertheless, in the years 1979–1985 some irregularities were observed, and also an arrest in the downward trend of DDT residues. The results of the investigations of animals and food products in Poland, and also of fish from the southern part of the Baltic Sea, unequivocally indicate that at the end of the 1970s and in the early 1980s, some DDT was still being used in Poland. It is difficult to say if it was exceptional and with special permission, or illegal. The mean of both α- plus γ-HCH found in adipose tissues of slaughtered and game animals is much below 30 µg/kg, and of hexachlorobenzene below 10 µg/kg. Furthermore the residue levels of ΣHCH and HCB during the past 10 years are nearly the same, without any clear trends.

Organophosphorus Insecticides

The monitoring program of organophosphorus insecticides includes the determination of the residue of 14 compounds (dichlorphos, diazinon, dimethoate, formothion, methyl parathion, pirimiphos ethyl, pirimiphos methyl, fenithrotion, malathion, chlorfenvinphos, demeton-s-methyl, thiometon, phenthion, and phosalone) in vegetables and fruits, and in tissues of slaughtered animals. None of the meat samples analyzed so far contained any dectable amounts of organophosphorus insecticides, while in the case of vegetables, and especially glasshouse vegetables, and

Table 3. Residue Levels of Σ DDT in Adipose Tissue of Cattle and Swine in Poland (ug/kg lipid wt)[7-12]

	Cattle		Swine	
Year	n	mean and range	n	mean and range
1969				2200
1970		920		1500
1971		470		1700
1972		420		970
1973		380		850
1974		150		630
1975		110		520
1976				390
1977		140		350
1978		160		370
1979		240		290
1980		200		250
	10	160 (59–660)	10	370 (300–530)
1979–81	2074	140 (10–2100)	3036	180 (10–1900)
1981	104	86 (4.6–710)	160	140 (<1.0–1300)
		110		150
1982	110	55 (3.3–260)	150	93 (7.0–420)
		110		210
1983	90	86 (10–740)	155	150 (16–760)
		130		220
1984	79	74 (8.3–830)	209	130 (1.0–1400)
		70		120
1985	65	130	350	180 (9–2900)
		80		140
1986	97	86 (11–740)	353	98 (1.4–810)
		70		100
1987	93	64 (6.7–270)	360	84 (3.8–1000)
1988[a]	27	53 (11–140)	159	75 (4.2–730)

[a] = For 6 months.

Table 4. Residue Levels of Σ DDT in Adipose Tissue of Game Animals in Poland (ug/kg lipid wt)[10,12-14]

	States		Roe-deer		Wild boar	
Year	n	Mean and range	n	Mean and range	n	Mean and range
1980					10	620 (180–1000)
1981	8	48 (trace–160)	3	86 (55–120)	55	320 (5.1–3400)
1982	5	60 (6.4–200)	5	48 (6.3–140)	50	150 (10–440)
1983	13	40 (18–130)	8	52 (7.2–110)	91	170 (32–610)
1984	15	40 (8–80)	15	63 (17–260)	84	250 (24–1300)
1985	45	28 (6–94)	24	59 (9–110)	146	250 (40–2100)
1986	57	18 (2.7–47)	36	53 (8.5–190)	177	310 (19–4300)
1987	56	44 (6.4–680)	41	69 (5.9–270)	141	340 (21–2300)
1988[a]	2	14 (11–17)	5	14 (5–45)	56	580 (82–3000)

[a] = For 6 months.

fruits, some samples were contaminated. Nevertheless the number of samples containing residues at a level exceeding the tolerance limits is small.

Carbamate and Pyrethroid Insecticides

Carbaryl, propoxur, carbonfuran, pirmicarb, cypermethrin, delthametrin, fenvalerate, and permethrin residues are monitored in vegetables and fruits only. The number of samples containing detectable amounts of carbamates or pyrethroids is very small. Until now, none of the vegetable or fruit samples contained residues of carbamates and pyrethroids at levels exceeding the tolerance limits.

Table 5. The Residue Levels of Σ DDT in Muscles of Herring from the Baltic Sea (ug/kg wet wt)[15-43]

Year	Western areas	n	Baltic proper	n	Northern areas	n
1965			230 (94–300)	9		
1967					260 (150–420)	4
1966–68			400 (12–1300)	18		
1968	350 (220–1200)	23				
1969	170 (40–1200)	130	4000 (1800–5500)	4		
1970	410	16	1200 (590–3300)	67	220	15
1968–70			1100 (150–5300)	111	530 (65–2500)	295
1969–71			1300 (110–7700)	527		
1971	2600 (280–7000)	37	460 (200–640)	3		
1972	550 (n.d.–4500)	98	1400 (100–5600)	11	380 (150–670)	192
1973	540 (n.d.–4000)	106	710 (n.d.–3200)	46	510 (98–1800)	50
1972–74			1000 (300–2500)	200		
1974	520 (70–2800)	~157	690 (300–1400)	25	190 (28–830)	155
1974–75			1000 (800–3300)	~100		
1975	340 (30–1250)	234	340 (50–3300)	355	230 (18–850)	258
1976	190 (30–340)	67	450 (100–2000)	107		
1977	150 (30–1500)	213	680 (30–2200)	598	410 (250–540)	366
1978	190	20	190 (140–480)	109	100 (16–2800)	93
1979	69 (36–99)	39	360 (51–700)	275	14 (9–18)	60
1980–81			320 (150–560)	339	260 (20–600)	377
1981	60 (17–86)	50	150 (15–870)	351	14 (5,3–27)	140
1982			56	10	140	7
1982–83	20 (<20–190)	24	<20 (<20–70)	23		
1983			240 (59–1100)	197		
1985			150	190		
1986			180 (130–280)	122		

n.d. = Not detected.

Fungicides

Organomercurials

The production and application of organomercurials for dressing of cereal seed started in Poland in 1954, and up to 1970 five formulations were registered (Table 6). There are no precise data on the quantity of mercury used for agricultural purposes in Poland. For example, 30 tons of mercury was utilized to synthesize organomercurials in 1970, while 208 tons was used for other purposes (Table 7). It has been estimated that in 1956–1970 a total of 217 tons of mercury was used for dressing seed, e.g., about 1 g of mercury per hectare yearly. However, in the years 1969–1970, it was nearly 1.66 g of mercury per hectare per year, assuming that the whole quantity produced in a given year was used in that year. The quantities of mercurials used in Poland for seed dressing did not have much of an impact on the environment. In the years 1971–1976 organomercury formulations were successively withdrawn from production. Nevertheless, some quantities of organomercurials were still used in Poland in the early 1980s. The investigations carried out in Poland on the degree of mercury residues in plants and animal tissues allowed us to determine that in general the pollution with mercury in Poland does not present toxicological hazards. The levels of mercury found in muscles and organs of slaughtered and game animals in Poland in the 1980s could be considered as low or very low (Tables 8 and 9).

In Poland the organomercury seed dressing agents were successively replaced with nonmercury fungicides. In Table 10 are summarized data on fungicides registered and applied in Poland and in other Eastern European countries in the year 1984. Because of many problems of economic and technical nature, withdrawal of mercury compounds for dressing seed gradually took place. Only Poland almost bans the use of organomercurials. Also the organomercurial formulations containing hexachlorobenzene and pentachloronitrobenzene are becoming unpopular. The best substitute products for mercury are compound formulations containing sulfur, manganese, thiuram, carboxine, carbendazin, and other benzimidazoles, mancozeb and Cu-oxine.

Table 6. Organomercury Compounds used in Poland as Seed-Dressing Agents in Agriculture

Compound	Phenylmercury acetate (PMA)
Trade names	Zaprawa nasienna R, pylista; Poland
	Zaprawa nasienna R. zawiesinowa; Poland
	Zaprawa nasienna RG, pylista; Poland
	Ceresan, West Germany
Notes	1954 – first produced (volume: about 500t/year)
	1971–74 – withdrawal (a mixutre with Lindane)
	1980 – withdrawal
	1954–1979 – probably about 12500t were produced (used)
	Mercury content – 2.4%
Compound	Ethylmercury P-toluene sulfonanilide
Trade names	Zaprawa nasienna Uniwersalna, pylista; Poland
	Ceresan M. Ceresan M-DB; USA
Notes	1966 – first produced
	1975 – withdrawal
	1966–71 – about 1300t were used
	Mercury content – 0.8%
Compound	Methylmercury dicyandiamide
Trade names	Zaprawa nasienna płynna 0.8; Poland
	Panogen, Panogen 8; Sweden
	Panogen 15, USA
Notes	1971 – first produced
	1975 – withdrawal
	1971–75 – probably about 350t were produced (used)
	Mercury content – 0.8%
Compound	Methoxyethylmercury chloride (MEMC)
Trade names	Ceresan Universal Nassbeize, West Germany
Notes	Imported
	1980 – withdrawal
Compound	Methoxyethylmercury acetate
Trade names	Zaprawa nasienna płynna M 2, Poland
	Panogen M, Panogen Metox; Sweden
Notes	Potentially used
	1975 – first used?
	1980 – withdrawal?
	Mercury content – 0.6–2.0%

Table 7. The Import and Consumption of Mercury in Poland in 1965–1972 (t)[44–48]

Year	1965	1969	1970	1971	1972	1973	1974	1975[a]	1976	1979
Import			235	270	203	252	208	260	357	228
Consumption										
Total			238	225	222	226	230	241	357	227.8
Chemical industry			200	184	183	186	194	209		182.3
Others			38	41	39	40	36	35		45.5
Agriculture[b]	9	20[c]	30	19	13	19	14	6		0

[a] = According to plan.
[b] = From the quota of the Chemical Industry.
[c] = 1969/71 about 23 t/year.

Dithiocarbamate Fungicides

The residues of thiuram, mancozeb, zineb, and maneb have been monitored in vegetables and fruits. In 1981 to 1985 quite a large percent of the vegetables contained detectable amounts of dithiocarbamates: from 18.5 to 30.6% in tomatoes, from 7.4 to 14.1% in cucumbers, and from

Table 8. Mercury in Muscles, Liver, and Kidneys of Swine and Cattle in Poland (ug/kg on a wet weight basis)[12,49-51]

Species and year	n	Muscles	n	Liver	n	Kidneys
Swine						
1972–75					831	44 (5–176)
					9[a]	(500–10800)
p.1976	30	5 (1–9)			30	50 (11–180)
1985–86	181	3 (<1–18)	181	5 (1–29)	180	8 (1–45)
1987	332	2 (<1–12)	355	4 (<1–19)	336	7 (1–130)
1988[b]	145	4 (<1–8)	143	6 (<1–17)	143	13 (3–66)
Cattle						
1972–75					819	18 (6–600)
p.1976	30	7 (4–14)			30	32 (13–58)
1979–81	565	7 (4–80)			875	17 (5–1800)
1985–86	68	2 (1–12)	69	5 (<1–16)	68	10 (2–54)
1987	92	1 (1–3)	92	3 (<1–19)	92	8 (2–29)
1988[b]	24	2 (1–4)	26	5 (2–10)	25	13 (4–43)

[a] = Samples excluded when calculated the mean.
[b] = For 6 months.

Table 9. Mercury in Muscles, Liver, and Kidneys of Game in Poland (ug/kg on a wet weight basis)[12,49]

Species and year	n	Muscles	n	Liver	n	Kidneys
Wild boar						
1985	2	9 (7–11)				
1986	100	4 (1–11)	2	41 (38–44)		
1987	168	4 (1–28)	43	14 (3–29)	10	42 (15–72)
1988[a]	42	6 (2–23)	17	24 (9–42)	17	56 (15–180)
Roe-deer						
1985	6	7 (3–13)				
1986	96	2 (<1–6)				
1987	149	1 (<1–10)	27	10 (2–36)	10	94 (31–260)
1988[a]	5	2 (1–2)				
Stag						
1985	2	17 (9–24)				
1986	74	2 (<1–6)	1	1	1	8
1987	102	1 (<1–10)	28	8 (2–17)	10	35 (16–54)
1988[b]	26	3 (1–7)	2	13 (2–23)	2	35 (2–65)
Elk						
1987	2	<1 (<1–<1)				
Aurochs						
1987	1	<1	1	1	1	5

[a] = For 6 months.

10.5 to 35.3% in the samples of lettuce. Some samples of tomatoes (0.4–5.3%), cucumbers (0.3–1.0) and lettuce (1.5–4.3) contained the residues of dithiocarbamates at a level exceeding the tolerance limit of 2.0 mg CS_2/kg.

Benzimidazole/MBC/Derivatives

The residues of carbendazin, benomyl, and thiophanate-methyl have been monitored in glasshouse vegetables, mushrooms, and fruits. Only a small number of the samples examined in 1981–1985 contained any detectable amounts of benzimidazoles. The tolerance limit (0.1 mg/kg) was exceeded in 2% of the glasshouse vegetables, 9% of mushrooms, and 3% of fruits.

Herbicides

The monitoring studies on residues of herbicides have been carried out in Poland since 1981, and they include only vegetables. So far, the residues of 12 (linuron, prometryn, metaxon, diquat,

Table 10. The Active Ingredients of Seed-Dressing Formulations Used in Eastern Europe (Situation in 1984)[a]

Active agent	Number of registered formulations						
	Bulgaria	Czecho-slovakia	East Germany	Poland	Romania	Hungary	USSR
Mercury compounds	1	2	7	0	2	4	1
Copper compounds	1	2	0	1(3)[b]	1	3	0
Thiuram	1	3	1	6(1)[b]	3	1	4
Mancozeb, Maneb	0	0	1	1(1)[b]	0	1	0
Hexachlorobenzene (HCB)	0	0	1	0	0	0	1
Pentachloronitro-benzene (PCNB)	0	0	0	1	1	0	1
Captan	1	1	4	1	0	2	0
Benzimidazoles	1	4	4	1(3)[b]	1	4	2
Carboxine	1	2	1	3(2)[b]	2	1	2
Triadimanol	1	0	0	0	0	1	1
Iprodion	0	1	1	0(1)[b]	0	0	0
Winchlozoline	0	0	1	0	0	0	0
Metaxyl	1	0	0	0	0	2	0

[a] = Mercury compounds remained in use also in France, Great Britain, and the Benelux countries.
[b] = Under registration.

paraquat, dinoseb, trichloroacetic acid, desmetryn, aziprotryn, lenacil, atrazine, and simazine) most frequently used herbicides have been determined. None of the samples analyzed so far contained any detectable residues of herbicides.

REFERENCES

1. **Rocznik Statystyczny,** 30, Główny Urzad Statystyczny, Warszawa, 1970.
2. **Rocznik Statystyczny,** 36, Główny Urzad Statystyczny, Warszawa, 1976.
3. **Rocznik Statystyczny,** 40, Główny Urzad Statystyczny, Warszawa, 1980.
4. **Rocznik Statystyczny,** 46, Główny Urzad Statystyczny, Warszawa, 1986.
5. **Rocznik Statystyczny,** 49, Główny Urzad Statystyczny, Warszawa, 1989.
6. **Bakuniak, E., and Moszczyńska, W.,** Directions in the development of plant protection in Poland in comparison to world tendencies during the last 25 years, Mat. XXV Sesji Nauk, *Inst. Ochr. Rośl. Poznań,* 27, 1985, (in Polish).
7. **Juszkiewicz, T., Stec. J., Niewiadowska, A., and Posyniak, A.,** Pesticide residues in animal tissues in the light of 11-year studies, Mat. XX Sesji Nauk, *Inst. Ochr. Rośl. Poznań,* 212, 1980, (in Polish).
8. **Juszkiewicz, T., and Niewiadowska, A.,** Residues of pesticides and polychlorinated biphenyls in tissues of animals, milk, eggs and environment in the light of 15-year-old studies *Med. Wet.,* 40, 323, 1984, (in Polish).
9. **Falandysz, J., and Falandysz, J.,** Residue levels of organochlorine pesticides in adipose tissue of swine and cattle from the northern part of Poland, 1980–1983, *Roczn. Państw. Zakl. Hig.,* 36, 215, 1985, (in Polish).
10. **Falandysz, J., and Falandysz, J.,** Organochlorine pesticides and polychlorinated biphenyls in adipose tissue fat of slaughtered and game animals from the northern region of Poland, 1984, *Roczn. Państw. Zakl. Hig.,* 37, 487, 1986 (in Polish).
11. **Falandysz, J., and Centkowska, D.,** Organochlorine pesticides and polychlorinated biphenyls in adipose fat of slaughtered animals from the northern region of Poland, 1986 *Bromat. Chem. Toksykol.,* 22, 176, 1989 (in Polish).
12. **Falandysz, J.,** unpublished data, 1988.
13. **Falandysz, J., and Centkowska, D.,** Organochlorine pesticides and polychlorinated biphenyls in adipose fat of game animals from the northern region of Poland, 1986 *Bromat. Chem. Toksykol.,* 22, 182, 1989 (in Polish).
14. **Falandysz, J., and Falandysz, J.,** Residues of organochlorine pesticides in adipose tissue of wild boars, roe-deer, stags, and elk, *Bromat. Chem. Toksykol.,* 19, 98, 1986 (in Polish).

15. **Andersen, K. S.,** Pesticider i danske levnedsmidler 1976–78, *Publikation nr.,* 46, 1980.
16. **Andersson, O., Linder, C-E., and Vaz, R.,** Levels of organochlorine pesticides, PCBs and certain other organochalogen compounds in fishery products in Sweden, 1976–1982, *Vär Föda,* 36 (Suppl. 1), 1984.
17. **Danish Marine Monitoring Methods and Data,** ΣDDT, dieldrin, PCB and mercury in fish from marine areas, 1968–1970, Part I, National Agency of Environmental Protection, 1977.
18. **Dybern, B.I., and Jensen, S.,** DDT and PCB in fish and mussels in the Kattegat-Skagerrack area, *Meddelande fran Havsfiskelaboratoriet, Lysekil, Nr.,* 232, 1978.
19. **Falandysz, J.,** Organochlorine pesticides and polychlorinated biphenyls in herring from the southern Baltic, 1981, *Z. Lebensm. Unters. Forsch.,* 179, 20, 1984.
20. **Falandysz, J.,** Organochlorine compounds and metals in muscle tissue of some Baltic fish, 1981–1983, *Roczn. Państw. Zakl. Hig.,* 36, 447, 1985 (in Polish).
21. **Falandysz, J.,** Organochlorine pesticides and polychlorinated biphenyls in herring from the southern Baltic, 1983, *Z. Lebensm. Unters. Forsch.,* 182, 131, 1986.
22. **Falandysz, J.,** unpublished data, 1986.
23. **Huschenbeth, E.,** Zur Speicherung von chlorierten Kohlenwasserstoffen im Fisch, *Arch. Fisch Wiss.,* 24, 105, 1973.
24. **Huschenbeth, E.,** Uberachung der Speicherung von chlorierten Kohlenwasserstoffen im Fisch, *Arch. Fisch. Wiss.,* 28, 173, 1977.
25. **Huschenbeth, E.,** Zur Kontamination von Fischen der Nord- und Ostsee sowie der Unterelbe mit Organochlorpestiziden und polychlorierten Biphenylen, *Arch. Fisch Wiss.,* 36, 269, 1986.
26. **International Council for the Exploration of the Sea,** Studies of the pollution of the Baltic Sea, *Coop. Res. Rep. No. 63,* 1977.
27. **Jensen, A.,** Harmful substances in fish and shellfish, The Danish monitoring program for 1979 and 1980 for the area covered by the Helsinki Convention, 1982.
28. **Jensen, S., Johnels, A.G., Olsson, M., and Otterlind, G.,** DDT and PCB in herring and cod from the Baltic, the Kattegat and the Skagerrak, *Ambio Spec. Rep.,* 1, 71, 1972.
29. **Jensen, S., Johnels, A.G., Olsson, M., and Otterlind, G.,** DDT and PCB in marine animals from Swedish water, *Nature (London),* 224, 247, 1969.
30. **Lipka, E., and Doboszyńska, B.,** Studies on DDT in fish of the Baltic Sea. I. DDT content of the fresh fish, *Bromat. Chem. Toksykol.,* 11, 167, 1978 (in Polish).
31. **Linko, R.R., Kaitaranta, J., Rantamaki, P., and Ernonen, L.,** Occurrence of DDT and PCB compounds in Baltic herring and pike from the Turku Archipelago, *Environ. Pollut.,* 7, 193, 1974.
32. **Luckas, B., Wetzel, H., and Rechlin, O.,** Ergebnisse der Trenduntersuchungen von Ostseefischen auf ihren Gehalt an DDT und seinen Metaboliten, *Acta Hydrochim. Hydrobiol.,* 8, 167, 1980.
33. **Luckas, B., and Lorenzen, W.,** Zum Vorkommen von chlororganisch Kusten Schleswig-Holsteins, *Dt. Lebenem. Rdsch.,* 77, 437, 1981.
34. **Maier-Bode, H.,** Untersuchingen uber den Insektizidgehalt von Ostseefischen, *Schr. Reihe. Ver Wass. Boden. Lufthyg.,* 34, 57, 1971.
35. **Miettinen, V., Verta, M., Erkomaa, K., Jarunen, O.,** Chlorinated hydrocarbons and heavy metals in fish in the Finnish coastal areas of the Gulf of Finland, *Finnish Fish. Res.,* 6, 77, 1985.
36. **Moilainen, R., Pyysalo, J., Wickstrom, K., and Linko, R.,** Tissue trends of chlordane, DDT and PCBs concentrations in pike (Esox lucius) and Baltic herring (Clupea harengus) in the Turku Archipelago, northern Baltic Sea, for the period 1971–1982, *Bull. Environ., Contam. Toxicol.,* 29, 334, 1982.
37. **Norén, K., and Rosén, G.,** Halter av klorpesticider och PCB fisk fran svenska vattenmraden, *Vår Föda,* 28 (Suppl. 1), 1976.
38. **Orbaek, K.,** Pesticider i danski levnedsmidler 1982–83, Publikation nr 107, Statens Levnedsmiddelinstitut, 1985.
39. **Paasivirta, J., and Linko, R.,** Environmental toxins in Finnish wildlife. A study of time trends of residue contents in fish during 1973–1978, *Chemosphere,* 9, 643, 1980.
40. **Perttila, M., Tervo, V., and Parmanne, R.,** Age dependence of the concentrations of harmful substances in Baltic herring (Clupea harengus), *Chemosphere,* 10, 1019, 1982.
41. **Roots, O. and Peikere, E.,** Polikloreeritud bifenuulide ja kloororgaanide pestitsudide Sisalduccd Laanemere Kaladed, *Keemia,* 27, 193. 1978.

42. **Tervo, V., Erkomaa, K., Sandler, H., Miettinen, V., Parmenne, R., and Aro, E.,** Contents of metals and chlorinated hydrocarbons in fish and benthic invertebrates in the Gulf of Bothnia and in the Gulf of Finland in 1979, *Aqua Fenn.,* 10, 42, 1980.
43. **Westöö, G., and Norén, K.,** Klorpesticid- och polyklorbifenyl-halter i fisk faangad i svenska vatten eller salufoerd i Sverige 1967–1970, *Vår Föda,* 22, 93, 1970.
44. **Baltic Marine Environment Protection Commission — Helsinki Commission.** Progress Reports on Cadmium, Mercury, Copper and Zinc. Baltic Sea Environment Proceedings No. 24, 1987.
45. **Bojanowska, A.,** The persistence of mercury compounds after the application of mercuric pickling, *Biul. Inst. Ochr. Rośl. Poznań,* 41, 103, 1971 (in Polish).
46. **Byrdy, S., and Fulde S.,** The influence of mercuric seed-dressing on the level of the environment contamination and the plans of chemical industry seed-dressing production, *Biul. Ins. Ochr. Rośl. Poznań,* 52, 197, 1972 (in Polish).
47. **Smart, N.A.,** Use and residues of mercury compounds in agriculture, *Residue Rev.,* 23, 1, 1968.
48. **Szprengier, T.,** Mercury in the environment — sources and contamination, *Proc. 1st Natl. Conf. Effects of Trace Elements Pollutants on Agricultural Environment Quality,* Pulawy, Poland, May 4 to 6, 1978, 73 (in Polish).
49. **Falandysz, J., and Gajda, B.,** Mercury content in the muscles, liver and kidneys of slaughtered animals and game in Northern Poland in 1985–1986, *Roczn. Państw. Zakl. Hig.,* 39, 113, 1988 (in Polish).
50. **Juszkiewicz, T.,** Residues of xenobiotics in animals, milk and environment in the light of a 13 year surveillance, in *Veterinary Pharmacology and Toxicology,* Ruckebusch, Y., Toutain, P.L., and Kortiz, G.D., Eds., MTP Press Ltd., Lancaster, 1983, 641.
51. **Juszkiewicz, T., Stec, J., Niewiadowska, A., and Szprengier, T.,** Contamination of the habitat and animals with chloroorganic insecticides, chlorobiphenyls and mercury, *Biul. Ins. Ochr. Rośl. Poznań,* 55, 195, 1975 (in Polish).

CHAPTER 27

Contributions of the World Health Organization to Pesticide Safety

Gary J. Burin

The World Health Organization, primarily through the International Programme on Chemical Safety, coordinates a number of activities that encourage pesticide safety. These activities include training programs in pesticide toxicology and safe use and the development of a variety of documents that are targeted to the governments, health officials, scientists, and individuals using pesticides. The Joint FAO/WHO Meeting on Pesticide Residues (JMPR) is the primary international body that considers the hazard of pesticide residues in food. The history and function of JMPR are discussed and the role of the JMPR is contrasted to a national regulatory agency.

OVERVIEW

The International Programme on Chemical Safety (IPCS) was established in 1980 by the signing of a Memorandum of Understanding between the United Nations Environment Program, the International Labour Organization, and the World Health Organization. Most of the activities related to pesticide safety at the World Health Organization are coordinated by the IPCS. Among the various activities of WHO concerning pesticides are the following.

Health and Safety Guides

These short guides, written for pesticides and industrial chemicals, give brief descriptions of toxicity, symptoms of poisoning, environmental concerns, antidotes, recommendations for safe handling, and protective clothing. They are written for the educated layman who has a need for this information. Summaries of toxicity and safe handling recommendations that are even more concise are produced in the form of International Chemical Safety Cards.

Environmental Health Criteria Documents

These provide detailed reviews of the scientific literature for certain widely used chemicals in the areas of toxicology, epidemiology, environmental fate, and toxicity. They are reviewed and approved by a committee of experts who meet for approximately 1 week to examine a given chemical. Although they do not set acceptable levels for pesticide residues in food, the Environmental Health Criteria document provides recommendations for the safe use of a chemical. It also identifies areas where further testing is needed to assess safety.

Poison Information Monographs

These are short documents that briefly review toxicity (acute, subchronic, and chronic) with a focus on the management of acute poisonings by pesticides, pharmaceuticals, industrial chemicals, and natural products. Their primary audience is intended to be medical professionals involved with the treatment of poisonings.

WHO/FAO Data Sheets on Pesticides

These guides describe the toxicity and safe handling of pesticides used in public health and include descriptions of the signs and symptoms of poisoning, emergency treatment, and recommendations for regulatory authorities. Their audience is intended to be public health officials and government authorities who must use pesticides to control disease vectors.

Training Programs

A variety of training programs are coordinated by WHO in the areas of pesticide safe use, toxicology, and risk assessment. These programs are often aimed at government officials and health professionals who must manage the use of pesticides in developing countries.

Joint FAO/WHO Meeting on Pesticide Residues

The reports and monographs arising from these joint Food and Agriculture/World Health Organization meetings are intended for use by governments and the Codex Alimentarius Commission in controlling dietary exposure of pesticide residues. JMPR recommends maximum residue limits (MRLs) and acceptable daily intakes (ADIs) for pesticides based on a review of all relevant toxicology, epidemiology, and residue chemistry data for a pesticide.[1] The MRL represents the maximum residue of a pesticide and its metabolites of concern, which may be present on crops with the use of good agricultural practices. The ADI, which is based upon the evaluation of toxicology data, represents the intake of a pesticide that could be ingested daily over a lifetime without appreciable risk. The ADI is applicable to the parent compound and those metabolites that occur in the experimental animal. The function of JMPR is discussed below.

THE JMPR AND CODEX ALIMENTARIUS COMMISSION

Austria first invited experts from various European countries to Vienna to discuss international standards for chemicals in food in the early 1960s. Out of this arose the Codex Alimentarius Commission (Codex), which establishes food standards for foods in international commerce. The purposes of Codex are "(a) to protect the health of consumers and to ensure fair practices in food trade; (b) to promote coordination of all food standards work undertaken by international, governmental and nongovernmental organizations; (c) to determine priorities and initiate and guide the preparation of draft standards through and with the aid of appropriate organizations; (d) to finalize standards and, after acceptance by governments, (e) to publish them in a Codex Alimentarius either as regional or as worldwide standards."

Many aspects of food safety are therefore considered by the Codex and its various subsidiary bodies. Nutrition, labeling, hygiene, and irradiation are important concerns in addition to the safety of chemicals that may appear in food. Among the various chemicals considered are food additives, veterinary drugs, and industrial contaminants, as well as pesticides. Today, more than 130 countries are members of the Codex and have an opportunity to comment on Codex recommendations concerning pesticides. The primary group dealing with pesticide food residues is the general subject committee, the Codex Committee on Pesticide Residues (CCPR). This group meets annually in The Hague in the Netherlands to consider recommendations for maximum residue limits (MRLs) and their implications for international trade, set priority for JMPR review, and consider problems related to pesticide residues in international trade. One objective of this Committee is to harmonize MRLs between countries and thus facilitate trade.

The first joint meeting of the Food and Agriculture Organization and World Health Organization Panel of Experts concerning pesticides took place in 1961. The topic of the meeting was "Principles governing consumer safety in relation to pesticide residues." In 1963 and then yearly from 1965, joint meetings have been held to discuss individual pesticides and the basic principles for the evaluation of their hazard. These sessions have been conducted under the title "Joint FAO/WHO Meeting on Pesticide Residues." The site of the meeting alternates yearly between Rome and Geneva. More than 200 pesticides have been evaluated since the first meeting. JMPR recommended MRLs and ADIs are sent to the CCPR for consideration. The MRLs that are accepted by the CCPR are placed into a step procedure for adoption by the Codex Alimentarius Commission.

The pesticide manufacturers are not members of JMPR, Codex, or CCPR. However, they are represented at Codex meetings by their trade organization (GIFAP, Groupe Internationale des Associations Nationales des Fabricants de Produits Agrochemique) and they sit in on CCPR meetings to facilitate discussions regarding individual chemicals and the process of industry/governmental interaction. Representatives of industry do not participate in JMPR sessions. However, JMPR may adjourn to discuss the toxicology or residue chemistry of an individual chemical with scientists representing the manufacturer of a pesticide. The majority of data generated in the areas of pesticide toxicology and residue chemistry are sponsored by pesticide manufacturers. Government research institutions and university laboratories also generate data.

The JMPR recommendations for ADIs for pesticide residues in food, MRLs and certain use restrictions are made by a panel of internationally recognized experts. Although the meetings are considered joint FAO/WHO meetings, the WHO Secretariat coordinates the toxicology aspects of the meeting and the FAO secretariat coordinates the residue aspects of the meeting. Chairmen, vice chairmen, and rapporteurs are selected by the panels to assist in reaching concensus opinions and the preparation of reports containing recommendations. WHO Temporary Advisers are selected by the Secretariat to prepare working papers on individual compounds that will serve as the starting point for discussion. A critique of each working paper is prepared by a Member who also proposes an evaluation. The entire JMPR, consisting of Members, Secretariat, and Temporary Advisers, reviews each working paper. After discussion and modification of the working paper and editing by the Secretariat, it is published in the form of a monograph by WHO.

The JMPR reports contain recommendations for Acceptable Daily Intakes based on No-Observed-Adverse-Effect-Levels (NOAELs) that have been identified in key studies and using safety factors that have been agreed on by the JMPR. Essential data requirements are identified as are data considered to be desirable. The key aspects of the toxicity of the pesticide are reviewed in the JMPR report and the toxicology and biochemical studies that were reviewed are summarized in the JMPR monograph (called "Evaluations"). Separate monographs are published in the areas of toxicology and residue chemistry.

CONTRAST OF JMPR WITH NATIONAL AUTHORITIES

Unlike a regulatory body such as the United States Environmental Protection Agency (U.S. EPA), the decisions of the JMPR are not enforceable by law. They are recommendations to

Member States, which may or may not be accepted by those member states, and to the Codex Alimentarius Commission. The following illustrate some of the important differences between the JMPR and a regulatory authority such as the U.S. EPA:

Requirements for a Change in Use

The U.S. EPA can require very specific changes in the use pattern of a pesticide in the United States. The changes that would influence the pattern of pesticide residues may involve limitations in the crops for which a pesticide can be used, limitations in the rates of application, or geographic restrictions. There may also be restrictions in how treated commodity can be used, e.g., prohibition of the feeding of treated corn to cattle. Specific recommendations are made by JMPR but implementation is left to the discretion of the individual member states.

Cancellation/Suspension

In response to a perceived health or environmental hazard or due to the failure to obtain necessary data from the manufacturer, the U.S. EPA can choose to cancel or suspend some or all uses of a pesticide. A cancellation allows a pesticide to be used during the period of 1 to 2 years that may be required for an administrative hearing. A suspension requires that use immediately cease, even if an administrative hearing has not yet occurred. A suspension is made only in the case of an imminent hazard to the public health. JMPR cannot, of course, prohibit a pesticide from being used in agriculture. However, they have recommended that pesticides not be used where residues may result in food.

Requirements for Data

Regulatory authorities can require that data be submitted either prior to the use of a pesticide or as a condition to continued use. The JMPR can determine that specific data are required to establish an ADI but cannot require data to be submitted as a condition for the use of a pesticide.

Laboratory Audits

Governmental authorities can audit the work of toxicology testing laboratories to examine the raw data that serve as the basis of reports submitted in defense of a pesticide product. WHO does not audit studies, although they can encourage individual governments to undertake such an audit.

THE BASIS FOR A JMPR EVALUATION

Generally, an extensive battery of tests is performed on pesticides used in agriculture, which include long-term studies in rats, mice, and dogs, a reproduction study in the rat, teratology studies in rats and rabbits (or mice), a metabolism study, a variety of acute studies, and mutagenicity studies in prokaryotes and eukaryotes by regulatory agencies.[2] A variety of other tests in the disciplines of ecological toxicity, environmental fate, and residue chemistry are also generally undertaken by the pesticide manufacturer (of these, only residue chemistry studies are considered by JMPR).

A review of all available toxicology data is undertaken with particular emphasis placed on human data, which may provide useful information on the sensitivity of the human species to manifestations of toxicity associated with the pesticide. Observations regarding toxicity in humans often are not available, especially for new chemicals. Those data that are available often consist only of reports of acute toxicity that have been observed in pesticide applicators, in workers involved with pesticide manufacture, or in cases of suicide or accidental exposure. Much emphasis is thus placed on the available animal data for the prediction of the safety of pesticide residues in human food. The lowest NOAEL is identified in the animal studies. This exposure level, expressed in mg/kg body weight/day, is then divided by a safety factor that is intended to account for the potential greater sensitivity of humans (compared to the test species) and the variation in sensitivity that is expected to occur within the human species.[3] In 1954, a safety factor of 100 was proposed to be used in those cases where data were not available to more precisely define the inter- and intraspecies differences in sensitivity. This factor is still commonly used when human data are not available. The factor may be increased to account for an unusually great concern for the observed effect, uncertainty in the NOAEL, or a lack of confidence in the overall data base. Intake of a pesticide in food may then be compared with the ADI proposed by JMPR to determine whether a dietary hazard has resulted (or will result) from the use of a pesticide. The ADI can also be compared with exposure through drinking water to determine the extent of hazard resulting from contamination of water by a pesticide.

The International Programme on Chemical Safety has published a document that outlines the approach to data evaluation and risk assessment used by the JMPR. This document, which has undergone extensive review by past members of the JMPR and by other international experts, is entitled "Principles for the Toxicological Assessment of Pesticide Residues in Food."[4] It discusses the JMPR approach to risk assessment in areas such as carcinogenicity, developmental and reproductive toxicity, neurotoxicity, mutagenicity, and other forms of toxicity. A similar document has been published that discusses the risk assessment of food additives and contaminants.[5]

The approach that is taken to assess the risk posed by pesticide residues varies somewhat among governments and international bodies. One area of difference is the interpretation of animal studies showing the induction of neoplasia that are used in determining the risk posed by pesticide residues. Some governments develop a cancer potency estimate based on the 95% confidence interval of the upper bound of the slope of the dose–response. This approach is predicated on the assumption that any increase in exposure will result in some increase in risk. The JMPR, on the other hand, will generally assume that cancer caused by nongenotoxic carcinogens, like that of most other forms of toxicity, has some threshold level below which there will be no increase in cancer risk. A safety factor can then be applied to the NOAEL for carcinogenicity to reach an ADI.

REFERENCES

1. **Vettorazzi, G., and Radaelli-Benvenuti, B. M.,** *International Regulatory Aspects for Pesticide Chemicals,* Vol. II, CRC Press, Boca Raton, FL, 1982.
2. **US EPA,** *Health Effects Test Guidelines,* EPA 560/6–82–001, U.S. Environmental Protection Agency, Washington, D.C., 1982.
3. **Klaassen, C. D.,** General Principles of Toxicology in *Casarett and Doull's Toxicology: The Basic Science of Poisons,* 3rd ed., Klaassen, C. D., Amdur, M. O., and Doull, J., Eds., Macmillan, New York, 1986.
4. **International Programme on Chemical Safety,** *Principles for the Toxicological Assessment of Pesticide Residues on Food,* World Health Organization, Geneva, 1987.
5. **International Programme on Chemical Safety,** *Principles for the Safety Assessment of Food Additives and Contaminants in Food,* World Health Organization, Geneva, 1987.

CHAPTER 28

World Policy in the New Environmental Age

David E. Alexander

INTRODUCTION

This chapter takes stock of progress in the conservation and management of the global environment during the 20 years that elapsed between the United Nations Conference on the Environment (Stockholm, 1972) and the Conference on Environment and Development (Rio de Janeiro, 1992). It also examines the status and prospects of environmentalism in the wake of the Rio conference.

At present, humanity appears to be entering a new phase of its relationship with nature. In many parts of the world environmental consciousness has become an established force of both political and economic significance. As environmentalism comes of age, strategies are being reassessed, aims reformulated, and the tactics used to manage natural resources, hazards, and living species reexamined. This has resulted in some progress toward global conservation and sustainable development, but the environmentalists still have some very serious obstacles to tackle.

The roots of modern environmentalism are diffuse.[1] These are some of them:

1. The work of inspired luminaries has been a guiding force: St Francis of Assisi, George Perkins Marsh, Vladimir Vernadsky, Pierre Teilhard de Chardin, Barbara Ward, and Max Nicholson are six names selected from a potentially very long list.
2. The conquest of near space on the one hand has allowed us to observe and monitor the depredations and changes occurring in the global environment, and on the other hand portrays the world as a small, finite, and very fragile dynamic system.
3. Cumulative evidence from repeated and sustained scientific work has shown incontrovertibly that worrying trends are appearing in the state of the atmosphere, biosphere, hydrosphere, and lithosphere.
4. Political and economic leaders have to some extent been forced to face up to the fact that not all of the environmental costs of development can be ignored, hidden, or postponed, and that some are substantial obstacles to growth at the present moment.
5. Unrestricted growth has been an object of both capitalism and traditional marxism, and this to a certain degree has forced environmentalists to politicize in order effectively to present an alternative strategy.

6. *Glasnost* and the breakdown of authoritarian socialism have revealed the scope of environmental depredation in countries of the former Eastern Bloc, which has facilitated a world view of such problems. Where the regimes still survive, as in China, the economic necessity of foreign trade and investment has laid the environmental imperative bare to international scrutiny.

According to some researchers, widespread misperception—or lack of vision—is a root cause of the current ecological crisis. The effects of misuse of the environment may be widely dispersed in space or time (thus appearing with a substantial lag), may not yield themselves up to scrutiny easily, and may thus be irreversible by the time they are perceived as problems. This syndrome appears in the research literature as the "tragedy of the commons,"—i.e., that human outlook is rarely based on the fact that the resources contemplated are finite and limited.[2]

WHAT IS THE NEW ENVIRONMENTAL AGE?

To begin with, concern about environmental problems is not a new phenomenon. For example, increased acidity of rain- and snowfall was observed in Europe as long ago as 1872. Indeed, severe pollution at a highly localized scale beset manufacturing production long before the industrial revolution.[3] The human impact on natural systems was ably evaluated as long ago as 1864 by George Perkins Marsh in his book *Man and Nature*.[4]

However, the first two or three decades after the Second World War constituted a period in which technology began to take effect and the exploitation of natural resources became, not merely profound, but also for the first time precisely monitored. It was a period in which the sustainability of world resources, especially the nonrenewable ones, first came to be questioned seriously and repeatedly. Moreover, the birth and development of ecological science revealed the complexity and persistence of impacts, which could evidently reach much further into the biosphere than simple cause and effect would indicate.

The *new environmental age*[5] represents the adaptation of the environmental movement to shifting parameters and a new set of problems. The oil crisis of 1973 sent the industrialized countries into a phase of economic retrenchment from which they have had great difficulty in recovering. Hence, many recent outbreaks of pollution and environmental contamination can be traced to lack of investment over the last 15 years in the measures needed to prevent leaks or purify waste products that have been discharged into the environment. Such is the result of one overriding but outdated concept: that environmental protection is a luxury, rather than a necessity, and that environmental costs are not part of normal profit-and-loss economics.

Accordingly, the new environmental age begins with the realization that resource utilization, pollution, damage to flora and fauna, and natural and technological hazards all exert costs which must inevitably be taken into consideration.[6] This has led to an intensification, and perhaps also a redefinition, of the debate on the relationship between development and conservation. Some degree of development is clearly necessary in order to provide the technological and monetary resources with which to carry out conservation. Too little conservation, on the other hand, and resources will become exhausted; hence the debate about sustainability and the neo-Malthusian reintroduction of the "carrying capacity" concept of the earth and its regions.[7]

This question has been treated with increasing sophistication recently. Growing integration of the world economy means that sustainability can no longer be defined in spatially discrete terms: many resources and assets can be transferred with relative ease from areas of surplus to those of deficit. Do we require conservation, renewal, transformation, or substitution of natural resources? To what extent should we aim merely to satisfy basic human needs (food, warmth, and shelter) rather than trying to serve more sophisticated desires embodied in rising expectations? The inclusion of ecological variables in economic models leads both to more rational development strategies and to consideration of their hidden costs.[8] Recent moves in this direction have been partly responsible for a general retreat (also amenable to characterization as a loss of fashionable status) from ''deep ecology,'' in which humanity is allowed relatively little opportunity to dominate nature, toward ''shallow ecology,'' in which anthropocentrism and technological solutions play a much greater role.[9]

The industrialized countries, with 30% of the world's population, are responsible for 90% of its pollution. In fact, the relationship between high per capita income and rate of generation and accumulation of wastes, many of which are directly or indirectly toxic, has never been clearer.[10] Consumerism in western societies and the residual effects of excessive central control in the former Eastern Bloc countries have both conspired to depoliticize the populations who are subject to them, while at the same time tending to draw a veil over the environmental damage that they are creating. The reaction has been the remarkably ubiquitous birth of Green politics as a minority force for socioeconomic change. But in the countries of the Third World and former Eastern Bloc, environmentalists are often only small and beleaguered minorities.[11] The attitude there is usually one of ''industrial growth at any price, and then tackle the environmental problems that this policy has created.'' While bearing in mind that many of these countries simply cannot afford to add to the costs of development, excessive use of resources without replacing them and uninhibited industrialization without adequate pollution control will both impose severe costs in the long-term. But under the duress of recession, developed countries tend to be reluctant to assume much of the responsibility for safeguarding the rest of the world's environment.

In summary, the new environmental age stems from a redefinition of the relationship between development and conservation in an age of rising technological sophistication. Increasing global interconnectivity and the transboundary nature of many environmental problems have led to the emergence of a global structure for tackling the issues at hand: the World Environment Centre was set up in 1974, and the World Commission on Environment and Development in 1983. The World Conservation Strategy is a joint initiative that has the World Wildlife Fund and United Nations Environment Programme at its core.[12] Further evidence of the new international order is given by the remarkable similarity in environmentalists' strategies, and the problems that they are facing, at the local level in diverse parts of the world.

WHAT ARE THE MAIN ENVIRONMENTAL PROBLEMS?

The world's principal environmental problems have crystallized into the following interconnected groups, which we must review in order to know what world policy needs to tackle.

Atmosphere

The composition of the earth's atmosphere is not fixed and has altered considerably with the evolution of life forms. However, human activity has caused it to change at a rate that may prove too fast for ecological systems to adapt to in the short term. Each spring the ozone (O_3)-rich layer of the atmosphere above Antarctica is partially depleted and there are fears that this protective stratum is becoming progressively impoverished.[13] The industrialized nations have started to eliminate the production of chlorofluorocarbons (CFCs), which are thought to be at least partially responsible for the loss of ozone, but, given that these compounds are by no means essential to industrial production, the rate of curtailment of their manufacture is unacceptably slow.

Complex models of fossil fuel consumption, economic growth, and governmental policies suggest that under optimistic conditions mean ambient temperature could stabilize at 1–1.75°C above 1960 levels by A.D. 2100, while pessimistic forecasts suggest a rise of up to 2.5°C if fossil fuel consumption is not curtailed.[14] Hence CO, CH_4, N_2O, CFC, and especially CO_2 accumulations in the atmosphere would limit solar reradiation. Melting polar ice would contribute to a sea level rise of 1.4–2.2 m; the Maldive Islands would virtually disappear, up to 30% of Bangladesh would be permanently submerged, and throughout the world populous low-lying areas would suffer physical and economic damage. It is, however, possible that these changes would provoke a climatic reversal leading to increased global cooling in relatively few decades, and hence the suite of problems faced by humanity would be different.

Lithosphere

The Worldwatch Institute of Washington, D.C., has estimated that 26,000 million tons of topsoil are eroded each year in excess of new soil formation.[15] Sediment discharge data for the world's major rivers compared with radiometrically dated lake and ocean cores (which indicate past rates of sediment accumulation) suggest that world erosion rates may have increased 260% in the last 3500 years, which is a clear indication of the impact on the land surface of agriculture and clearance of natural vegetation.[16] Mismanagement is undeniably the principal reason why 60,000 km^2 of land becomes desertified (unproductive) each year.[17]

Hydrosphere

The extent of toxic contamination of the hydrosphere is unknown at the world scale.[18] However, pesticides and heavy metals may have residence times of at least 10^3–10^4 years in the ocean basins, where they enter the food chain in concentrated form through biological magnification. Thirty-two American states have groundwater that is contaminated by a range of up to 50 pesticides. Moreover, throughout the industrial northern latitudes, lakes have acidified (in Norway, for example, the pH of rivers has dropped from 5.8 to 4.6–5.0).

Demand for water is high and increasing in the world's expanding megalopoli, to the extent that water tables have fallen in parts of Africa, China, India, and North and Central America as aquifer recharge rates are consistently exceeded by extraction. Moreover, it is widely being predicted that the next Middle Eastern war will concern water supplies rather than oil resources.[19]

Biosphere

Concepts of evolution, survival of the fittest, and genetic diversity will take on a new meaning as the rate of species extinction increases in response to mankind's exploitation of habitats, food sources, and the flora and fauna themselves. Before 1900 species were being eliminated artificially at a rate of perhaps only 25 per century. In 1974, a thousand species disappeared under human duress, but the present rate is closer to 7000 per year, such that one-fifth of all living species will disappear during the next two decades,[20] although the veracity of this forecast depends on whether the number of living species is as high as recent estimates suggest (15–20 million). Among at least 730 species of mammal threatened with extinction are 23 kinds of whale, and four species of rhinoceros. Although it is a step in the right direction, the Convention on International Trade in Endangered Species (CITES) has not solved the problem and is in any case perpetually vulnerable to loopholes, or even outright abrogation.[21]

In no place is the depletion of the genome more profound than in the tropics. Rainforests have survived in some form since the Cretaceous period, but are now disappearing at the rate of 100,000 km^2 per year,[22] with a consequently profound impact on carbon exchange, climate, surface water quality, soil stability, nutrient cycling, genetic diversity, and—through the general circulation—perhaps also on the thickness of the ozone layer at the poles. Population pressure is clearly one of the principal reasons for the expansion of slash-and-burn agriculture beyond sustainable limits. But this is not the only problem associated with the rainforest that has proven intractable to international mediation. In many cases, political conniving lies behind rank exploitation: the Balbina Dam in Brazil, for example, generates only 112 MW of electricity for the city of Manaus, but has flooded 2360 km^2 of rainforest, one-third of it less than 4 m deep.[23] In such instances political and economic power struggles are responsible for propagating a "hollow frontier" of development,[24] which involves a very short cycle of consumption without replacement, after which the virgin land rapidly becomes sterile and unproductive. These forces seem often to be impervious to international pressure for better conservation.

Although genetic resources are being reduced by intentional or inadvertent exploitation, they can at least be managed with the help of biotechnology, which is defined as "the application of scientific and engineering principles to the processing of materials by biological agents to provide goods and services."[25] The field includes manipulation of recombinant DNA, artificially induced mutations, intensive breeding for artificial selection, use of pesticides or antibiotics to cause directional selection, and introduction of exotic species as a means of biological control.[26] Fears have been expressed, for example, about potentially uncontrollable multiplication of pathogenic microorganisms in favorable and extensive environments. The diffusion of the myxoma virus, and that of fungal pathogens associated with chestnut blight and Dutch elm disease, implies just how terrible the spread of a genetically engineered pathogen might be.

One other potential side effect of technology concerns possible manned exploration of Mars, which has a thin atmosphere that is potentially capable of supporting organic life. This raises the twin spectres of contamination of the Martian environment with terrestrial microorganisms and back-contamination of Earth with organisms from Mars, if such exist. However, the natural ecological controls on environmental proliferation tend to be strict, and organisms that are novel to a particular habitat or niche tend not to fare well. Yet the environmental impact of biotechnology is largely unknown and, as so few data are available, few attempts have been made to predict it.

WHAT ARE THE OBSTACLES TO THE SOLUTION OF THESE PROBLEMS?

Apart from the more obvious difficulties (such as insufficiency of funds, research, data, organizations, and appropriate technology), there are several particularly formidable obstacles to the solution of current environmental problems.

Materialism and Growing Ignorance

It is increasingly recognized that neither research nor implementation of environmental programs can take place without recourse to value systems and ethics.[27] On the one hand, there has been an enormous diffusion of environmental consciousness and to varying degrees a sustainable earth ethic, and this has at least partly shaped the present-day environmental movement. But on the other hand, the growth of consumerism has led to a retreat from "deep ecological" values. In developed countries individualism has replaced collective activity, responsibility has been overshadowed by rights, and social justice has become subjugate to market economics. The new generation is often found lacking in basic skills, ideological foundations, an appreciation of global issues, and appropriate breadth of education.[28] Under such conditions, which represent the strictures of purely economic rationality, it becomes difficult to teach concepts of value that have no monetary or tangible aspect. The paradox is that a society so rich in information flows should have become so superficial in its treatment of multifarious problems, while the environment remains an issue that defies adequate analysis from a sectoral or partial, rather than holistic, viewpoint.

Technocentric Approaches

It is widely recognized that technical development has been so far-reaching and rapid that society has been unable to evolve at a comparable rate and hence has had difficulty in assimilating the changes wrought by technology.[29] One present area of debate concerns the extent to which the development and use of technology is compatible with ecological conservation. Technology is both the potential enemy and possible friend of environmental management, in the sense that it may consume resources wastefully and permanently, yet it may also be capable of providing sophisticated solutions to environmental problems. Thus we may regard it as inherently separate from the values systems in which it is embodied. The more fundamental problem concerns the way in which technology is used—that is, the values and approaches employed by decision makers.

In the pursuit of technocracy, society has tended to place excessive reliance on data and quantification. When facts and numbers are treated as explanation in their own right, and when mankind attempts to assign numerical values to intangible factors, then ethics, objectives, and value systems become purely intrinsic and are susceptible to being ignored or taken for granted. Hence, *tactics* replace *strategy*, even though the necessary solutions to problems may be far more psychosocial than mechanistic.[30,31] Mechanistic approaches to environmental management, especially those in which quantification seems to guide the experimental design, tend to assume an unwarranted level of ecosystem stability. However, given the complexity of the environment, excessive ecocentrism may be as counterproductive as the stronger forms of technocentrism.

Economic Retrenchment and International Debt

The ups and downs of the world economy have caused the fortunes of environmentalism to fluctuate. As mentioned above, after the oil crisis of 1973 the developed countries tended to reduce their commitment to environmental protection in order to maintain economic growth. Meanwhile, environmental deterioration continued unabated. For example, acid deposition may have damaged half of the forest trees of Germany, especially in stands of Norway spruce, oak, and beech. Moreover, by 1980 the European Community was producing 480,000 tons of CFCs each year. In Athens, increasing car ownership coupled with the fact that 60% of vehicles do not meet basic exhaust emissions standards has led to a perennial smog problem. The question, therefore, is to what extent people will be willing to pay higher taxes and prices, and forego consumer benefits, and will companies be willing to sacrifice profits, in order to tackle problems like these?

Third World countries are often constrained to view environmental protection as a luxury that they cannot afford. Crippling debt burdens may prevent them from industrializing under their own impetus: hence, investment capital returns to the richer lending countries, and any residue is invested abroad because the home country offers little chance of economic success. Developed countries offer aid that is tied more to the involvement of their own companies and technical staff than projects that truly benefit the developing country targeted for assistance. Multinational companies, on the other hand, may be interested primarily in keeping their costs down by using cheap labor and facilities in the Third World, although there is evidence that some, at least, are sensitive to international criticism that they are exploiting the environment, not merely the workers. Yet in the poorer countries of the world, especially where population pressure is high, there may be moves to industrialize at any cost, and to put off consideration of environmental impacts. The problem is complex, as Blaikie[32] has noted:

> Thus environmental degradation is seen as a *result* of underdevelopment (of poverty, inequality and exploitation), a *symptom* of underdevelopment, and a *cause* of underdevelopment (contributing to a failure to produce, invest and improve productivity).

The "Implementation Shortfall"

It is not uncommon to find that proposals for environmental protection regulations, or even existing laws themselves, are conceptually good but are difficult to implement in practice. Examples of such situations include emissions standards that cannot be reached, safe disposal methods that are economically infeasible (or technically unattainable where large volumes of material are concerned), regulations that contain loopholes, and inspectorates that are too small and deprived of resources to be able to enforce compliance. The United States' Council on Environmental Quality, for instance, has been concerned that it is increasingly difficult to ensure compliance as the regulatory focus shifts from requiring initial investment in pollution control equipment to enforcing emissions standards over time; that the growth in the number of pollutants, pollution sources, and rules may overwhelm the resources available to monitor emissions and effluents adequately; that the high cost of monitoring prevents the inspection of sufficient sources to ensure adequate compliance with regulations; and that the quality of self-monitoring data varies widely.[33]

WHAT CAUSES HUMAN VULNERABILITY TO ENVIRONMENTAL RISK?

A clear upward trend can be observed in the human toll of natural and technological disasters.[34,35] Seveso, Bhopal, Three Mile Island, Chernobyl, Nevado del Ruiz volcano, the Armenian and Mexican earthquakes, and many other tragedies emphasize the environment as a source of risk, and not merely a habitat. Several vital questions are posed by such events.

First, is the vulnerability of human systems rising faster than risk reduction can tackle it? Evidence tends to be patchy, but the answer may well be "yes." For example, it is estimated that each year one thousand people move into the vicinity of the San Andreas fault-line in California. Under such circumstances, earthquake risk is a function of the presence of people and built structures in seismically active zones: yet it is unusual for there to have been a prior risk–benefit analysis. Over the next 40 years world population will double, and vulnerability per person will have to be halved if present levels of overall environmental security are to be maintained; yet there is no sign that any such goal will be reached.[36]

Second, to what extent is ecologically sustainable growth an accepted strategy nowadays? Governments that flagrantly ignore ecological precepts are likely to incur the wrath of electors, or at least of the international community. But *prima facie* compliance or concern may mask indifference or inability to act. Thus Scandinavia has protested vigorously to Western Europe, and Canada likewise to the United States, about the long-distance airborne transportation of sulfur dioxide pollutants that cause acid deposition. But the responses have often been dismissive, as was Britain's refusal to join the "Thirty Percent Club" of western nations committed to achieving timely reductions in sulfur emissions. Although the Chernobyl nuclear accident forced international reappraisal both of radiation tolerance levels and of past negligence in nuclear weapons testing, power generation, and fuel reprocessing, there is still too little international agreement on transboundary pollutants, including isotopic ones.

Third, new problems are emerging regarding the containment of the by-products of modern technology. For example, since it was discovered in the 1940s, cloud seeding has made rather erratic progress as a means of reducing the power of hail storms and hurricanes, and of producing rain during droughts. The former Soviet Union claimed to have mastered the technique of seeding hail-producing convective storms in order to reduce crop damage, but the State of Pennsylvania has legally banned all experiments with cloud seeding over its territory, because the transformation of latent heat expenditure in storms is judged unpredictable enough to cause a risk of more violent and less desirable atmospheric effects.[37] In another context, hazardous chemical by-products of Western European and North American manufacturing processes have often found a permanent home very cheaply at uncontrolled dump-sites in places like West Africa, or have been relegated to "ghost ships" that no port will admit.[38]

In the light of these considerations, much thought needs to be given to the following questions:

1. What role should existing and future technology play in the amelioration of environmental impacts? How can technological inputs to this process be integrated with social, economic and political ones: in other words, how can interdisciplinary communication be improved?

2. What can—and must—be learned from intercultural comparisons, especially between the vulnerability levels of the richer and poorer nations of the world?
3. Can environmental problems be treated as separate from other negative impacts on human life, such as unemployment, poverty, and war?

First, let us consider disciplinary constraints. As the complexity of modern science and engineering has increased, so the breadth of training has narrowed. Hence, specialists tend to bring to the environmental field an expertise that is somewhat constrained by the boundaries of the disciplines in which they have been nurtured. At the least, experts may show a tendency to view this highly interdisciplinary field from the very particular viewpoint determined by their own professional education. Thus, many physical and natural scientists are not at home amid the shifting sands of policy, environmental perception, social investigations, and economics. Likewise, social scientists may have a poor appreciation of physical realities, prediction possibilities, and engineering achievements. This mitigates against developing and disseminating a balanced overview of particular issues in which the demands of the *problem* replace those of the *discipline,* or, in other words, in which problems are solved by applying explicitly interdisciplinary methodology.[39]

One of the most vital tasks of the institutions that promote applied environmental science should be to concentrate on the applicability of technology to particular situations (its probable effectiveness at reducing the human and economic toll exacted by misuse of the environment). But it is equally vital to ensure that such efforts not become a mere technology fair, especially as it is potentially easier for a government to spend money on a highly tangible asset, such as pollution control apparatus, than on a much more diffuse investigation of what net benefit such investment may actually bring. We also require careful evaluation of how technology can be transferred successfully to places that lack the foreign exchange, trained manpower, and spare parts to make it function. Furthermore, there is still a dearth of information on how social and economic systems cope with environmental risks and impacts. Rather than being an absolute shortage of published results, it is an overconcentration of research on places—especially in North America—that are not necessarily representative of other cultures.

Cultural differences apart, it is clear that true reductions in environmental risk can be achieved only by examining the *hazardousness of place*[40]—that is, by adopting a broad systems approach that deals with multiple rather than single risks. Although valuable progress has been made, much of the developed world (from which the advances in such knowledge mainly emanate) is insulated against the most serious multiple impacts, which it does not understand well. Such effects are to be found, instead, among the countries in which civil or international conflict coexists with natural or technological hazards. For example, it is difficult and perhaps ultimately unproductive to try and determine the extent to which the refugee problem in Sub-Saharan Africa stems from drought, economic policies, or armed conflict.[41] Indeed, the Third World easily falls prey to a ''poverty–repression–militarization'' cycle,[42] into which environmental hazards must be inserted if a true picture of vulnerability is to be obtained.

WHAT ARE THE EMERGING SOLUTIONS?

Despite these problems, some positive signs have emerged.

Grass Roots Environmentalism

The impact of pollution and the depredation of resources tends to be progressive and cumulative. One can only hope that the same is true of the environmental movement. In this respect, the signs are encouraging. For example, residents of the Lake Baikal area banded together to fight the construction of a wood processing and pulp mill that would have polluted the lake very seriously, despite the employment opportunities that this plant represented.[43] Moreover, concerned about deteriorating water quality, transboundary air pollution, overexploitation of farmland, and overuse of pesticides, Green Parties have made substantial progress in various European countries where ecology has long been absent from the political agenda.[44] Initiatives and developments such as these are occurring in diverse countries throughout the world, although the environmental movement is in clear need of international sustenance in developing countries.

There is often an implicit assumption in environmentalism that one should seek to maintain the equilibrium—or at least the dynamic equilibrium—of ecological systems. We may question the wisdom of this. Nature itself does not necessarily follow such a strategy. For example, the momentous global changes associated with fluctuation in climate (variations in isostasy, in sea level, in the distribution of flora and fauna, etc.) appear abnormal only in the very restricted timespan of human perception.[45]

International Organization

Clearly, many environmental problems are either global in scope or distributed without respect to international boundaries. One of the most well-known examples of this is the explosion in 1986 at the Chernobyl nuclear reactor, near Kiev in the Ukraine, which increased radioactivity levels about 50 times in parts of Norway and Sweden.[46] In a different context, Bangladesh has to contend with the fact that 92.5% of the catchment area of its rivers, and 90% of their discharges, occur outside its national jurisdiction, in India, Nepal, Bhutan and Tibet. Hydrological management of the Ganges-Brahmaputra delta area is thus a fully international problem with ramifications all the way to the Himalayan Mountains.[47]

There is no lack of data on international environmental problems, especially as remote sensing has allowed recent environmental change to be monitored systematically on a world scale. Some have suggested, however, that what is needed is an international environmental planning discipline,[48] that is capable of providing a theoretical framework for the analysis of information on disparate local environments. The work of the UNESCO "Man and the Biosphere" Programme and of the Scientific Committee on Problems of the Environment (SCOPE) is helping to remedy this deficiency, and the enormous increase in the quantity, sophistication, and organization of famine relief shows the way toward increasing global collaboration. In short, it seems that, in the midst of the information age, the new world order that is emerging will probably include an environmental component, under the duress of increasing global degradation of the atmosphere, soils, waters, and vegetational resources. It remains to be seen whether chronic world population pressure will hold environmentalism in check, through advocating the need for more unbridled industrialization, or will facilitate it, through emphasizing the need to produce habitable environments wherever possible.

The Responsibility for Environmental Degradation

During the early 1970s the wiser developed countries began to institute the "polluter pays principle", in which those who are responsible for environmental degradation are charged with putting it right or contributing to the funds assigned to that end.[49] Subsequent economic recession caused a return to ideas of "economic necessity" and a slackening of any pollution controls that might limit investment and growth. But now there are signs that the "polluter pays principle" is returning to favor: gradual moves to reduce SO_X and NO_X emissions are a good example of this, as further deterioration of atmospheric quality may prove economically catastrophic if not checked. It has also been suggested[50] that there will soon be a thriving business in environmental monitoring, pollution control, and purification technology, which may begin to turn the tables on the "economic versus ecological" controversy.

Environmental Auditing

Recent developments in the United States, Canada, the Netherlands, and Germany suggest that it may be time to add a new level between monitoring and decision-making with respect to the environment. According to Schaeffer et al.,[51] "the term *audit* denotes the purposeful act of characterizing environmental attributes so as to understand a particular environment for the purposes of promoting more effective management." It may also signify ascertainment of compliance.[52] Hence, the method involves summarizing the functioning, or the functionality, of the environment, or of any attempts to modify or regulate it (and thus is similar to but broader in scope than environmental impact assessment). The creation of a structure for research and analysis that integrates the natural aspects of environment with positive, regulatory issues is a significant gain for environmentalism.

THE EARTH SUMMIT

The United Nations Conference on Environment and Development (UNCED) is a recently established, ongoing forum for international negotiation on environmental matters. Its first major achievement was the so-called "Earth Summit," a conference held in Rio de Janeiro in June 1992 and attended by the representatives of at least 166 national governments.

Reaction to the achievements of the Rio meeting has been decidedly mixed; and debate has varied from vociferous to mute (*Science* magazine, for instance, devoted only one page to the convention during the following 2 months). Some commentators have argued that merely to have held such a conference testifies to the importance of the environment on the global agenda and the seriousness with which ecological problems are now regarded. Others have branded the Summit a debacle, at which the multinational companies, through the World Bank, finally established the dominance of the form of international exploitation known as "free-market environmentalism."[53] At least, the conference demonstrated clearly that scientific environmentalism, which depends on the analysis of data, and political environmentalism, which depends on the judgments inherent in policy formulation, have very different agendas.[54]

It has been argued that a great opportunity was lost when the United States failed to take a leadership role comparable to that which it had played at Stockholm 20 years previously.[55] The American desire to protect "economic interests" rather than sign the Biodiversity Treaty was regarded as a move to prolong the exploitation of genetic resources in Third World countries, especially tropical ones. Yet notable advances in conservation have been achieved by the United States at more modest levels, both at home and abroad (such as the bilateral genetic exploitation treaty signed by Washington and Costa Rica).[56]

Amidst the polemics and controversies, it is evident that the world's expectations of the Earth Summit were too high. The industrialized nations did not commit a sufficient amount of their gross domestic product to environmental conservation, but who would expect them to when compelled by a severe recession to search for the easiest means of increasing employment and business profits? Yet the conference did end with a startling disparity between what was required of industrialized countries and what was expected of developing nations:

> Its secretariat provided delegates with materials for a convention on biodiversity but not on free trade; on forests but not on logging; on climate but not on automobiles. Agenda 21—the Summit's "action plan"—featured clauses on "enabling the poor to achieve sustainable livelihoods" but not on enabling the rich to do so; a section on women but none on men.[57]

The danger in this startling paradox is that by its own weakness UNCED may have condemned itself to irrelevance. The solution may be a radical restructuring of the entire United Nations organization in order to reflect better the diversity of its own membership. The vociferous complaints made by countries like India about the attitude of the developed nations seem to have scarcely been listened to; yet the fact that they were made in a continuing global forum may signify the early stages of a "new world environmental order." The Rio Summit has at least clarified some of the issues and the official positions of the main protagonists on the world stage.

CONCLUSION

Material progress has been made at least partly at the expense of the world's natural environment. As impacts are in many cases persistent, progressive, and cumulative, what started out as local effects have now risen to the status of global problems. The present age is characterized by vastly increased information flows and growing international involvement. So far, the challenge of exploiting these characteristics to restrict the negative impact of technology, stem the loss of resources, and reverse trends in pollution has not been met and progress is sluggish. But there are signs that the ecological ethic is becoming more prominent and more acceptable in decision making. In view of this, it should be borne in mind that the resolution of the world's environmental problems is a question of ideology applied to policy and organization, rather than simply a matter of applying more and more technology.

REFERENCES

1. **Pepper, D.,** *The Roots of Modern Environmentalism,* Croom-Helm, Beckenham, Kent, 1984.
2. **Hardin, G.,** The tragedy of the commons, *Science,* 162, 1243, 1968.
3. **Goudie, A. S.,** *The Human Impact on the Natural Environment,* 3rd ed., Blackwell, Oxford, 1990.
4. **Marsh, G. P.,** *Man and Nature, or, Physical Geography as Modified by Human Action,* Lowenthal, D., Ed., Belknap Press, Harvard University Press, Cambridge, MA, 1965.
5. **Nicholson, M.,** *The New Environmental Age,* Cambridge University Press, Cambridge, MA, 1987.
6. **Hartwick, J. M., and Oleweiler, N. D.,** *The Economics of Natural Resource Use,* Harper & Row, New York, 1986.
7. **Brown, B. J., Hanson, M. E., Liverman, D. M., and Merideth Jr., R. W.,** Global sustainability: Toward definition, *Environ. Manage.,* 11(6), 713, 1987.
8. **Carpenter, R. A., and Dixon, J. A.,** Ecology meets economics: A guide to sustainable development, *Environment,* 27(5), 6, 27, 1985.
9. **Pepper, D.,** *ibid.*
10. **World Resources Institute,** *World Resources, 1990–91,* Oxford University Press, Oxford, 1990.
11. **Nicholson,** *op. cit.,* 185.
12. **Ayala, H.,** New view of environmental health: Toward an environmental research policy for international strategies, *Environ. Manage.,* 11(2), 141, 1987.
13. **El-Hinnawi, E., and Hashmi, M.,** *The State of the Environment,* Butterworth, Sevenoaks, Kent, 1987.
14. **Bolin, B.,** Changing climates, in *The Fragile Environment,* Friday, L., and Laskey, R., Eds., Cambridge University Press, Cambridge, 1989, 127.
15. **Brown, L. R.,** and collaborators, *State of the World 1990,* W.W. Norton, New York, 1990.
16. **Judson, S.,** Erosion of the land, or what's happening to our continents? *Am. Scient.,* 56, 356, 1968.
17. **United Nations Conference on Desertification,** *Desertification: Its Causes and Consequences,* Pergamon, Oxford, 1977.
18. **El-Hinnawi, E., and Hashmi, M.,** *op. cit.,* 41.
19. **Hillel, D. J.,** *Out of the Earth: Civilization and the Life of the Soil,* The Free Press, New York, 1991, 236.
20. **Wilson, E. O.,** Threats to biodiversity, in *Managing Planet Earth: Readings from Scientific American,* Scientific American edition, W.H. Freeman, New York, 1990, 49.
21. **World Resources Institute,** *op. cit.,* 135.

22. **Myers, N.,** The future of the forests, *The Fragile Environment,* Friday, L., and Laskey, R., Eds., Cambridge University Press, Cambridge, MA, 1989, 22.
23. **Fearnside, P. M.,** Brazil's Balbina Dam: Environment versus the legacy of the Pharaohs in Amazonia, *Environ. Manage.,* 13(4), 401, 1989.
24. **Blaikie, P.,** *The Political Economy of Soil Erosion in Developing Countries,* Longman, London, 1985, 35.
25. **Bull, A. T., Holt, G., and Lilly, M. D.,** *Biotechnology: International Trends and Perspectives,* Organization for Economic Co-operation and Development, Paris, 1982.
26. **Levin, S. A., and Harwell, M. A.,** Potential ecological consequences of genetically engineered organisms, *Environ. Manage.,* 10(4), 495, 1986.
27. **Barbour, I. G.,** *Technology, Environment and Human Values,* Praeger Scientific, New York, 1980.
28. **Lemons, J.,** The need to integrate values into environmental curricula, *Environ. Manage.,* 13(2), 133, 1989.
29. **Barbour, I. G.,** *ibid.*
30. **Conacher, A.,** Environmental problem-solving and land-use management: A proposed structure for Australia, *Environ. Manage.,* 4, 391, 1980.
31. **Miller, A.,** Technological thinking: Its impact on environmental management, *Environ. Manage.,* 9(3), 179, 1985.
32. **Blaikie, P.,** *op. cit.,* 9.
33. **Palmisano, J.,** Environmental auditing: Past, present, and future, *Environ. Auditor,* 1(1), 7, 1989.
34. **Berz, G.,** Losses in the range of US $50 billion and 50,000 people killed: Munich Re's list of major natural disasters in 1990, *Natural Hazards,* 5(1), 95, 1992.
35. **Harriss, R. C., Hohenemser, C., and Kates, R. W.,** Our hazardous environment, *Environment,* 20(7), 6, 38, 1978.
36. **Coburn, A. W., Pomonis, A., and Sakai, S.,** Assessing strategies to reduce fatalities in earthquakes, *International Workshop on Earthquake Injury Epidemiology for Mitigation and Response,* Johns Hopkins University, Baltimore, MD, 1989.
37. **Carswell, S., and McBoyle, G. R.,** Analysis of state laws in weather modification: An update, *Bull. Am. Meteorol. Soc.* 64, 471, 1983.
38. **Alexander, D. E.,** Pollution, policies and politics: The Italian environment, in *Italian Politics: A Review, Vol. 5,* Catanzaro, R., and Sabetti, F., Eds., Francis Pinter, London, 1991, 90.
39. **Alexander, D. E.,** Applied geomorphology and the impact of natural hazards on the built environment, *Natural Hazards,* 4(1), 57, 1991.
40. **Hewitt, K., and Burton, I.,** *The Hazardousness of a Place: A Regional Ecology of Damaging Events,* Research Publications No. 6, Department of Geography, University of Toronto Press, Toronto, 1971.
41. **Timberlake, L.,** *Africa in Crisis: The Causes and Cures of Environmental Bankruptcy,* Earthscan, Washington, D.C., 1984.
42. **COPAT,** *Bombs for Breakfast,* Committee on Poverty and the Arms Trade, London, 1981.
43. **Pryde, P. R.,** in *Resource Conservation and Management,,* Miller, G.T., Ed., Wadsworth, Belmont, CA, 1990.
44. **Biorcio, R., and Lodi, G.,** *La sfida verde: il movimento ecologista in Italia,* Liviana, Padua, Italy, 1987.
45. **Houghton, R. A., and Woodwell, G. M.,** Global climatic change, *Sci. Am.,* 260(4), 18, 1989.
46. **Flavin, C.,** *Reassessing Nuclear Power: The Fallout from Chernobyl,* Worldwatch Institute, Washington, D.C., 1987.
47. **Brammer, H.,** Floods in Bangladesh: I—geographical background to the 1987 and 1988 floods, *Geographic. J.,* 156(1), 12, 1990.
48. **Ayala, H.,** *ibid.*
49. **MacNeill, J.,** Strategies for sustainable economic development, in *Managing Planet Earth: Readings from Scientific American,* Scientific American edition, W.H. Freeman, New York, 1990, 119.
50. **DocTer,** *European Environmental Yearbook,* 2nd ed., Institute of Environmental Studies, Milan, 1991.
51. **Schaeffer, D. J., Kerster, H. W., Perry, J. A., and Cox, D. K.,** The environmental audit. I. Concepts, *Environ. Manage.,* 9(3), 191, 1985.
52. **Palmisano, J.,** *ibid.*
53. **The Ecologist,** The Earth Summit debacle, *The Ecologist,* 22(4), 122, 1992.

54. **Moghissi, A. A.,** Who speaks for the environment? *The Environmentalist,* 18(4), 329, 1992.
55. **Alm, A. L.,** U.S. retreat at the Earth Summit, *Environ. Sci. Technol.,* 26(8), 1503, 1992.
56. **Stone, R.,** The Biodiversity Treaty: Pandora's box or fair deal? *Science,* 256(5064), 1624, 1992.
57. **The Ecologist,** *ibid.,* 122.

INDEX

A

Accipiter nisus (sparrowhawk), 66
Accumulators, 53–55
Acetaminophen, 94, 130
Acetone, 85
Acetylcholinesterases, 66, 106
Acid phosphatase, 113
Activity variables, 38
Acute fish lethality test, 156–157
Acute toxicity testing, 152
Adipose tissue, DDT levels in, 250
Agricultural chemicals
in Asia and Oceania, 20–21
modeling vapor drift and bioconcentration in foliage, 11–15
pesticide levels in food, 247–254
Air pollution
biomonitors, 43, 46
global problems, 265–266
lichens sensitive to, 43
pesticide saturation concentration calculation, 12
sentinel organisms, 41
Albumin, protein adducts, 132, 133
Alcaligenes, 236
Alcohol dehydrogenase, 72
Aldehyde dehydrogenase, 71
Aldicarb, 12
Aldrin, 249
Alkylation, serum proteins, 133
Allium cepa (onion), 13
Allochthonous species, 39
Alternative research and testing methods, 145–157
alternatives to animal experiments, 146–149
ecotoxicology, 150–157
Alveolitis, pathogenesis of, 203–205
Aminotransferases, 108
Ampicillin, 131, 132
Anaptychia ciliaris, 43
Anethum graveolens (dill), 13
Angiosarcoma, liver, 166
Animal studies
alternatives to, 146–149
arsenic in drinking water, 167–169
extrapolation to human toxicity, 70
radionuclide inhalation, 194
Anthracene, 115
Anthracosis, 195–196
Antibiotics, 130–132, 266
Anticoagulant rodenticides, 108
Aphids, 66
Arachlor, 12
Arctic region, organochlorines in, 21, 23–26
Arsenic, cancer risks, 163–174
animal studies, 167–169
comparison with other environmental exposures, 173
dose-response analysis, 169
epidemiological studies, 209
exposure assessment, 171
hazard identification, 164–169
human studies, 164–167
micronutrient value, 173
risk characterization, 171–173
thresholds, 169–171
Arthrobacter, 236
Arthropod biomonitors, 46, 47
Asbestos, 209

Asia and Oceania, fate of persistent organochlorines, 19–27
Aspirin, 130
Assays, alternative methods, 145–157, see also Alternative research and testing methods
Assemblages, 48
Atmospheric transport and deposition
 global environmental problems, 265–266
 in Mediterranean, 5–6
 organochlorines, global distribution, 23, 24
ATP-TOX, 154
Atrazine, 239–241, 254
Auditing, environmental, 272
Autochthonous species, 39
Activity variables, 40
Aziprotryn, 254

B

Bacteria
 alternatives to animal experiments, 148
 gut flora and absorption of toxic substances, 72–73
 microbial decontamination approaches, 235–245
 as sentinel organisms, 41
Barbiturates, 66, 85, 88, 89, 95
Bees, as sentinel organisms, 41
BEIR IV model, 227–228
Benomyl, 253
Benzene biodegradative pathways, 237
Benzene hexachloride, fate of inhaled toxic substances, 200
Benzimidazoles, 247, 253, 254
Benzo[a]anthracene, 241
Benzo[a]pyrene
 biodegradation pathways, 241
 cytochrome P-450 induction, 95
Benzo[a]pyrene metabolites, 66, 67
Bile, excretion of chemicals in, 72
Bile acids, 93
Bioaccumulation, fish species used in assays, 76
Bioassays, 45–48, see also Biomarkers; Biomonitoring
Biochemical biomarkers, 105–109
 blood enzymes, 108
 enzyme induction, 107–109
 enzyme inhibition, 106–107
 metabolic pathways, 108
Biocoenotic level biomonitoring, 38, 40
Bioconcentration, 40
 in foliage, evaluative models, 11–15
 HCBs in Mediterranean, 7
Biodegradation, microbial decontamination approaches, 235–245
Biological monitoring, see Biomonitoring
Biomagnification, 56

Biomarkers, 30–33, 155–156
 advantages and limitations, 31–32
 biochemical, see Biochemical biomarkers
 classification, 30–31
 interpretation, 33
 molluscan lysosomes, 111–120
 selection and prioritization, 32–33
Biomonitoring, 29–35, 37–57
 advantages and limitations, 31–32
 biomarkers, 30–33
 biomass and microcosms, 45–48
 concept of, 30
 detectors, 44–45
 environmental pollution as social dilemma, 29–30
 indicator species, 42–44
 integrated study, 33–34
 interpretation of data, 33, 37–38
 selection and prioritization, 32–33
 sentinel species, microcosms and, 39, 41–42
 typology, 38–39
Biosphere, global environmental problems, 266–267
Biotechnology, 241–245, 266
Biotic indices, 40, 48–53, 57
 defined, 48–49
 examples, 49–53
Biotransformations
 human toxicity studies, 69–73
 microbial decontamination approaches, 235–245
Birds
 blood enzymes, 108
 ergosterol fungicides and, 107, 108
 esterases, 106
 persistent organochlorines in, 20
Blackfoot disease, 166
Bladder cancer, arsenic and, 169, 170, 172
Blood cells
 arsenic in, 173
 cytochrome P-450, 107–108
Blood enzymes, 108
Bluegill sunfish (*Lepomis macrochirus*), 34, 76, 77
Body size, sampling strategy, 55
Breast milk, persistent organochlorines in, 20
Breath test, cytochrome P-450, 97–98
Brodifacoum, 108
B-type esterases, 106
Butylate, 12
Butyrylcholinesterase, 106

C

Cadmium toxicity, 69–70
Caffeine, 94
Calcium homeostasis, heavy metal effects, 142
Calibration of model data, 3

Cancer risk, see Carcinogens
Captan, 254
Carbamates
　fate of inhaled toxic substances, 200
　fish toxicity, 77
　persistence of, 236
　pesticide levels in food, 247–254
Carbaryl, 250
Carbendazin, 253
Carbofuran, 250
Carbon disulfide, 87, 89
16 α-Carbonitrile, 88
Carbophenothion, 108
Carboxylesterases, 106
Carcinogens
　arsenic, see Arsenic, cancer risks
　covalent interactions of, 89
　epoxide hydrolases and, 67
　human papilloma virus and, 189–190
　phenoxy herbicides, 179–185, 213–217
Case-control studies
　epidemiological approaches, 209, 210
　phenoxy herbicides, 214
Ceftriaxone, 131, 132
Cell injury
　biomonitoring approaches, 40
　mollusc lysosomes in, 111–112
Chemical saturation concentration, 12
Chlorcyclizine, 95
Chlordanes
　cytochrome P-450 induction, 95
　in marine environment, 19–27
Chlorfenvinphos, 249
Chlorinated dioxins, see 2,3,7,8-Tetrachlorodibenzo-*p*-dioxin
Chloroaromatics, see also Organochlorines
　biodegradation pathways, 238, 239
　evaluative models, HCBs and PCBs, 4–8
3-Chlorobenzoate, 238
Chlorofluorocarbons, 265
Chlorthal-dimethyl (DCPA), 12–15
Cholesterol, 93
Cholinesterases, 106
Cimetidine, 86–87
Clam (*Mercenaria mercenaria*), 118
Classical system theory, 38
Clements and Weaver, phytometers, 43, 56
Clofibrate, 85, 88, 107
Clotting factors, 108
Coastal region, fate of persistent organochlorines in, 19–22
Cobalt, 87
Codex Alimentarius Commission, 258–259
Coenzymes, 71
Cohort studies, 209, 210, 214
Community analysis, 49, 51, 52
Comparative metabolism, 65–67
Confounding factors
　cancer studies, 165, 166
　epidemiological studies, 208

Copper
　and gill changes of mussels, 135–142
　lysosome sequestration of, 113, 115
　seed dressing formulations, 254
Corynebacterium, 236
Crassostrea gigas, see Pacific oyster
Crop rotation, 13
Cross-sectional approach, 209, 210
Crotalaria spectabilis, 89
Cupermethrin, 250
Cyclosporin A, 94
Cymoxanil, 12
Cypermethrin, 12, 66
Cysteine adducts, 130
Cytochromes P-450
　assessment of induction
　　in vitro, 96–97
　　in vivo, 97–98
　biochemical bioassays, 107–108
　dichloromethane metabolism, 70–71
　microsomal monooxygenases, 65–66
　role in drug metabolism and toxicity, 81–90
　　drug interactions, inhibition and stimulation, 86–88
　　heme oxygen activation, 83–84
　　mechanisms of action, 82–83
　　multiple forms of, 84–86
　　xenobiotic toxicity, 88–90
Cytochrome P-450 monooxygenase, 96
Cytoskeleton
　mussel gills, copper-induced changes, 135–142
　protein adducts, 130

D

Daikon (*Raphanus sativus*), 13
Daphnia, 76, 156–157
DCPA, 12–15
pp'-DDMY, 108
DDT and metabolites
　accumulators, 54
　and blood enzymes, 108
　and cytochrome P-450, 95, 107
　food levels in eastern Europe, 247–254
　in marine environment, 19–27
　in Mediterranean, 4, 5, 7
　persistence of, 236
Debrisoquine, 86
Deforestation, 266
Deltamethrin, 250
Demeton-*p*-methyl, 249
Desmetryn, 254
Detectors, 40, 44–45
Detoxification
　fate of inhaled toxic substances, 200
　mechanisms of, 65–67, see also Cytochromes P-450
Dexamethasone, 85, 88, 95

Diazepam, 94
Diazinon, 66, 249
Dibenzodioxins, see 2,3,7,8-Tetrachlorodibenzo-p-dioxin
Dibenzofurans, cytochrome P-450 induction, 95
Dichloromethane, 70–71
Dichlorphos, 249
Dichlorvos, 12
Dieldrin, 66, 249
Diethylnitrosamine (DEN), 168
Dill (*Anethum graveolens*), 13
Dimethoate, 249
Dimethylarsonate (DMA), 168, 170, 171
2,3-Dimethylnaphthalene, 115
Dinoseb, 254
Dioxins
  cancer risk, see 2,3,7,8-Tetrachlorodibenzo-p-dioxin
  and cytochrome P-450, 95, 107
Diphenacoum, 108
Diphenyloxazole, 87
Diquat, 253
Diversity, 50, 51
DNA, 33–35
  adduct formation, 129, 155
  biomarkers, 155
  biotechnology, 266
  covalent interactions of carcinogens, 89
  fate of inhaled toxic substances, 200
Dose-response curves, 47–48, 169
Dosimetry, ecotoxicity testing, 152
Drinking water, arsenic in, 163–174, see also Arsenic, cancer risks
Drug interactions, and cytochromes P-450 system, 86–88
Drug metabolism, role of cytochromes P-450 in, 81–90

E

Early warning systems, 41, 56
Earth Summit, 272–273
Ecological fallacy, 209
Ecological groups, 42–44
Ecological realism, 57
Economics, world, 268
Ecotoxicology
  alternative testing methods, 150–157
    limitations of current procedures, see 152
    potential uses, 153–157
    purposes of, 151–152
  marine environment, fate of persistent organochlorines, 25–27
Eicosanoids, 93
Electrophoresis, protein adducts, 130–131
Embryo tests, 154
Endangered species, 266
Endotoxin, 203

Environmental auditing, 272
Environmental decontamination, microorganisms for, 235–245
Environmental epidemiology, 207–210
Environmental realism, 38
Environmental restoration, 29–30
Environmental variables, 3
Enzymes, 33, see also Biochemical biomarkers; Cytochromes P-450; Metabolism; Monooxygenases
  human toxicity studies, 69–73
  induction of cytochrome P-450 monooxygenase system, 93–98
    in vitro, 96–97
    in vivo, 97–98
  lysososmal, see Molluscs, lysosomes
Epidemiological studies
  phenoxy herbicides, 213–217
  radon, 221–228
Epidemiology, environmental, 207–210
Epithelial cells, virally induced changes, 189–190
Epoxide hydrolases, 66–67, 94
Epoxyeicosatrienoic acids, 93
Equilibrium fugacity models, 7
Ergosterol fungicides, 107
Erythrocyte acetocholinesterase, 106
Erythromycin, 88, 94, 95
Esterases, 66, 106
Estrogen, 117, 118
Ethanol, 85, 94, 95, 107
Ethics, environmental, 266
Ethylene, 41, 87
Evaluative environments, 3
Evaluative models in field situations
  HCBs and PCBs in Mediterranean, 4–8
  tributyltin in harbor, 8–11
  vapor drift and reconcentration in plants, 11–15
Extinction, 266

F

Fathead minnow, 76, 77
Fatty acids, 93
Felopidine, 94
Fenitrothion, 249
Fenvalerate, 250
Field studies
  calibration of models, 3
  comparison with laboratory studies, 47–48
  evaluative models in, see Evaluative models in field situations
  extrapolation from laboratory data, 106
  fish as biomarkers, 33–34
Filamin, 135
Fish
  as biomonitors, 33–34, 46
  cytochromes P-450, 95, 107

DDT in, 249, 251
metabolism of xenobiotics, 75–78
as sentinel organisms, 41
Fish lethality test, 156–157
Flavobacterium, 235, 236
Fluoride, sentinel organisms, 41
Food
  pesticide levels in, 247–254
  WHO safety recommendations, 258
Formaldehyde, 71, 97–98
Formothion, 249
Formyl chloride, 70–71
FRAME Toxicity Committee, 154
Fugacity models, 9–10, 13
Fungicides
  and cytochrome P-450, 107
  persistence of, 236
  pesticide levels in food, 247–254

G

Genetic polymorphism of drug oxidation, 85–86
Genotoxicity, 89, 148
Geographic studies, 209
Global distribution, organochlorines, 19–27
β-Glucuronidase, 113, 117
Glucuronyl transferases, fish, 78
Glutamate oxaloacetate aminotransferase, 108
Glutathione
  conjugation of xenobiotics, 76, 81, 82
  copper-induced gill damage in mussels, 136, 137, 141
Glutathione S-transferases, 94
  dichloromethane metabolism, 70–71
  fish, 78
Glutathionyl adduct of dichloromethane, 70–71
Golden orfe, 76
Grass roots environmentalism, 270–271
Guppy, 76
Gut microflora, and absorption of toxic substances, 72–73

H

HCB, see Hexachlorobenzene
Heavy metals, see Metals
Heliothis virescens (tobacco budworm), 66
Heme oxygenase, 87
Hemoglobin adducts, 155
HEOM, 66
Hepatic microsomal monooxgenases, 65–66
Heptachlor, 249
Herbicides
  biodegradation pathways, 238
  persistence of, 235–236
  pesticide levels in food, 247–254
  TCDD and cancer risk, 179–185, 213–217

Herpes simplex virus type 2, 190
Hexachlorobenzene
  evaluative models, 4–8
  seed dressing formulations, 254
Hexachlorocyclohexanes, in marine environment, 19–27
Hexagenia limbata (mayfly), 156
n-Hexane, protein adducts, 130
2,5-Hexanediol/hexanedione, 90
Hexobarbitone, 88
Holistic approach, 38
Hot spots, inhaled plutonium, 197
Human studies, 69–73
  alternatives to animal experiments, 148
  arsenic in drinking water, 164–167
Hydrocortisone, 118
Hydrolase latency, 116
Hydrosphere, 266
Hygiometers, 43, 56
Hypogymnia physodes, 43

I

Imazalil, 107
Immunochemistry
  copper-induced gill damage in mussels, 135–142
  cytochrome P-450, 96, 107
  esterases, 106
  protein adduct detection, 129–133
Inadvertent residues, 15
Index of Atmospheric Purity (IAP), 40–50, 53
Index of toxiphoby, 50
Indicator species, 42–44
Individual level biomonitoring, 38, 40
Infraindividual level biomonitoring, 38
Inhalation exposures
  arsenic, 164–167, 169, 170, 172
  fate of toxic substances, 199–201
  pathogenesis of pulmonary alveolitis, 203–205
  plutonium oxide, 193–198
  radon, 221–228
Insect and mite biomonitors, 46, 47
Insecticide resistance, monooxygenases and, 66
Insecticides
  biodegradation pathways, 238
  persistence of, 235–236
Intestinal microflora, 72–73
Intrinsic properties, 3
In vitro tests, 148, 150, 154–155

J

Japanese quail, 108
JMPR (Joint FAO/WHO Meeting on Pesticide Residues, 258–261

K

Keratinocytes, papilloma virus and, 189–190
Kidney cancer, arsenic and, 165, 167–170, 172
Kohlrabi (*Brassica oleracea*), 13

L

Laboratory data
  comparison with field data, 47–48
  extrapolation from, 106
LC_{50} tests, 156
LD_{50} values, 152
Lecanora, 43
Lenacil, 254
Lepomis
  auritus (redbreast sunfish), 34
  macrochirus (bluegill sunfish), 34, 76, 77
Lepraria incana, 43
Lethality testing, 152
Level II fugacity model, 9–10
Lichens, 43–44, 46–53
Lindane, 200, 248
Linuron, 253
Lipid peroxidation, 93, 136
Lipids, molluscan lysosomes, 113–114, 116, 119
Lipofuscin, 113, 114, 117, 119
Lipophilicity and lipid solubility
  and cytochrome P-450, 87–88, 107–108
  cytochrome P-450 monooxygenase system and, 93
  DDT levels in livestock, 250
  fate of inhaled toxic substances, 199
  persistence of xenobiotics, 236
Lithosphere, 266
Littorina littorea, 113, 117, 119
Liver, 34
  arsenic and cancer risk, 165–166, 168, 170, 172
  detoxification and circulation in, 72
Liver enzymes, see also Cytochromes P-450; Monooxygenases
  epoxide hydrolases, 66–67
  fish, 78
Livestock, DDT levels, 250
Lobaria pulmonata, 43
Lung
  inhalation of radionuclides and, 193–198
  pathogenesis of pulmonary alveolitis, 203–205
Lung cancer
  arsenic and, 164–167, 169, 170, 172
  epidemiological studies, 209
  radon and, 221–228
  TCDD exposure and, 185
Lymphatics, lung, 195–197
Lymphomas, phenoxy herbicides and, 179–185, 213–217
Lysosomes, mollusc, 111–120

M

Malathion, 66, 249
Malignant lymphomas, 179–185, 213–217
Malondialdehyde, 136, 137, 139, 141
Mancozeb, 252, 254
Maneb, 252, 254
Manomethylarsonic acid (MMA), 170, 171
Mapping, 56
Marine environments, evaluative models
  HCBs and PCBs in Mediterranean, 4–8
  tributyltin in harbor, 8–11
Marine snails (*Littorina littorea*), 113, 117, 119
Mathematical models, 47, 57, 147
Mayfly (*Hexagenia limbata*) larva, 156
Mediterranean, evaluative models, HCBs and PCBs, 4–8
Mediterranean Action Plan (MAP), 4, 8
MED POL model, 6–8
Melipax, 248
Membrane integrity, 119
Membrane permeability, 116
Mephenytoin, 94
Mercenaria mercenaria (clam), 118
Mercurial fungicides, 251–254
Metabolism
  biochemical biomarkers, 108
  comparative, 65–67
  enzyme induction, 93–98
  fate of inhaled toxic substances, 199
  in fish, 75–78
  human toxicity studies, 69–73
  microbial biodegradation, 235–245
Metallothionein, 33
Metals
  accumulators, 54
  copper-exposed mussels, gill changes of, 135–142
  evaluative models, tributyltin, 8–11
  and hepatic drug metabolism, 87
  lysosome sequestration of, 113
  and mollusc lysosomes, 119
Metapyrone, 87
Metaxon, 253
Metaxyl, 254
Methanol, human toxicity mechanisms, 71–72
Methoxychlor, 54, 248, 249
3-Methylcholanthrene, 85, 88, 95, 107
2-Methylnaphthalene, 115
Methyl parathion, 249
Microbiotests, 154
Microcosms, 40–42, 45–48
Micronutrients, arsenic as, 173

Microorganisms
  environmental decontamination strategies, 235–245
    biodegradation pathways, 237–241
    engineered organisms, 241–245
  intestinal flora, 72–73
  as sentinel organisms, 41
Microsomal bioassay system, 46
Microsomal monooxygenases, see Monooxygenases
Microtox, 156–157
Microtubules, copper-induced changes in mussel gills, 135–142
Mixed-function oxidase system, 155
Mixtures of chemicals, ecotoxicity testing, 152
Models, see also Evaluative models in field situations
  epidemiological, 207–210
  mathematical, 47, 57, 147
  multiplicative risk, 227–228
Molecular-biochemical level biomonitoring, 33, 40
Molecular-organic level, biomonitoring, 38
Molluscs
  copper-exposed mussels, gill changes of, 135–142
  lysosomes, 111–120
    cell injury, 111–112
    cellular consequences of damages to, 118, 119
    pathological reactions, 112–118
Monocrotaline, 89
Monohydroxyeicosatetraenoic (HETEs) acids, 93
Monooxygenases, 65–66, see also Cytochromes P-450
  biochemical bioassays, 107–108
  fish, 76
Multinational corporations, 268
Multiple-species biomonitoring systems, 38
Multiple-way table, 39, 40
Mumerobates rostrolamellatus, 46, 47
Mussels (*Mytilus*), 119
  copper-exposed, gill changes, 135–142
  lysosomes, 114
Mussel watch, 53–54
Mutagenesis, 46, 47
Mytilus, 118, 119
  galloprovincialis, 135–142
Myzus persicae, 66

N

NADP-cytochrome P-450 reductase, 83
Naphthalene, 241
Naphthoflavone, 107
α-Naphthoquinone, 87
α-Naphthyl thiourea, 125–127
Nifepidine, 94

Nitrapyrin, 12
Nivanol, 94
No Adverse Effects Levels (NOAEL), 261
Nocardia, 236, 240
Non-Hodgkins lymphomas, phenoxy herbicides and, 179–185, 213–217
No observed effect level (NOEL), 70
Norbormide, 125–127
Nylander, hygiometers of, 43, 56

O

Oak Ridge Air Quality Index (ORAQUI), 48
Oak Ridge effluents, biomonitoring, 34
Ocean, see Marine environment
Oceania, fate of persistent organochlorines, 19–27
Onions (*Allium cepa*), 13–14
Oocytes, mussel, 118
Organization level, 55
Organochlorines
  biodegradation pathways, 238, 239
  and cytochrome P-450, 107
  evaluative models, HCBs and PCBs, 4–8
  food levels in eastern Europe, 247–254
  in marine environment
    coastal region, 19–22
    ecotoxicology, 25–27
    open ocean, 21, 23–25
Organomercurials, 251–252
Organophosphates
  and esterases, 66, 106–107
  fate of inhaled toxic substances, 200
  fish toxicity, 76, 77
  persistence of, 236
  pesticide levels in food, 247–254
Osmoregulation, copper-induced gill damage in mussels, 139
Oxidative stress, 113, 136, 142
Oxygen activation, cytochrome P-450, 83–84
Oysters
  as accumulators, 54
  lysosomes, 114
  tributyltin effects, 8–11
Ozone levels, 265

P

Pacific oyster (*Crassostrea gigas*), 8
PAHs, see Polychlorinated aromatic hydrocarbons
Papillomavirus, 189–190
Paraquat, 254
Parathion, 87, 249
Parmelia species, 43
Parsley (*Petroselinum crispum*), 13–15
Pathogens, 266
PCBs, see Polychlorinated biphenyls

Penicillin, 130, 132
Penicillolyated serum albumin, 130
Pentachlorobenzene (PCNB), 254
Pentachlorophenols, 238
Periwinkles, 114, 117, 119
Permethrin, 250
Pesticides
 biodegradation pathways, 238
 fate of inhaled toxic substances, 200
 global problems, 266
 levels in food, 247–254
 modeling vapor drift and bioconcentration in foliage, 11–15
 persistence of, 235–236
 resistance to, monooxygenases and, 66
 TCDD and cancer risk, 179–185
 World Health Organization safety programs, 257–261
Petroleum products
 and lysosomes, mollusc, 115
 and mollusc lysosomes, 118
Pharmacokinetic studies, cytochrome P-450, 96
Phenacetin, 94
Phenanthrene, 115, 117
Phenobarbital, 85, 89, 95
Phenobarbitone, 66, 88
Phenoxyherbicides
 cancer risks, 179–185, 213–217
 persistence of, 236
Phenthion, 249
Phenylbutazone, 95
Phenytoin, 95
Phosalone, 249
Photobacterium phosphoreum, 154
Physicochemical properties
 alternatives to animal experiments, 147
 and environmental fate, 3
Phytometers, 45–48
Piperonyl butoxide, 66
PIVKA levels, 108
Plants
 bioassays, 45–48
 evaluative models of vapor drift, 3, 11–15
 as sentinel organisms, 41
Plasma enzymes, 108
Plasma proteins, 130, 132, 133, 156
Plasmids, degradative, 241–245
Plutonium oxide inhalation, 193–198
Policy
 Earth Summit, 272–273
 emerging solutions, 270–272
 environmental problems, 265–267
 global, 263–273
 human vulnerability to environmental risk, 269–270
 new environmental age, defined, 264–265
 obstacles to solutions, 267–269
 World Health Organization, 257–261
Polybrominated biphenyls, 95

Polychlorinated aromatic hydrocarbons (PAHs)
 biodegradation pathways, 240–241
 biomarkers, 155
 cytochromes P-450 and, 65, 88, 95, 107
 and lysosomes, molluscan, 113, 115–118
 membrane permeability, 116
Polychlorinated biphenyls (PCBs)
 biodegradation pathways, 240
 and cytochrome P-450, 107
 and lysosomes, mollusc, 115
 in marine environment, 19–27
 persistence of, 236
Polychlorinated dibenzo-*p*-dioxins (PCDDs), see 2,3,7,8-Tetrachlorodibenzo-*p*-dioxin
Population, 266
Population level biomonitoring, 33, 38
Porphyrin, 81, 82
Predictive value
 ecotoxicity testing, 151
 field versus laboratory studies, 48
Pregenolone, 88
Preprothrombin, 108
Preputial gland, 117, 118
Primicarb, 249
Primiphos ethyl and methyl, 66, 249
Priority pollutants, persistence of, 235–236
Prochloraz, 107
Prometryn, 253
Propoxur, 250
Propranolol, 94
Protein adducts, 129–133, 155
Protocatechuate, 238, 239
Protococcus viridis, 43
Protoheme, 82
Protoporphyrin IX, 82
Pseudomonas, 235–237, 240, 244
 putida, TOL plasmid, 244
Pyrethroids, levels in food, 247–254
Pyridine-2-aldoxime methiodide, 106

Q

Quantitative structure-activity relationships (QSAR), 147, 150, 153, 156–157

R

Radionuclide inhalation, 193–198
 animal effects, 194
 anthracosis, 195–196
 dose distribution in lungs, 194–195
 hot spots, 197
 lymphatic system and, 195–197
 nonstochastic effects, 198
 parenchymatous changes, 196
 radon, 221–228

Radon
  characteristics of, 221–222
  exposure assessment, 222
  indoor, 223–225
  miners, 222–223
  nonpulmonary malignancies, 225–226
  risk estimates in miners and general population, 227–228
  smoking and, 226–227
Rainforest, 266
Ramalina species, 43
Raphanus sativus (daikon), 13
Recombinant DNA, 266
Redbreast sunfish (*Lepomis auritus*), 34
Red killifish, 76
Regulation
  World Health Organization programs, 257–261
  world policy, 263–273
Remediation, 29–30
Research methods, alternative, see Alternative research and testing methods
Resistance, pesticide, 66
Respiration
  copper-induced gill damage in mussels, 139
  sentinel organisms, 41
Retinoids, 93
Retinol, 155
Richness, 50, 51
Rifampicin, 95
Rodenticides, 108
  pesticide levels in food, 248
  toxicity evaluation, 125–127
Rotational crops, 13

S

Salmo gairdneri (trout), 76, 77
Salmonella bioassay, 46
Sampling strategy, 54–55
Schiff bases, 130
Seasonal variation, sampling strategy, 55
Seawater, see Marine environment
Selective toxicity, 65–67
Selenium-binding protein, 130
Semivolatile organochlorinated hydrocarbons (SOCs), see Hexachlorobenzene; Polychlorinated biphenyls
Sentinel species, 39–42
Serum enzymes, 108
Serum proteins
  adduct detection, 130
  adducts, 130, 132, 133
  biomarkers, 156
Sex, 55
Simazine, 254
Single-species biomonitoring systems, 38
SKF 525-A, 87
Skin cancer, arsenic and, 164–165, 169

Society
  pollution as problem of, 29–30
  value systems and environmental ethics, 266
Soft-tissue sarcomas, phenoxy herbicides and, 179–185, 213–217
Soil carryover, pesticides, 13
Soil erosion, 266
Sparrowhawk (*Accipiter nisus*), 66
Spartein, 87
Species diversity, 50, 51
Species endangerment and extinction, 266
Species richness, 50, 51
Spurious peaks, 4
Starling (*Sturnus vulgaris*), 106
State variables, 38, 40
Steroid hormones, 93
  and cytochrome P-450, 88
  and membrane stability, 118
Stevioside, 72–73
Sturnus vulgaris (starling), 106
Styrene oxide, 133
Suicidal inhibitors, 87, 89
Sulbactam, 131, 132
Sulfur dioxide, sentinel organisms, 41, 53
Supercooled vapor pressure, 13
Synergy
  biomonitoring considerations, 55
  field versus laboratory studies, 48
Synthetic topology, 39, 40

T

2,4,5-T, 179, 238
TCDD, see 2,3,7,8-Tetrachlorodibenzo-*p*-dioxin
Technocracy, 267–268
Terbutryn, 12
2,3,7,8-Tetrachlorodibenzo-*p*-dioxin (TCDD)
  cancer risks, 179–185, 213–217
  and cytochrome P-450, 88, 107
  cytochrome P-450 induction, 95
  epidemiological studies, 209
  metabolism of, cytochrome P-450 and, 85
Tetrachloroterphthalate (chlorthal-dimethyl; DCPA), 12–15
Tetrahydrofolate coenzymes, 71
Thiometon, 249
Thiophanate-methyl, 253
Thiuram, 252, 254
Thivers (indicator species), 42–44
3T3 cells, human papilloma virus and, 189–190
Tobacco budworm (*Heliothis virescens*), 66
Tolbutamide, 94, 95
Tolerance, 50
TOL plasmid, 241–245
Topology, synthetic, 39, 40
Topsoil, 266

Toxaphene, 248, 249
Toxi-Chromotest, 154
Toxicity, see also Ecotoxicity; Metabolism
  fish species used in assays, 76
  human toxokinetics, 69–73
  selective, 65–67
Toxic necrosis, 89
Toxicology, see Ecotoxicology
Toximeter, 43, 56
Toxiphoby, index of, 50
TOXKITS, 157
Tradescentia stamen hair test, 46
Triazine, 235
Tributyltin, evaluative models, 8–11
Trichloracetic acid, 254
Trimethylbenzene, 239
Troleantromycin, 88
Trophic transfer, temperature and, 54
Trout (*Salmo gairdneri*), 76, 77
Turbulence, 3, 13
Typology of biological changes, 38–39

U

United Nations Conference on Environment
    and Development, 272–273
Urinary tract cancer, 165, 167–170, 182
Usnea ceratina, 43

V

Vapor drift, evaluative modeling, 11–15
Vapor pressure, 12, 13
Vietnam veterans, 183, 184, 213, 215
Viruses, human papilloma virus-induced
    changes, 189–190
Vitamin K antagonists, 108
Volatilization potential, pesticides, 11–14

W

Warfarin, 108
Water, global environmental problems, 266
Water Quality Index (WQI), 48
World Health Organization, 257–261
World Policy, 263–273

X

Xylenes, 237–238

Z

Zebra fish, 76
Zineb, 252
Zoxazolamine, 88